Physics of High-T_c Superconductors

Physics of High-T_c Superconductors

J. C. Phillips

AT&T Bell Laboratories
Murray Hill, New Jersey

Published by arrangement
with AT&T

ACADEMIC PRESS, INC.
Harcourt Brace Jovanovich, Publishers
Boston San Diego New York
Berkeley London Sydney
Tokyo Toronto

ACADEMIC PRESS, INC.
1250 Sixth Avenue, San Diego, CA 92101

United Kingdom Edition published by
ACADEMIC PRESS INC. (LONDON) LTD.
24–28 Oval Road, London NW1 7DX

Library of Congress Cataloging-in-Publication Data

Phillips, J. C.
Physics of high-T_c superconductors.

Bibliography: p.
Includes index.
1. High temperature superconductors.
I. Title.
QC611.98.H54P47 1989 537.6′23 88-34278
ISBN 0-12-553990-8

Printed in the United States of America
89 90 91 92 9 8 7 6 5 4 3 2

Preface

"I have no data yet. It is a capital mistake to theorize before one has data. Insensibly one begins to twist facts to suit theories, instead of theories to suit facts."

— A Scandal in Bohemia, from The Adventures of Sherlock Holmes, by Sir A. Conan Doyle.

Condensed matter chemists and physicists have written thousands of research papers in less than two years on the subject of high-temperature superconductivity, and the literature has become as forbidding as the darkest jungle. The aim of this book is to cut a well-marked path through this jungle, not by listing every paper, but rather by highlighting what I judge to be the important experiments. To understand these experiments one also needs a few simple ideas, and I have mentioned these where necessary. The reader will soon come to realize that my primary emphasis is on the materials themselves. Some of my more mathematical colleagues have seized upon high-T_c superconductors as the ideal area for testing ideas from subjects seemingly very remote from materials science, ranging from quark confinement to liquid He^3. Perhaps much can be learned from these analogies, but here the reader will find them discussed only very briefly. The new world of materials requires, I believe, its own ideas, and I have tried to indicate where these may by found in the context of high T_c superconductivity.

In writing this book I have drawn heavily upon the support of my colleagues at AT&T Bell Laboratories. I am especially grateful to B. Batlogg for access to his computerized preprint library, and to R. Matula for assistance in literature searches. Discussions of materials problems with G. Fisanick, L. Schneemeyer and D. W. Murphy have been most helpful, and correspondence with Profs. M. L. Cohen and C. Kittel of the Univ. of California, Berkeley, has provided useful perspectives. Most important of all, this book could never have appeared without the cheerful and accurate manuscript preparation provided by Mrs. A. E. Bonnell.

Contents

III. New Materials *60*

IV. New Theory *118*

V. Isotope Effect *185*

Appendices

I. Old Materials

1. General Classifications

There are three general patterns in high-T_c superconductors which are widely observed. First, the transition temperatures of elements or compounds containing transition elements are generally higher than those containing simple s-p metals. This is generally explained by the larger d-band densities of states $N(E_F)$ in the former compared to the s-p states of the latter. Secondly, binary compounds can have significantly higher transition temperatures than elements. A simple explanation for this is statistical: there are more binary compounds than elements. If this reasoning is correct, then still higher temperatures should be found in true ternary compounds (not pseudo-binaries). Until recently this has appeared not to be the case. However, it is important to realize that only 7200 true ternary compounds are known, compared to 15,000 binaries,[1] so that this argument may be correct. In fact, as we shall see, the higher T_c, the more unstable the material, and among 100,000 possible ternaries probably so far the ones that are more easily made have been preferentially prepared.

The compound $(La, Sr)_2$ CuO_4 can be regarded as a pseudoternary and it has $T_c > 30K$. It is one of the new materials to be discussed in Chapter III. Clearly, while statistical arguments alone favor higher maximum T_c's in ternaries than in binaries, and in binaries than in elements, some other guiding principles are needed if the search is to be successful.

The third general principle which is often mentioned is high symmetry, especially cubic symmetry. This principle is less useful in practice than might appear. This is because it is generally much easier to test a material for superconductivity than it is to determine its crystal structure. Over a period of three decades B. T. Matthias and coworkers, for example, examined representative samples of virtually all binary alloys for superconductivity, many of which contained compounds of unknown structure. Unless a $T_c > 1$ K was found, no efforts were made to determine these crystal structures.

A careful review of known T_c's has been made by P. Villars and the results are listed in Appendix C. Generally speaking for the elements $T_c < 10$ K while for 60 compounds $T_c > 10$ K. The former region includes altogether 575 superconductors with $1\ K < T_c < 10$ K, and this region is called the physical region, in contrast to the chemical region where $T_c > 10$ K. In his monumental studies of the crystal structures of 22,000 compounds and alloys Villars found that with three elemental coordinates he was able successfully to separate according to crystal structure 97% of the binaries and 95% of the ternaries. Using these same coordinates he finds that in the physical region $T_c < 10$ K a scatter-shot (random) plot results, but in the chemical region ($T_c > 10$ K), the known high-T_c materials are isolated in three small islands. The details of Villars' analysis are also presented in Appendix C and they serve as the basis for organizing the discussion of materials in this book. However, his discussion is quite technical and involves concepts at the cutting edge of materials science, so the general reader may wish to follow the presentation here in the text which is more conventional and requires no specialized knowledge.

2. Lattice Instabilities in Elemental, Binary and Pseudobinary Superconductors

According to the BCS theory (Appendix B), the simplest expression for T_c involves the density of states at the Fermi energy, $N(E_F)$, the average electron-phonon interaction V, the Debye temperature θ_D and an average Coulomb repulsion energy μ. In terms of these parameters T_c is given by

$$T_c = 1.14\,\theta_D \exp[-(\lambda-\mu)^{-1}] \tag{2.1}$$

where $\lambda = N(E_F)\overline{V}$ and $\theta_D = \hbar\omega_D/k_B$, and $\hbar\omega_D$ is an average phonon energy.

A great deal of effort has been expended both experimentally and theoretically on elemental and binary superconductors to test the validity of Eqn. (2.1). The functional form of this equation, with its exponential singularity, makes the results for T_c very sensitive to λ and μ, and less sensitive to θ_D. Unfortunately only θ_D can be measured accurately in the normal state. However, the general theory discussed in Chapter II shows that increasing λ destabilizes the lattice and decreases ω_D. At the same time in pseudobinary alloys where the electron density ρ varies slowly with composition, μ (which depends primarily on ρ and is considerably smaller than λ for high T_c materials) is also slowly varying. This means that even without having complete, detailed, accurate and controlled quantum-mechanical calculations of $\overline{V}$ and $N(E_F)$, such as those discussed in Chapter II, one can connect T_c to λ by measuring $\omega(\mathbf{k})$ by neutron scattering. We shall next examine such data qualitatively, deferring quantitative and critical discussion until the next chapter.

a. Elements

The non-transition or s-p elements with the highest T_c's include the tetravalent $Z=4$ metals Sn(T_c = 3.7 K) and Pb (T_c = 7.2 K). With increasing valence λ increases more rapidly than μ as both $N(E_F)$ and $\overline{V}$ increase. For $Z>4$ covalent bonds form and $N(E_F)$ decreases rapidly. The lighter tetravalent elements, C, Si and Ge have covalent tetrahedral structures and are not metallic. Under high pressure Si and Ge become metallic and superconducting

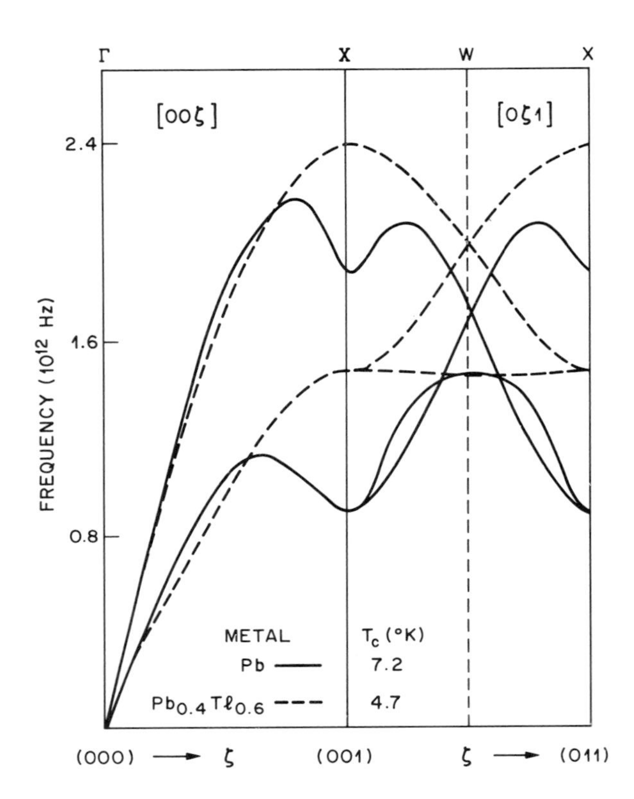

Fig. I.1. Comparison of lattice vibration dispersion $\omega(\mathbf{k})$ in Pb ($T_c = 7.2$ K) and $Pb_{0.4}Tl_{0.6}$ ($T_c = 4.7$ K). Note the dramatic softening for **k** near (001) in Pb which is associated with a 50% increase in T_c relative to $Pb_{0.4}Tl_{0.6}$ (from ref. 2).

with values of T_c in good agreement with theoretical predictions, as discussed in the next chapter.

Superconductivity was observed very early in Pb ($T_c = 7.2$ K), and large single crystals of this material are easily prepared, so that this system has provided very accurate data. Neutron scattering data[2] showing $\omega(\mathbf{k})$ for wave vector **k** along a few symmetry directions are shown in Fig. 1 for Pb and a $Pb_{0.4}Tl_{0.6}$ alloy ($T_c = 4.7$ K). Note that near $\mathbf{k} = 0$ the two dispersion curves are very similar, but that there is a drastic softening for both longitudinal and transverse modes in the pure metal for **k** near (001), in units of $2\pi/a_1$ where a is the cubic lattice constant. This softening reflects a substantial enhancement of the electron-phonon interaction and it occurs near **k** = (001) because of a special property of the nearly free electron energy bands of Pb, which is discussed in the next chapter, and is called Fermi surface nesting. The incipient lattice instability ($\omega^2 \longrightarrow 0$) in Pb is immediately evident in Fig. 1 and it is noteworthy that it is associated with shortwave-length phonons with out-of-phase vibrations of (100) planes, which are atomically the most open and lease dense.

The value of T_c in $Pb_{1-x}Tl_x$ alloys have also been carefully studies,[3] and the results are sketched in Fig. 2. As the electron-phonon interaction decreases and $\omega(001)$ increases with x, there is a drastic drop in T_c which is roughly proportional to $T_c(0) - ax^{1.5}$. The correlation between lattice stiffening and reduction of T_c has been observed in almost all careful studies of superconducting metals with high T_c's, including other s-p metals such as

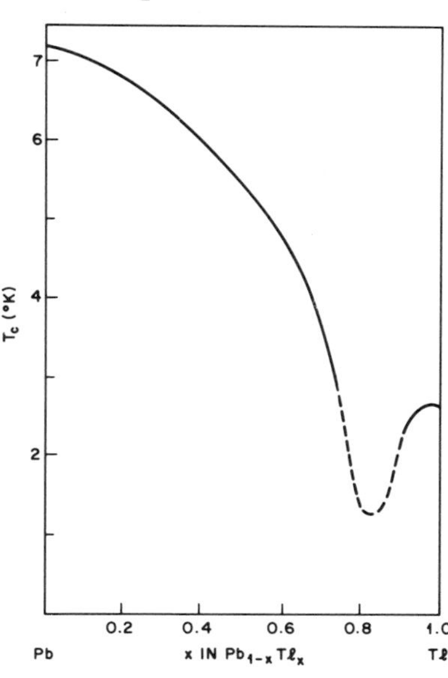

Fig. I.2. Variation with composition of T_c in $Pb_{1-x}Tl_x$ alloys (from ref. 3).

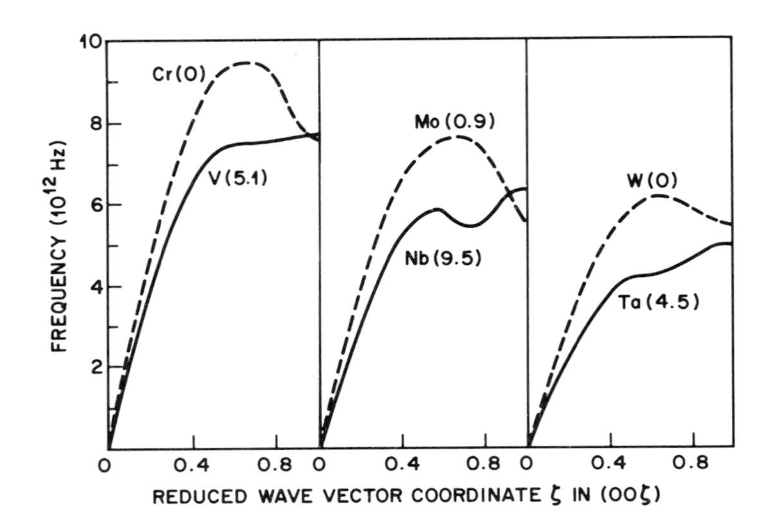

Fig. I.3. Longitudinal acoustic modes in the [001] direction for Groups V and VI transition metals. The values of T_c are indicated and the correlation of T_c with lattice softening is obvious (from ref. 4).

Sn, In and Hg. Even without microscopic lattice dynamical and electronic energy band calculations materials physics common sense tells us that these correlations provide strong evidence for supposing that electron-phonon interactions are the basic mechanism responsible for high-temperature superconductivity.

Among transition elements the highest T_c's are found in groups V and VI, and here again there is an excellent correlation[4] between phonon softening and T_c enhancement as shown in Fig. 3. Here the phonon softening again occurs for **k** along the (00δ) direction with δ near 0.7. Careful studies of this dip in Nb_xMo_{1-x} alloys showed[5] that for $x = 0.25$ the dip disappeared completely, but that a faint dip reappeared near $\mathbf{k} = \mathbf{H}$ at $x = 0$, corresponding to

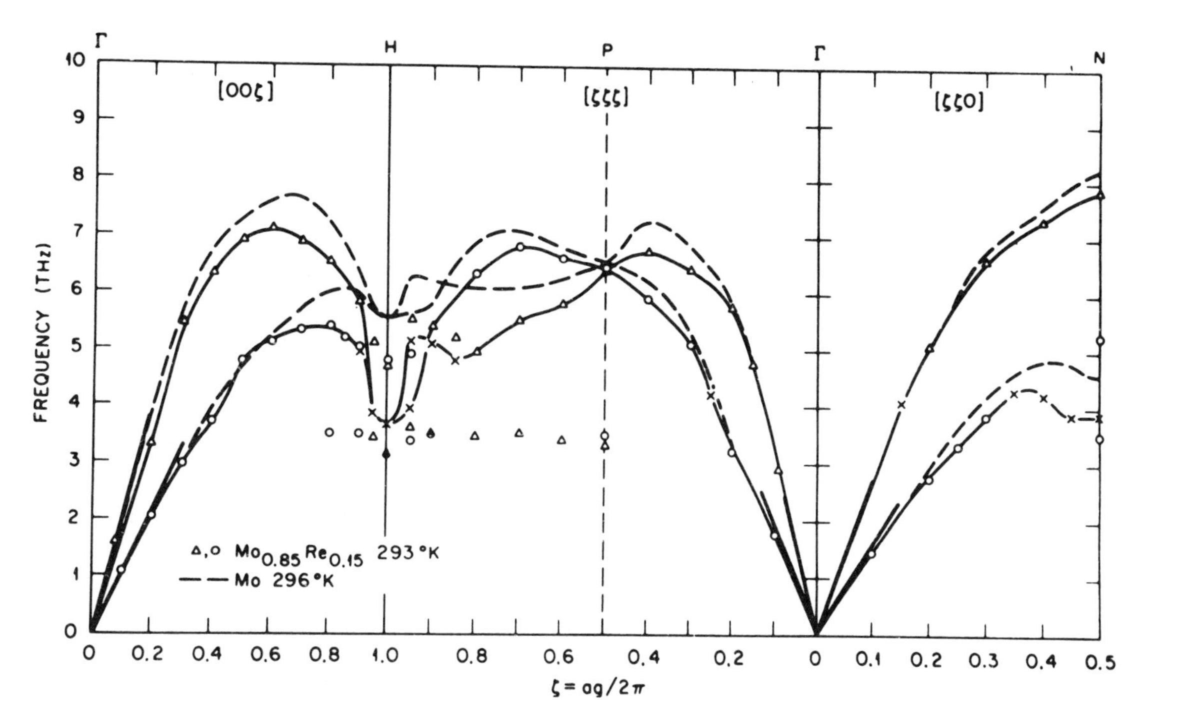

Fig. I.4. Comparison of "dispersion curves" (including localized phonon resonances) in $Mo_{0.85}Re_{0.15}$ ($T_c = 10°K$) with Mo($T_c = 0.9°K$). The gigantic enhancement of T_c with only 15% Re substitution is associated with the large dip in the transverse acoustic $\omega(\mathbf{k})$ for $\mathbf{k}$ near $\mathbf{H} = (001)$. This instability leads to phase separation for larger Re substitution, and it suggests partial ordering of Re substituents with off-center relaxation at Re contents of 15% (ref. 6.).

$T_c < 0.1$ K for $x = 0.25$ and $T_c = 0.9$ K for $x = 0$. Similarly, when Mo is alloyed with Re, Smith et al. showed[6] that $\omega(\mathbf{H})$ was indeed anomalous, with a reduction by almost a factor of two for $Mo_{0.85}Re_{0.15}$, as shown in Fig. 4. The decrease continues up to $Mo_{0.75}Re_{0.25}$, where $T_c = 10$ K. For greater concentrations of Re, T_c increases further to 12 K as a metastable phase embedded in a multiphase sample is formed, which is further evidence for the critical rôle played by lattice softening. Similar T_c enhancement followed by a phase instability is found in Zr_xNb_{1-x}, whereas for Mo_xNb_{1-x} a single-phase substitutional alloy is formed for all values of x with T_c decreasing as x increased.

A special comment should be made here on Fig. 4. For vibrational studies it is generally convenient to separate electronic effects from mass effects by substituting atoms belonging to the same period, so that mass changes are small. This is not the case for Re_xMo_{1-x} allys, but these materials were studied just because T_c changes from <0.1 K at $x = 0$ to >10 K at $x=$ only 0.15. In Fig. 4 we notice a localized vibrational mode with $\omega(\mathbf{k})=$ const. for $\mathbf{k}$ near $\mathbf{H}$. This is a localized resonance, and because $m(Re) > m(Mo)$, the appearance of this resonance is expected on general grounds. In dynamical systems ordinarily, however, the continuum modes are *repelled* by the localized modes. This is not the case here: instead there is a dramatic softening of the continuum modes $\omega(\mathbf{k})$ for $\mathbf{k}$ near $\mathbf{H}$. It is likely[7] that the softening and the increase in T_c are both associated with formation of σ phase precursor clusters with composition near $Re_{0.6}Mo_{0.4}$, because in evaporated thin films formed in this phase T_c reaches 15 K.

b. Binary and Pseudobinary Compounds

The best-known and most studied binary and pseudobinary compounds belong to AB and A_3B structure types (usually called NaCl or B-1, and A-15 respectively).

These types form separate islands on Villars' quantum structure diagrams (Appendix C). These islands contain other structure types with $T_c > 10$ K, but for simplicity the discussion here is confined to these two types including their pseudobinary alloys.

The first superconductor with a transition temperature T_c in the pumped liquid hydrogen region[8] was NbN ($T_c = 15$ K), as was found in 1941, in the NaCl structure.

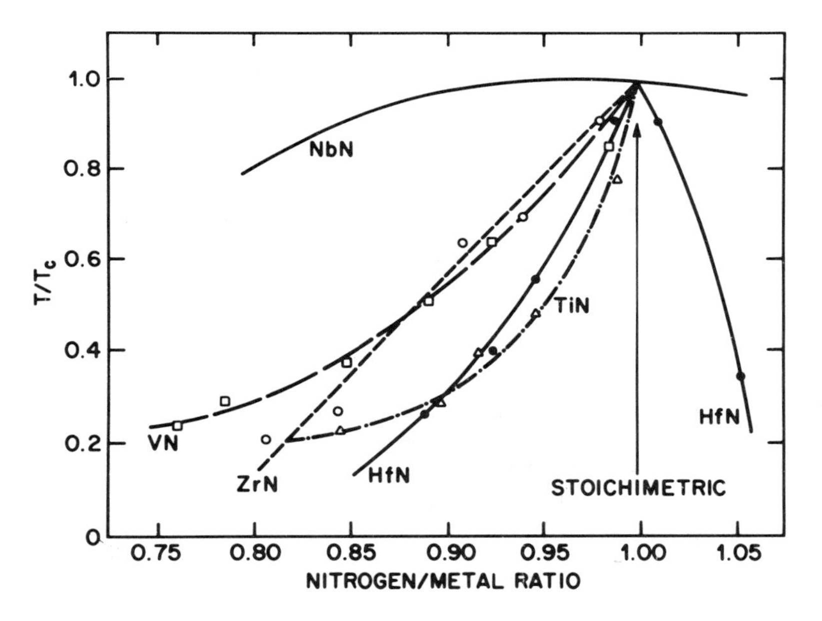

Fig. I.5. Effect of deviation from stoichiometry on T_c, various nitrides. Notice that NbN, with the largest T_c, is the least effected by vacancies or antisite defects. In examining such figures, it is important to realize that the ideal 1:1 stoichiometry, even if it is attained, does not necessarily mean a perfect lattice, as there may be compensating vacancy and antisite defect concentrations (from ref. 8).

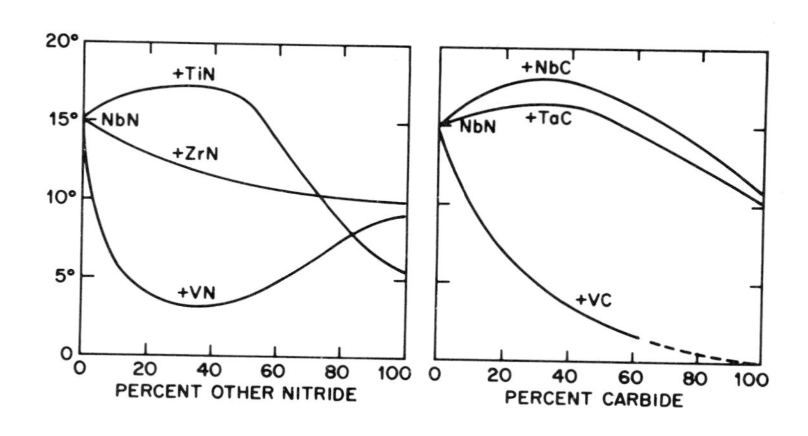

Fig. I.6. Effect on T_c (NbN) of replacing Nb or N by Ti, Zr, or V, or C, TaC and VC (ref. 7).

Superconducting borides, carbides, oxides and phosphides also occur in this structure.

Lattice instabilities in this family manifest themselves in many properties which make these materials difficult to handle experimentally, such as brittleness, high melting temperatures T_m, and a tendency to decompose with anion vacancy formation and reduction of T_c. These compounds can be stable in the NaCl structure with anion vacancy concentrations as high as 20% or 30%, but these vacancies rapidly reduce T_c in general. The situation for the nitrides is shown in Fig. 5, where it is seen that except for NbN_{1-x}, increasing $|x|$ rapidly rapidly decreases T_c. Note that this Figure shows T_c/T_{co} and that NbN also has the highest T_c in this family (group B in Appendix C).

The special properties of NbN led to many alloying studies in which either Nb or N were replaced by nearby transition metals, or C, or both.[7] The results are shown in

Fig. 6, which indicates that $NbN_{0.7}C_{0.3}$ achieves a T_c of 18 K. Such pseudobinary solid solutions were found to fall into two distinct classes, I and II. Class I included TC and TN, with T = Nb, Ta, T_i, Zr, and Hf, which class II contained T = V. With intraclass mixing T_c increases or decreases slowly, while with interclass mixing T_c decreases rapidly.

According to a chemical analysis[9] of heats of formation, only V among these T elements has or is expected to have a smaller heat of formation ΔH_f than T = Nb. This leads us immediately to a simple explanation of the two-class regularity which puzzled Hulm and Blaugher,[7] namely,

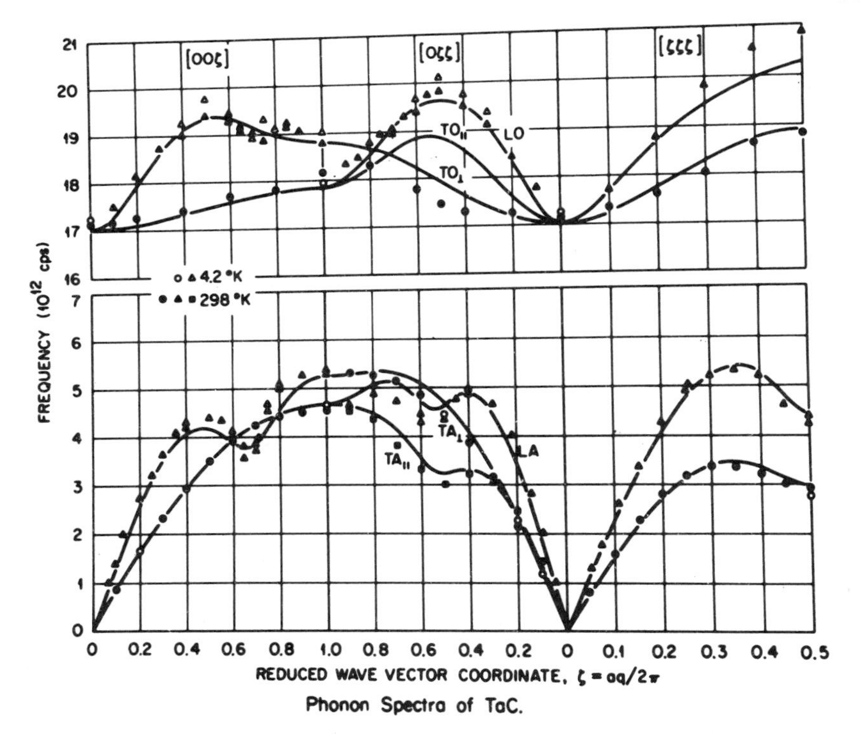

Fig. I.7. Lattice softening in TaC, $T_c = 10$ K. Anomalies near **k** = (0,0,0.7), (0,0.5,0.5) and (0.5,0.5,0.5) in the acoustic branch are identified by comparison with ZrC($T < 1$ K), whose lattice vibration dispersion curves are shown in refs. 4 and 6.

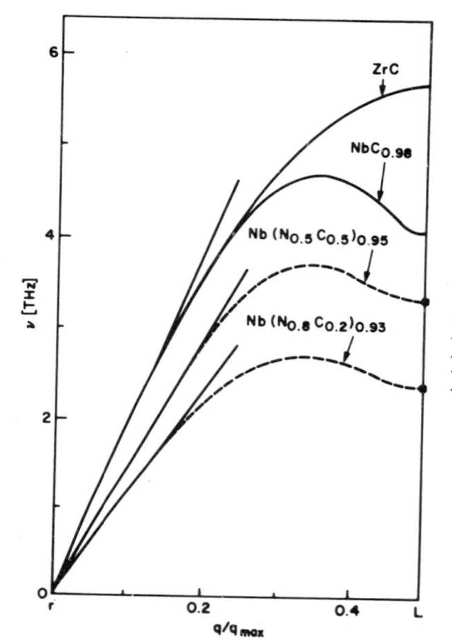

Fig. I.8. Dispersion of TA phonons in the ($\zeta\zeta\zeta$) direction, as derived in Ref. 9, for alloys of $NbC_{0.98}$ with $NbN_{0.91}$.

qualitatively T_c increases as $|\Delta H_f|$ decreases and the lattice less stable. This explains the general trend in Class I up to T = Nb. However, in class II, for T = V, we have $|\Delta H_f|$ so small that large anion vacancy concentrations are formed in inter-class alloys and T_c decreases drastically. This point is discussed further in Chapter II.

Neutron studies of $\omega(\mathbf{k})$ and soft phonons in TC and TN compounds are greatly hampered by the difficulties associated with preparing single crystals large enough to scatter the beam effectively. So far this difficulty has been surmounted only for NbC and TaC (T_c = 10 K), which shows[6] characteristic softenings, for example, near

k = (0,0,0.7), (0, 0.5, 0.5) and (0.5, 0.5, 0.5) which are absent from the vibrational spectra of ZrC ($T_c < 1$ K). The anomalies in TaC are shown in Fig. 7. The results[4] for NbC ($T_c = 11$ K) are similar to those for TaC.

Because large single crystals have not been grown for NbN ($T_c = 15$ K), $\omega(\mathbf{k})$ has not been measured directly. On $NbN_{1-x}C_x$ powders the phonon density of states and the electronic and lattice specific heat have been measured, however and these show[10] even larger lattice softenings than are found for NbC and TaC. This is shown in Fig. 8, where the dispersion of the transverse acoustic mode (as inferred from specific heat data) is shown. At the zone boundary the TA frequency drops by about a factor of two from $NbC_{0.98}$ ($T_c = 11$ K) to $Nb(C_{0.2}N_{0.8})_{0.95}$ ($T_c = 17.5$ K). This again shows the strong correlation between T_c and lattice softening.

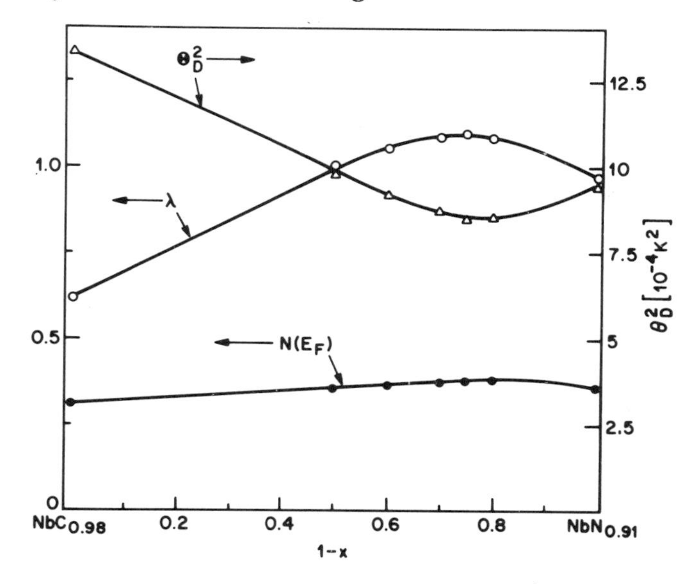

Fig. I.9. Variations of θ_D and microscopic electronic parameters λ and $N(E_F)$ in $NbC_{0.98}$ - $NbN_{0.91}$ alloys (ref. 10).

For those interested in electronic properties the effects on θ_D, $N(E_F)$ and λ, quantities which enter Eqn. (2.1), are shown in Fig. 9. Note that $N(E_F)$ varies slowly with composition as $NbC_{0.98}$ is alloyed with $NbN_{0.91}$. According to Fig. 6, T_c reaches its maximum value just when θ_D in Fig. 9 reaches its minimum value! This is because the minimum in θ_D corresponds to a maximum in the average electron-phonon interaction $\overline{V}$, and this maximum in $\overline{V}$ produces a maximum in λ, as shown in Fig. 9. The variation of λ is much more important than the variation of θ_D, because of the way λ enters the exponential in Eqn. (2.1).

Among the intermetallic superconductors the most favorable group (denoted by group A in Appendix C) is that based on A_3B compounds in the cubic A-15 structure. This structure contains chains of closely spaced A atoms running parallel to the cubic axes, with the cavities between the chains occupied by B atoms. Six binary compounds in this structure have $T_c > 17$ K, with the favorites being V_3Si (discovered in 1953) and Nb_3Sn (discovered in 1954).[8] The high T_c's in all these materials are greatly depressed by small deviations from stoichiometry, unlike NbN in Fig. 5. As a result efforts to raise T_c in this family were frustrated for many years. The highest T_c's near 23 K were obtained for Nb_3Ge films stabilized by traces of O impurities.[11] Quite recently rapidly quenched and oxidized Nb-Ge-Al films were reported[11] to become superconductive with an onset temperature $T_{co} \sim 44$K (broad transition) or 30K (narrow transition, film composition close to Nb_3Ge). The data resemble in many respects much of the data reported for the Cu oxide high-T_c superconductors.

One of the unusual features of A_3B or A-15 compounds that distinguishes them from the AB or B-1 compounds is that in the B-1 case mode softening occurs

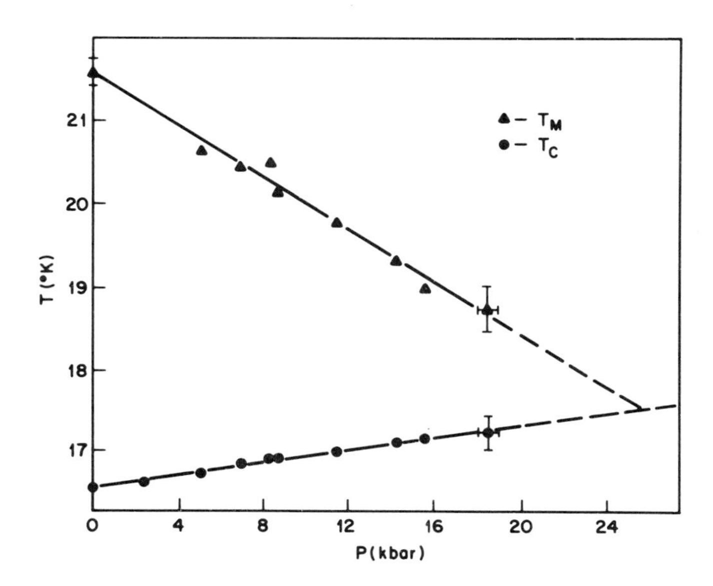

Fig. I.10. Pressure dependence of the Martensitic lattice transformation temperature T_M and T_c of a V_cSi single crystal (ref. 12).

at short wave lengths for both acoustic and optic modes, but in the A-15 case the optic mode softening (which is very large) occurs also at $\mathbf{k} = 0$. These optic modes couple to the (110) transverse acoustic modes and produce a large softening of the (110) transverse sound velocity which in turn can generate a cubic-to-tetragonal phase transition. This phase transition occurs at low temperatures and hence is called a Martensitic transition. The better the sample, the more likely it is to undergo the transition, which is favorable for T_c. However, T_c decreases rapidly in the tetragonal phase. Thus ideally one prefers samples in which the Martensitic transition temperature T_m is equal to the superconducting transition temperature T_c, so that one is just on the knife edge of an unstable lattice which has not actually gone unstable. This point is illustrated in Fig. 10 for V_3Si as a function of pressure.[12] An increase in T_c is observed up to P = 20 kbar, with the projected coincidence between T_m and T_c occurring at 24 kbar.

The close connection between the Martensitic transformation and optic mode softening was studied by Raman scattering.[13] The lattice instability is demonstrated by the width of the Raman band, which increases until the cubic-to-tetragonal phase transformation takes place. The observed 30% depression of the optic mode frequency corresponds to a large reduction of the (110) TA sound velocity which has been measured directly and by neutron scattering.[14]

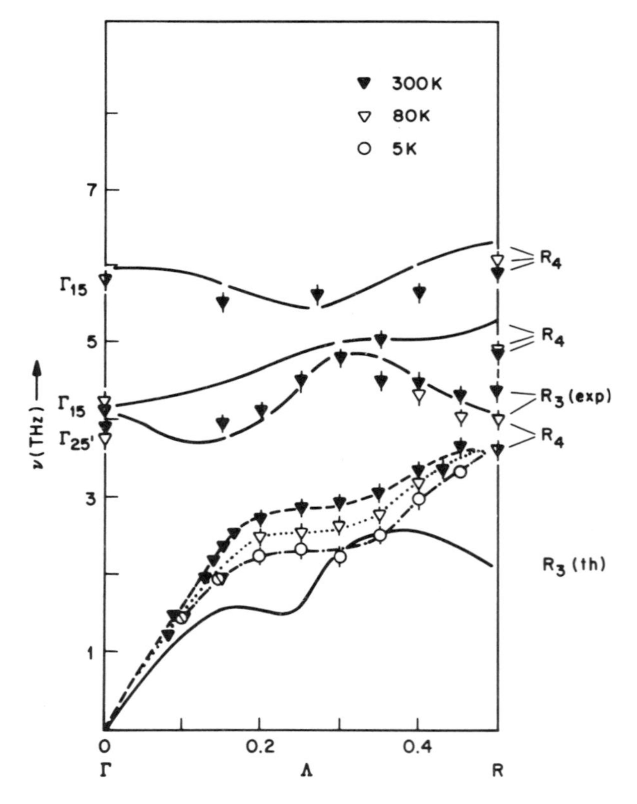

Fig. I.11. Phonon dispersion curves for Nb_3Sn for the $\Lambda = (\varsigma, \varsigma, \varsigma)$ direction for different temperatures compared to Weber's theory (solid curves) (ref. 15).

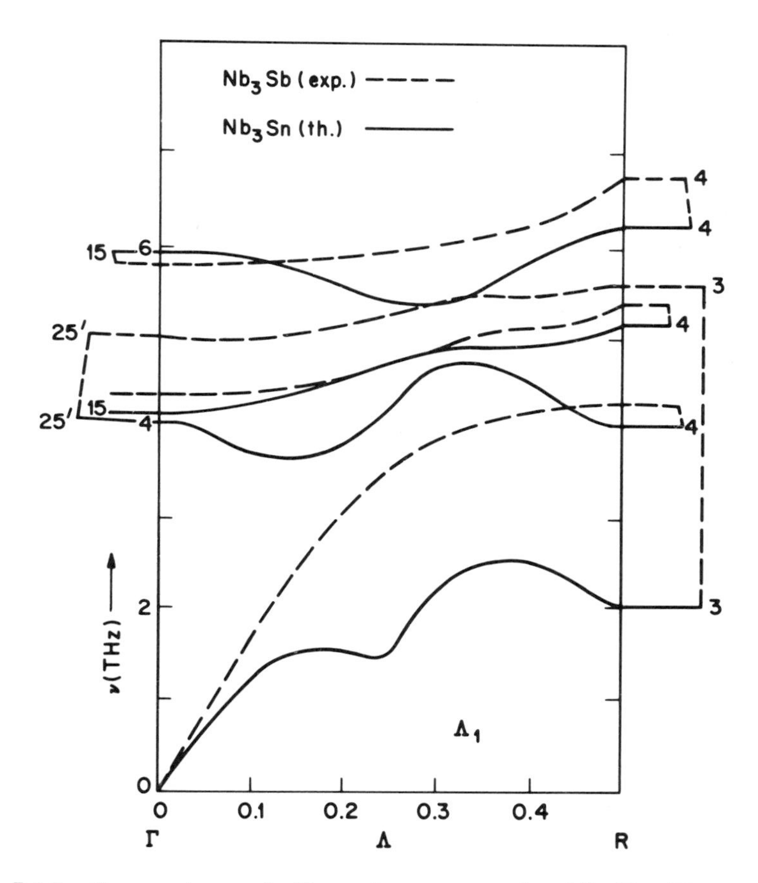

Fig. I.12. Comparison of dispersion curves for Nb_3Sb ($T_c < 1$ K) and Nb_3Sn ($T_c = 18$ K). Note the depression of $\omega(R_3)$ by almost a factor of 3 from Nb_3Sb to Nb_3Sn (ref. 16).

Once again direct measurement of $\omega(\mathbf{k})$ by neutron scattering[15] was hampered by the availability of only five single crystals of Nb_3Sn with a total volume less than 0.05 cm^3. After extensive normal-mode analysis the data shown in Fig. 11 for the $(\varsigma, \varsigma, \varsigma)$ direction were obtained.[15] The instability near $\varsigma = 0.3$ agrees qualitatively with the theoretical softening near $\varsigma = 0.2$. The theory will be discussed in more detail in Chapter II. However, the shift and broadening of experiment relative to theory observed in

Fig. 11 could well be the result of tetragonal domain formation, i.e., atomic relaxation associated with precursors to the cubic-to-tetragonal transition which is observed in V_3Si.

To appreciate the lattice instability in $Nb_3Sn(T_c = 18$ K) shown in Fig. 11, it is useful to have a control, i.e., to compare the dispersion curves of Nb_3Sn with those of $Nb_3Sb(T_c < 1$ K). This is done[16] in Fig. 12. If we compare Figs. 11 and 12, we see substantial softening of the optic and acoustic modes in Nb_3Sn compared to Nb_3Sb in selected regions. Note also that in Nb_3Sn the transverse acoustic mode softening is strongly temperature-dependent and is associated probably with the incipient Martensitic phase transition.

The examples of $\omega(\mathbf{k})$ which we have discussed illustrate the basic point that high T_c materials show pronounced lattice softening, but only in relatively small regions of $\mathbf{k}$ space and for certain specific modes. With polycrystalline or powder samples only the density of phonon states $N(\omega) = F(\omega)$ is accessible by inelastic neutron scattering (strictly speaking one measures $N(\omega)$ weighted by neutron scattering strengths for each mode, but the weighting is usually a slowly varying function of ω). Representative results[17] for $G(\omega)$, the weighted $N(\omega)$, for five A_3B compounds in the A-15 structure are presented in Fig. 13. The phonon energies have been scaled to make the transverse acoustic peak frequencies all nearly equal. For V_3Ga, V_3Ge and Nb_3Sn the soft modes produce a peak near 15 meV which is quite obvious and this is the spectral analogue of the mode softening seen earlier in $\omega(\mathbf{k})$.

It would seem from Fig. 13 that one cannot observe mode softening in the vibrational spectra of V_3Si and Nb_3Al. As is so often the case in searching for a subtle effect, its absence in some cases merely reflects the limitations of one particular technique. For polycrystalline

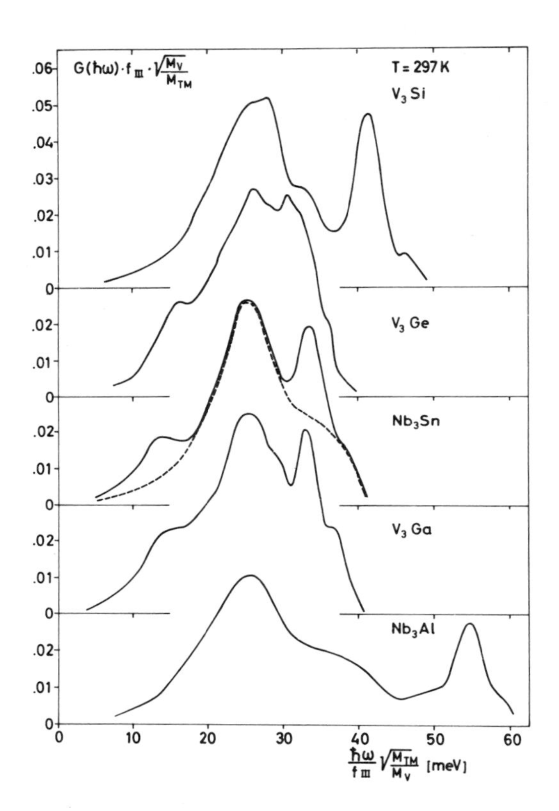

Fig. I.13. Data from ref. 17 showing $G(\omega)$ as measured by inelastic neutron scattering for A_3B (T_c, K) compounds: V_3Si (17), V_3Ge (11), Nb_3Sn (18), V_3Ga (17) and Nb_3Al (14).

samples tunneling is a more sensitive technique than neutron scattering, because tunneling measures $\alpha^2 N(\omega)$, where $\alpha(\omega)$ is the average electron-phonon coupling strength at each frequency ω. Moreover $\lambda = N(E_F)V$ is related directly to the tunneling characteristic by the sum rule,

$$\lambda = 2\int \alpha^2(\omega)F(\omega)\omega^{-1}d\omega \tag{2.2}$$

so that soft phonon modes with very strong electron-phonon coupling are more likely to show up in the tunneling spectra.

In practice tunneling experiments, especially on high-T_c superconductors which are intrinsically unstable, encounter serious sample problems at the tunnel junction both in terms of deterioration of the superconductor close to the surface and in terms of forming a uniform, thin tunneling barrier. However, since 1980 these problems have been overcome in several beautiful experiments on both A-15 materials and the B-1 material NbN. The dramatic sensitivity of $\alpha^2F(\omega)$ to increases in T_c and λ in Nb_3Al, which could not be seen in $G(\omega)$ in Fig. 13, *are* seen[18] in Fig. 14 with tunneling. Incidentally, this figure illustrates

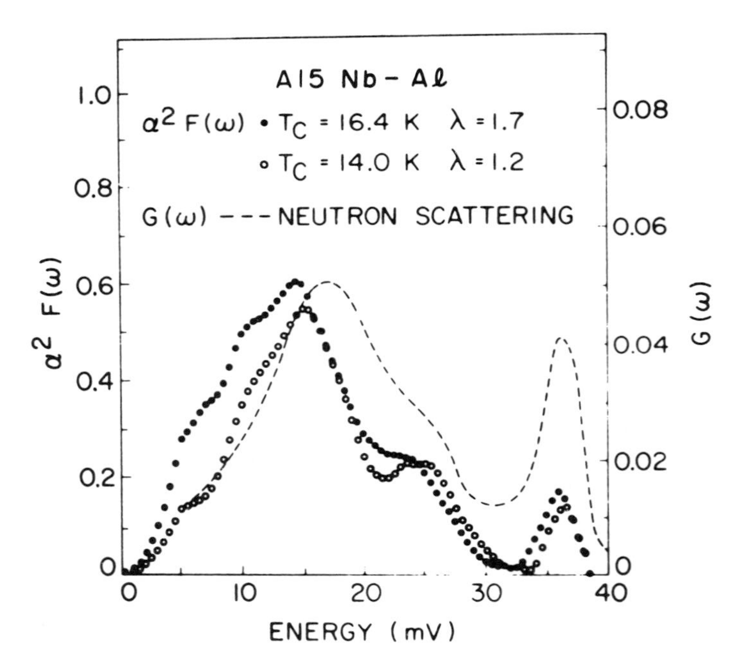

Fig. I.14. The electron-phonon spectral functions α^2 $F(\omega)$ for two Nb-Al junctions with E_g/kT_c of 3.6 and 4.4, are compared with $G(\omega)$ from Fig. 13 (ref. 18).

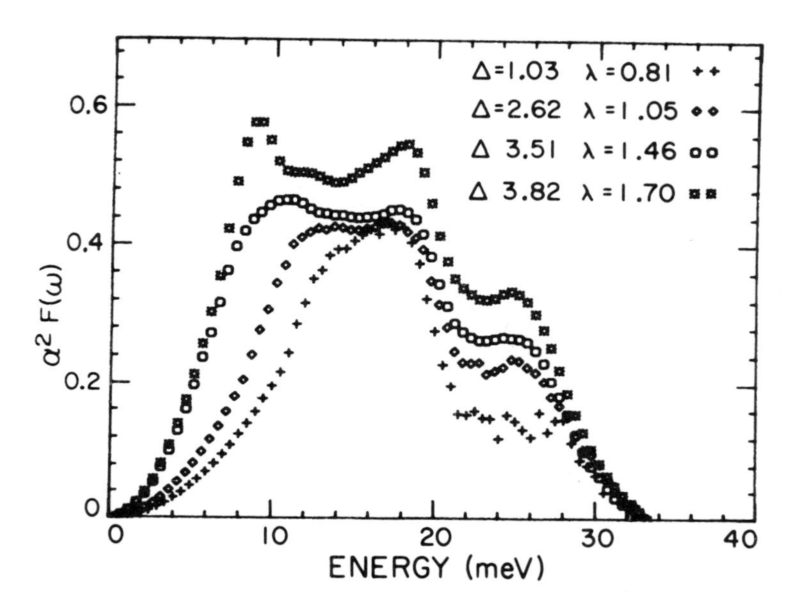

Fig. I.15. Similar to Fig. 14 for four Nb-Ge samples with T_c's of 7.0, 16.8, 20.1 and 21.2 K, repsectively. Note the strengthening and movement to lower energies of the phonon peak near 8-14 meV (ref. 19).

an interesting bonus which can be achieved by tunneling. The samples are evaporated thin films, and two spectra are shown. The one with weaker coupling has $T_c = 14$ K, as in bulk Nb_3Al (Fig. 13), while the second film has $T_c = 16.4$ K. It is often possible with soft materials to achieve higher T_c's with thin evaporated films than with bulk samples, because (for example) the thin film may be quenched by deposition into a more metastable configuration than the bulk grains.

Tunneling on "Nb_3Ge" samples with a wide range of T_c's is shown[19] in Fig. 15. The enhancement of T_c with the growth of the soft mode near 8 meV is quite striking. Finally, data[19] on NbN reveal that the acoustic modes are exceptionally soft in this material, as can be seen in Fig. 16.

This is partly due to the large difference in mass between Nb and N, and a true soft mode is not so evident in this spectrum as in the A-15 spectra just examined. In fact the interesting feature[19] of this spectrum is the large value of α for the high energy peak. This peak makes a significant contribution to λ in spite of the factor ω^{-1} in (2.2), and it shows how strong the electron-phonon coupling can be for first-period elements like N and O. These data on $\alpha^2F(\omega)$ should be compared to the actual acoustic mode dispersion curves[20] $\omega(\mathbf{k})$ for $NbN_{0.84}$ which are shown in Fig. 17. Note that a drastic softening occurs for antiphase vibrations of (100) lattice planes, $\mathbf{k} = \mathbf{X} = (100)$.

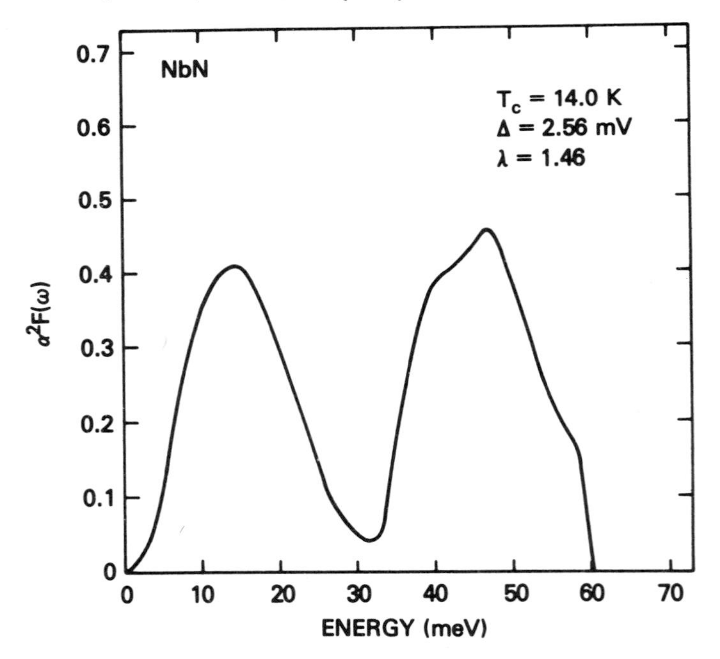

Fig. I.16. As in Figs. 14-15, for NbN with $T_c = 14$ K. This is *less* than the bulk value (17 K) and this is probably the reason a soft mode has yet to be resolved (ref. 19).

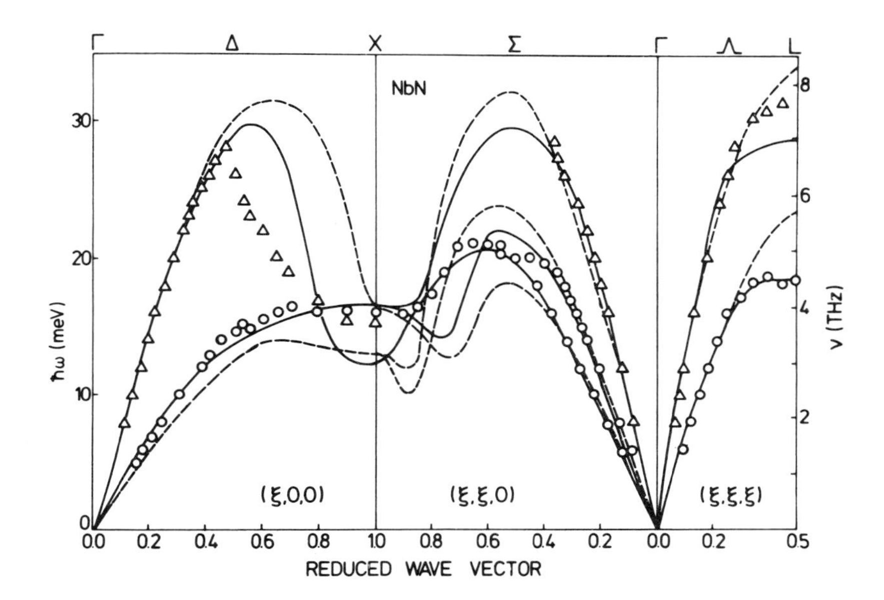

Fig. I.17 Acoustic phonon dispersion curves on NbN. Experimental values for $NbN_{0.84}$ are compared to two theoretical models, dashed lines for NbN and solid lines for actual stoichiometry (rigid band model) (ref. 20).

There is an important effect which has just barely been observed[21] by neutron scattering in $G(\omega)$ and by Raman scattering on high T_c (19.4 K, compare Fig. 14) Nb_3Al. This is mode softening: as T decreases from 300K to 10K, a soft mode at 14 meV broadens and shifts to 12 meV. Effects of this kind are expected to become more pronounced as T_c increases. (These comments were written before the phonon anomaly discussed in Fig. VI-12 was observed.)

REFERENCES

1. P. Villars, J. C. Phillips and H. S. Chen, Phys. Rev. Lett. *57*, 3085 (1986).
2. S. C. Ng and B. N. Brockhouse, Sol. State Comm. *5*, 79 (1967).
3. T. Claeson, Phys. Rev. *147*, 340 (1966).
4. H. G. Smith, Austral. J. Phys. *33*, 861 (1980).
5. R. M. Powell, P. Martel and A. B. Woods, Phys. Rev. *171*, 727 (1968).
6. H. G. Smith, N. Wakabayashi, and M. Mostoller, in Douglass II, p. 223. This is the abbreviation that will be used in future references to *Superconductivity in d- and f- Band Metals* (Ed. D. H. Douglass, Plenum, New York, 1976).
7. L. R. Testardi, Rev. Mod. Phys. *47*, 637 (1975).
8. J. K. Hulm and R. D. Blaugher, in Douglass I, p. 1. This is the abbreviation that will be used in future references to *Superconductivity in d- and f- Band Metals* (AIP Conf. Proc. *4*, Ed. D. H. Douglass, New York, 1972).
9. J. C. Phillips in Douglass I, p. 339.
10. P. Roedhammer, E. Gmelin, W. Weber and J. P. Remeika, Phys. Rev. *B15*, 711 (1977).
11. J. R. Gavaler in Douglass II, p. 421; T. Ogushi and Y. Osono, Appl. Phys. Lett. *48*, 1167 (1986).
12. C. W. Chu and L. R. Testardi, Phys. Rev. Lett. *32*, 766 (1974).
13. S. Schichtanz, R. Karser, E. Schneider and W. Gläser, Phys. Rev. *B22*, 2386 (1980).
14. J. D. Axe and G. Shirane, Phys. Rev. *B8*, 1965 (1973).
15. L. Pintschovius, H. Takei and N. Toyota, Phys. Rev. Lett. *54*, 1260 (1985).
16. W. Weber, Physica *126B*, 217 (1984).
17. B. P. Schweiss, B. Renker, E. Schneider and W. Reichardt, in Douglass II, p. 189.

18. J. Kwo and T. H. Geballe, Phys. Rev. *B23,* 3230 (1981).

19. K. E. Kihlstrom, R. W. Simon and S. A. Wolf, Physica *135B,* 198 (1985).

20. W. Weber, in ***Superconductivity in d- and f- Band Metals*** (Ed. H. Suhl and M. B. Maple, Academic Press, N.Y., 1980, hereafter referred to as Douglass III), p. 131.

21. P. Müller, R. Hackl, R. Kaiser, N. Nücker and A. Müller, Physica *135B,* 355 (1985).

II. Old Theory

1. Electron-Ion and Electron-Phonon Interactions in Simple Metals

Bardeen carried out the first rigorous calculation of electron-ion interactions in alkali metals.[1] He assumed that when an ion was displaced, the ion core and its potential moved rigidly and he calculated the screening of the displaced potential by the conduction electrons self-consistently. For the alkali metals this calculation is easy because the Fermi surface lies entirely inside the first Brillouin zone and the conduction electron spectrum is that of a nearly free electron gas. The interaction of the electrons with the ion cores is also especially simple for ion cores with positive charge $Z = 1$. Subsequently it was realized that the Bardeen method could easily be generalized to most s-p metals (such as Al, $Z = 3$) by utilizing for the self-consistent screening the well-studied dielectric function of a free electron gas. The interaction with the ion-core is then treated by the pseudopotential method,[2] which provides accurate and reproducible results without the use of adjustable parameters. The modern theory[3] using this approach is in good agreement with experiment for most non-transition metals. More recent work on non-transition metals has utilized improved pseudopotential form factors to achieve virtually exact agreement[4] with experimental values of λ and T_c in metals such as Al and this theory has accurately predicted[5] both new crystal structures and T_c in these structures for high-pressure phases of materials such as Si.

The evolution of the calculation of electron-phonon interactions from Bardeen[1] to Cohen et al.[5] represents probably the greatest accomplishment of modern computational physics. However, the technical details of either calculations or experiments are not the primary concern of this book. Our interest lies in using the results of these calculations on simple systems to develop insight into electron-phonon interactions which can be applied to the very complex materials which are high-T_c superconductors. The following discussion has been tailored to this purpose.

To obtain the electron-phonon coupling from self-consistent electron-ion interactions one must know the normal modes of vibration of the lattice. For phonons in crystals with simple structures propagating in directions of high symmetry these normal modes are determined by symmetry alone, but in general we confront *two* formidable problems, the electronic problem and the lattice vibrational problem. In principle very accurate self-consistent calculations of the total energy of the lattice in equilibrium and subjected to small displacements solve both the electronic and the lattice problem , but in practice the number of such small displacements needed increases like N^2, where N is the number of atoms in the unit cell. In such calculations based on a large number of energy differences, rounding errors rapidly become overwhelming. For the calculations to remain under control one needs internal checks on both the electronic and the lattice dynamical problems which are provided by other experiments. With this in mind we now examine the general formalism as embodied in current calculations of the electron-phonon interaction.

We assume that the material is a perfect crystal, and that all the one-electron Bloch functions $\psi_{n\mathbf{k}}(\mathbf{r})$ are known and are represented by the abbreviated symbol $|\mathbf{k}>$. We imagine displacing each ion at $\mathbf{r} = \mathbf{R}_l$ by an amount $\delta\mathbf{R}_l^\alpha$ for each phonon mode α, where the phonon modes α are also assumed to be known. The electron-phonon matrix element is

$$I_\alpha(\mathbf{k},\mathbf{k}') = \sum_l \epsilon_l^\alpha \cdot < \mathbf{k}' \,|\, \nabla V(\mathbf{r}-\mathbf{R}_l) \,|\, \mathbf{k} > \tag{1.1}$$

where $V(\mathbf{r}-\mathbf{R}_l)$ is the self-consistent electron-ion potential and $\epsilon_l^\alpha = \delta\mathbf{R}_l^\alpha / |\delta\mathbf{R}_l^\alpha|$. Next we determine the Fermi surface S in **k**-space for which $E_{n\mathbf{k}} = E_F$ and with it the density of states $N(E_F)$ given by

$$N(E_F) = [2/(2\pi)^3] \int \frac{dS_{n\mathbf{k}}}{|\nabla_\mathbf{k} E_{n\mathbf{k}}|} \tag{1.2}$$

The phonon vibrational frequencies are denoted by $\omega_\alpha(\mathbf{q})$ where $\mathbf{q} = \mathbf{k}' - \mathbf{k}$. The phonon amplitudes $\delta\mathbf{R}_l^\alpha$ are properly weighted in terms of the electron-phonon coupling g defined by

$$g_\alpha(\mathbf{k},\mathbf{k}') = (\hbar/2M_\alpha\omega_\alpha(\mathbf{q}))^{1/2} I_\alpha(\mathbf{k},\mathbf{k}') \tag{1.3}$$

where M_α is an effective mass for mode α. (For diatomic crystals with $M_1 << M_2$, M_α is nearly equal to the total/reduced mass for the acoustic/optic modes.) The tunneling definition (I.(2.2)) of λ becomes

$$\lambda = \sum_\alpha N(E_F) < I_\alpha^2 >_{\mathbf{k},\mathbf{k}'} / M_\alpha [\omega^2] \tag{1.4}$$

where $[\omega^2]$ is defined by

$$[\omega^2] = \int d\omega\, \omega \alpha^2(\omega) F(\omega) / \int d\omega\, \omega^{-1} \alpha^2(\omega) F(\omega) \tag{1.5}$$

and where $<A>_{\mathbf{k},\mathbf{k}'} =$

$$\int dS_{\mathbf{k},\mathbf{k}'} |\nabla_{\mathbf{k}} E_{\mathbf{k}}|^{-1} |\nabla_{\mathbf{k}'} E_{\mathbf{k}'}|^{-1} A(\mathbf{k},\mathbf{k}') / \int dS_{\mathbf{k},\mathbf{k}'} |\nabla_{\mathbf{k}} E_{\mathbf{k}}|^{-1} \tag{1.6}$$

All the steps leading to (1.4) are straightforward in principle, but in practice many difficulties are concealed. Measurements are seldom made on ideal crystals, and because high-T_c superconductors are generally unstable, the materials often contain defects such as vacancies. If the defects are randomly distributed, then the various averages leading to (1.3), such as the definitions of the Bloch functions, may cause only small errors, providing local modes (not of the Bloch type) are not important either for the electronic states or the vibrational states. However, we saw in Fig. I.4 that local modes occur already even in the "simple" case of substitutional bcc $Mo_{1-x}Re_x$ alloys, so in analyzing theoretical results based on elegant formulae such as (1.3) these limitations should always be kept in mind.

Apart from material idealizations which enter (1.3), there are questions of computational precision. One test of the accuracy of λ as calculated for determining T_c is provided by the relationship between the effective mass m^* which enters the electronic specific heat through $N(E_F)$, as given by (1.2). In an isotropic one-band model.[3]

$$m^* = m_b(1 + \lambda_{ep} + \lambda_{ee}) \tag{1.7}$$

where λ_{ep}, the electron-phonon contribution, corresponds to λ in (1.3). When we compare λ_{ep} from (1.5) with λ in (1.3), we usually make two approximations. First, we neglect λ_{ee} compared to λ_{ep}. The electron-electron contribution λ_{ee} is known to be small in nearly free electron metals such as the alkalis and the alkaline earths, but in more covalent or more ionic materials this need not be the case. The second approximation is the isotropic one, which means that there is no difference between a single angular average, Eqn. (1.2), and a double average, Eqn. (1.4). Again this is a good approximation only for nearly free electron metals, which always have low T_c's. With these approximations good agreement between λ (as inferred empirically from T_c, or calculated from (1.3)) and λ_{ep} was obtained[3] for Be, Mg, Zn and Cd, the cases where λ is large enough ($\lambda \sim 0.3-0.4$) to be significant, yet the nearly free electron approximation is still valid.

Allen and Cohen made a careful, complete study of λ in the isotropic approximation in 20 s-p metals ranging from Na ($\lambda = 0.15$ $\mu = 0.16$, $T_c = 0$) to Pb ($\lambda = 1.3$, $\mu = 0.12$, $T_c = 7.6°$ K). They analyzed their results in terms of several obvious chemical trends originally noted by Matthias: T_c increases with ion core charge Z (larger rigid ion potential) and decreasing conduction electron charge density n (poorer self-consistent screening of the ion core by the conduction electrons). These trends can be derived from simple models of the Thomas-Fermi type, but quantitatively the simple models fail to predict λ accurately because the details of the ion-core pseudopotential form factors *are* important. For instance, for larger λ the T-F model gives λ scaling with $n^{-1/3}Z^{2/3}$ but the actual scaling is closer to $n^{-1/3}Z^{5/3}$.

The failures of the isotropic approximation are most serious for Ga (T_c(pred.) = 0, T_c(exp.) = 1.1° K) and Sn (T_c(pred.) < 8.2° K, T_c(exp.) = 3.7° K). From the point of view of high T_c's, the failure for Sn is interesting. White Sn has a complex crystal structure which is partially covalent (like Ge) and partially metallic (like Pb). The isotropic approximation treats Sn as fully metallic, so it is not surprising that the theory overestimates T_c. Incipient formation of covalent bonds reduces the density of states factor $N(E_F)$ in (1.4) and this effect is not included in the pseudopotential theory, which treats electron-ion scattering only to second order. Covalent-metallic and ionic-metallic mixed bonding are important in high-T_c superconductors, as we shall see later.

The isotropic approximation goes hand-in-hand with perturbation theory for the electron wave functions. With these approximations it is difficult to treat transverse vibrations accurately and hence to evaluate the relative weights of longitudinal and transverse phonon-electron interactions. These approximations can be avoided by carrying out completely self-consistent calculations in equilibrium and for polarized "frozen-in" phonon modes of small amplitude and short wave-length. So far the most careful calculation of this kind[4] has been carried out for Al (fcc crystal structure, nearly free electron energy bands, $T_c = 1.2° K$). In spite of the apparent simplicity of this metal as regards crystal and electronic structure, substantial variations of λ between longitudinal and transverse modes are found (typically a factor of two or three). Calculations of this kind are needed to treat covalent-metallic[5] and ionic-metallic binding. Simplified calculations of this type by Weber for transition metal compounds are discussed in II.3.

2. Simple Models

We saw in I.2 that strong electron-phonon interactions and high T_c's are associated with lattice instabilities, and that these instabilities are generally observable with sophisticated experimental techniques such as neutron scattering and superconductive tunneling. However, the instabilities usually involve only a small fraction of the phonons. This is not surprising, since the superconductive condensation energy is at most of order $(kT_c)^2/W$, where W is a band width of order eV, and this is quite small compared to kT_m, the order of the change in the vibrational energy of the lattice when all of the phonons become unstable and the lattice melts at $T = T_m$. Indeed we would not expect *a priori* to be able to find any simple models of lattice instabilities in high-T_c superconductors - or so the pessimist would reason.

The optimist would proceed somewhat differently. He would argue that the reasoning just given, while appropriate to most solids, might well fail for just those few materials with high T_c's, and he would look for simple models in just those cases. In fact two such simple models are available, one for the A-15 group and one for the B-1 group of T_3M and TM compounds, respectively, where T is a transition metal and M is a metalloid. We discuss these simple models here for two reasons: they are much easier to follow and to check than the more "rigorous" models, and they give us hope that such simple models may exist for the new ceramic superconductors discussed in later chapters. We will also see in the next section that for "merely" technical reasons, calculations of lattice instabilities and electron-phonon interactions are much more difficult for compounds containing transition metals than for simple metals.

The simplest group of high-T_c transition metal compound superconductors is the TM group with the cubic NaCl structure, where M = C, N or O and T is one of the four transition metals from the 3d and 4d series and Ti and V columns. This group is part of a larger group (denoted by B in Appendix C) which includes compounds in hexagonal and other structures, but with a simple model one discusses only the twelve compounds from the 4×3 TM cubic NaCl tableau. Only a few of these compounds have been analyzed by the more elaborate theoretical methods discussed in the next section, and as we saw in I.2, it has so far not proved possible to grow single crystals of most of these compounds, which usually have large anion vacancy concentrations even in the powder form.

One way of understanding the high-T_c vibrational anomalies observed in neutron or tunneling experiments is to assume that in the ideal structure near the normal equilibrium separation some transverse acoustic (TA) modes would have $\omega_{TA}^2(\mathbf{q}) < 0$ for certain wave vectors $\mathbf{q}$. This can occur because the repulsive ion-ion interactions are overscreened by strong electron-ion interactions. The dynamical instability associated with such overscreening can be removed in several ways depending on the crystal structure. For the A_3B compounds crystallizing in the A-15 structure which contains A chains, the A atoms can move off-center (random buckling), which has the effect of mixing the LA and TA modes and stabilizing the latter. For the TM compounds with actual composition TM_{1-x} and $x > 0$, atoms can relax near vacancies. However, another kind of relaxation is possible, that of the cubic lattice constant a. Both relaxations can be described as anharmonic, but relaxation of a can be inferred by studying a(n) in a series TM^n, where $M^4 = C$, $M^5 = N$ and $M^6 = O$. The advantage of this approach is that a(n) can be measured very accurately.[6]

The results obtained by plotting $T_c(TM^n)$ and a (TM^n) for T= Zr, Nb, Ti and V are shown in Fig. 1. One expects, in the absence of lattice instabilities, that a(n) will be a linear function of n (this is called Vegard's law), and this is approximately correct for T = Ti and V, where the T_c is lower. However, for T = Nb and Zr, where T_c is much higher, a(n) is strongly bowed. The magnitude of the bowing, denoted by $|a_2|$, is largest for T = Nb, and this series also has the highest T_c's; it is second largest for T = Zr, which has the second highest T_c's.

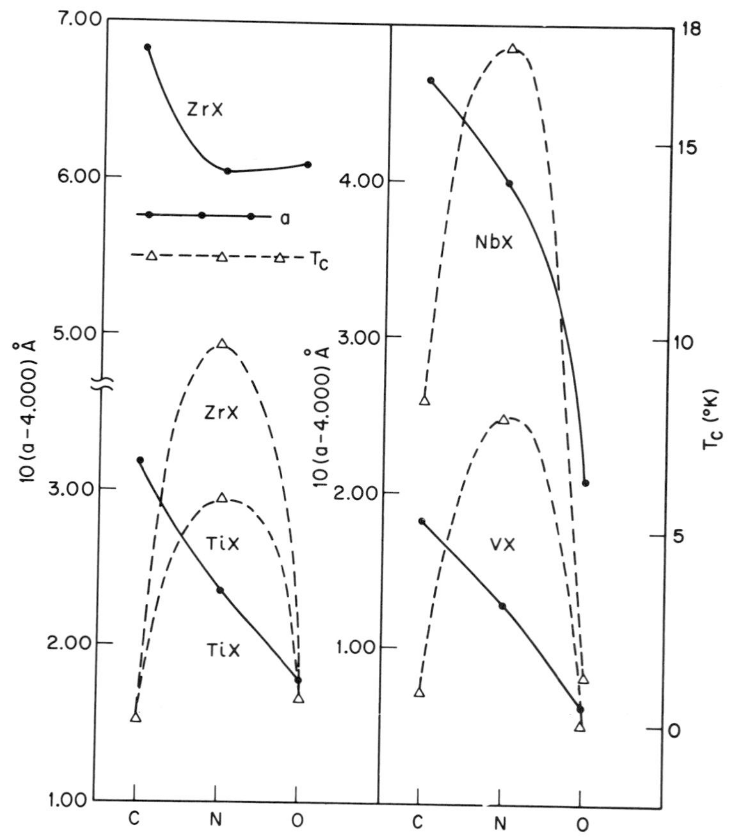

Fig. II.1. Lattice constants a(n) and $T_c(n)$ for TM^n compounds for $M^{(4-6)}$ = C, N and O. Note that the magnitude of the bowing of a(n) correlates well with maxima in T_c (ref. 6).

What about the sign of a_2? This correlates very well with the sign of dT_c/dP, and the origin of this correlation is easily understood.[7] In units of $10^{-2\circ}$K/kbar, dT_c/dP in NbN_{1-x} and ZrN are +0.4 and −1.7, respectively, and the NbN value might easily increase as $x \rightarrow 0$. From Fig. 1 we see that application of pressure should reduce the bowing in NbM^n, thereby making the material more unstable and increasing T_c, while the opposite is expected for ZrM^n, in agreement with experiment. It is important to realize that the δa bowing effects shown in Fig. 1 are of order $\delta a/a \sim 10^{-2}$, which is an accuracy that has barely been achieved in quantum-mechanical calculations of lattice constants for simple metals, and that such accuracies have yet to be reported in calculations of lattice constants of transition metal compounds.

Another simple model is the "peaked N(E)" electronic model for the A_3B compounds in the A-15 structure shown in Fig. 2 introduced by Friedel et al.[8] Note here the A atom chains, which in this model are supposed to be electronically decoupled from each other and from the B atoms. Such

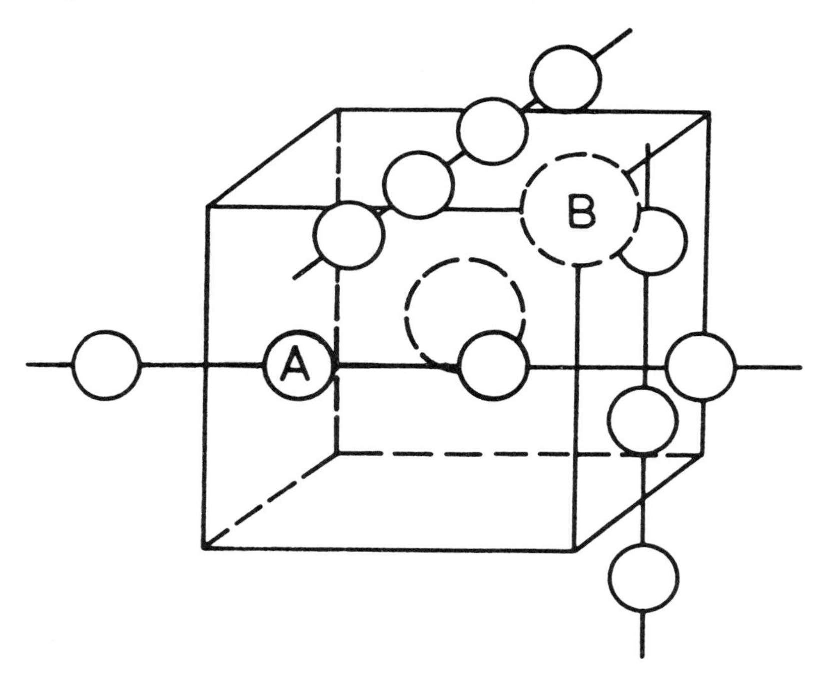

Fig. II.2. The A15 structure showing A-chains passing through the centers of cube faces with B atoms at cube corners.

one-dimensional structures are known to have singularities in the electronic density of states N(E) of the form N(E) $\sim (E_o - E)^{1/2}$. In the Friedel model it is assumed that such singularities exist and that $o < (E_o - E_F) < k\theta_D$, so that the effects of the attractive electron-phonon interaction are greatly reduced beyond the smaller energy range $(E_o - E_F)$. Then E_g and T_c no longer depend on θ_D and no isotope effect is predicted, in agreement with experiment for Nb_3Sn (V, 1) and with band structure calculations which show a peak in N(E) near $E = E_F$.

The three electronically decoupled one-dimensional chains in the Friedel model are described by three energy bands $E_\alpha = E_\alpha(k_\alpha)$, where $\alpha = (x, y, z)$. The Fermi surfaces of these three bands consist of parallel planes separated by $q = 2k_F$. These chains are unstable against the formation of Peierls charge density waves with periodicity q superimposed on the lattice periodicity $G = 2\pi/a$, where a is the chain periodicity. The Peierls instability is reflected in the ideal, fully decoupled model by the appearance of a cyclic (zero-frequency) vibrational mode, $\omega(q) = O$. In practice when large areas of the Fermi surface are nearly parallel (this is called Fermi surface nesting) one expects to observe very soft (nearly zero-frequency) vibrational modes essentially of the Peierls type. However, Fermi surface nesting is not the only mechanism which can produce soft modes, although it is the simplest one.

The band structure and lattice dynamical calculations described below have shown that many of the details of the Friedel model are not valid (such as the electronic decoupling of the chains), but the basic point of the model, which is that structural features (such as chains) can produce significant structure in N(E) on a scale comparable to $k\theta_D$, the Debye energy, remains valid and is likely to be important in many high-T_c superconductors. Peaks in N(E) near $E = E_F$ also often are accompanied by soft modes, but again one can have soft modes even in the absence of very large values of $N(E_F)$.

3. Transition Metals and Their Compounds

The pseudopotential method, which is so useful for s-p metals, begins with a free-electron gas and treats the interaction of plane waves with ions by perturbation theory. This method is not suitable for transition metals and their compounds. Here the d electrons are in states which resemble atomic states which hybridize only weakly with plane-wave like s-p states. Thus a different approach is required to calculate electron-phonon interactions in transition metals and their compounds, just as a different method is required to solve the wave equation itself in these materials.

The solution of the wave equation when d states are important uses for basis functions the augmented plane waves (APW) of Slater. Many of the approximations to the crystal potential which were used in earlier applications of the Slater method have been removed in modern approaches, such as the calculations of Mattheiss, which form the basis of the electron-ion interaction calculations which we now discuss. However, the degree of self-consistency and general flexibility in these calculations is still much less than that possible for s-p metals, as one would expect. While none of these technical difficulties constitute limitations in principle, in practice these limitations are at present unavoidable, and accordingly they should be kept in mind for the following discussion.

The general formalism for tight-binding calculations of electron-phonon interactions in transition metals and their compounds was developed by Varma and Weber.[9] They separated the dynamical matrix which determines lattice vibration frequencies into two parts,

$$\mathbf{D} = \mathbf{D}_1 + \mathbf{D}_2 \tag{3.1}$$

where $\mathbf{D}_1$ is short-ranged. Its range is about the same as the tight-binding or atomic orbital overlap terms which are used to fit the energy bands as obtained in self-consistent APW calculations, which typically include first and second-neighbor overlap. The term responsible for lattice instabilities is $\mathbf{D}_2$, which contains the changes in the Hamiltonian H and the atomic orbital overlap matrix S which are first order in the atomic displacements $\delta\mathbf{R}_l^\alpha$ and the gradients ∇_l treated by second-order perturbation theory. If the energy bands $\epsilon_\mathbf{k}$ vary rapidly in $\mathbf{k}$-space, then $\mathbf{D}_2$ can be large for widely separated atomic sites $\mathbf{R}_l$ and $\mathbf{R}_{l'}$.

In principle both $\mathbf{D}_1$ and $\mathbf{D}_2$ can be obtained from calculations of the total energy at equilibrium and for a sufficient number of configurations. In practice the vibrational dispersion curves $\omega^\alpha(\mathbf{q})$ have been used to describe $\mathbf{D}_1$ in terms of four adjustable parameters/inequivalent atom. For transition metals this has the advantage of combining ion-ion, ion-core and core-core interactions, which are separately large but which cancel each other to a great extent, into a single easily managed function. Here some of the d electrons become, in effect, part of a polarizable ion core.

There are two keys to the success of this scheme. The first is the extent to which the potentials contained in $\mathbf{D}_1$ really do not contribute to electron-phonon scattering. So far this assumption has been proved valid only for non-transition metals (such as Sn and Pb), which are treated fairly well by pseudopotential methods, and for early transition metals (such as Nb). This assumption has never been tested for highly polarizable metalloids (the most troublesome of which is anionic oxygen) or late transition metals in d^8 or d^9 configurations. The second point is that one needs a sufficiently simple unit cell that the number of adjustable parameters contained in $\mathbf{D}_1$ is small. These parameters are determined by fitting to observed vibrational spectra, and with a large number of atoms/unit cell and a correspondingly large number of adjustable parameters, the determination of the latter is usually ambiguous.

The term $\mathbf{D}_2$ can be written approximately as

$$D_2^{ij}(\mathbf{q}) = -\sum_{\mathbf{k},\mathbf{k}'} \frac{g^i(\mathbf{k},\mathbf{k}')g^j(\mathbf{k},\mathbf{k}')}{\epsilon_{\mathbf{k}} - \epsilon_{\mathbf{k}'}} \tag{3.2}$$

with $\epsilon_{\mathbf{k}'} < E_F < \epsilon_{\mathbf{k}}$ and

$$\mathbf{g}(\mathbf{k},\mathbf{k}') = \int (\mathbf{v}_{\mathrm{k}} - \mathbf{v}_{\mathbf{k}'}) u_{\mathbf{k}} u_{\mathbf{k}'} d\mathbf{r} \tag{3.3}$$

where $u_{\mathbf{k}} = e^{-i\mathbf{k}\cdot\mathbf{r}}\psi_{\mathbf{k}}$ and $\mathbf{v}_{\mathbf{k}} = \nabla_{\mathbf{k}}\epsilon_{\mathbf{k}}$. In general the overlap $u_{\mathbf{k}}u_{\mathbf{k}'}$ is large for $\mathbf{k}$ and $\mathbf{k}'$ close and in the same band, and $\mathbf{v}_{\mathrm{k}} - \mathbf{v}_{\mathbf{k}'}$ is large near a saddle point, where the energy denominator in (3.2) is small over a large $\mathbf{k}$ volume. For elemental transition metals Varma and Weber found that the magnitude of $\mathbf{D}_2$ was about half that of $\mathbf{D}_1$, that all the anomalies in $\omega^{\alpha}(\mathbf{q})$ came from $\mathbf{D}_2$, and all the anomalies in $\mathbf{D}_2$ came from saddle points within 0.5 eV of E_F. Note that most of the screening of the ion-ion interaction is short-range and is contained already in $\mathbf{D}_1$, and that only the long-range screening is contained in $\mathbf{D}_2$. The d-electron screening of the ion-ion interaction in transition metals is much more complete (of the order of 90% or more) in transition metals compared to s-p screening in simple metals.

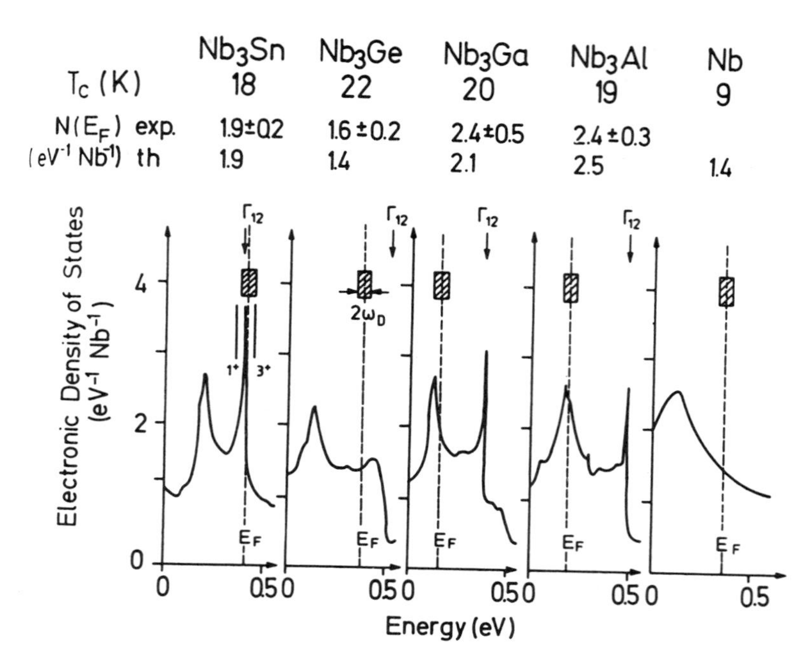

Fig. II.3. Electronic densities of states N(E), from ref. 13, for Nb and Nb_3M compounds. The region $E_F - \omega_D < E < E_F + \omega_D$, where ω_D is the Debye energy, is marked explicitly, as is the position of the two-fold degenerate Γ_{12} state which is often associated with a large peak in N(E) and which is important to the tetragonal Martensitic transition.

With this formalism Weber, Varma and Mattheiss have carried out by far the most extensive calculations of electron-phonon interactions in transition metals and their compounds, as illustrated in Figs. 3 and 4. These calculations show that the chemical trends in the compounds are (perhaps not too surprisingly) quite different from the trends in the elements. This point is important because several simplified formulae were developed in the early days by McMillan[10] relating T_c to λ, $N(E_F)$ and $< \omega^2 >$ and these formulae are still often used because of their algebraic convenience. However, McMillan's formulae

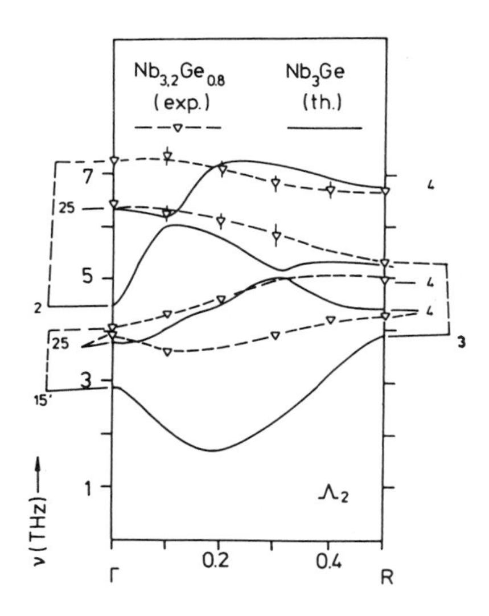

Fig. II.4. Comparison of lattice dispersion curves for $Nb_{3.2}Ge_{0.8}$ and Nb_3Ge, from ref. 14.

were based entirely on algebraic trends in elemental simple and transition metals ($T_c < 10°$ K) and they do not describe just those chemical effects associated with soft modes which lead to $T_c > 10°$ K in compounds.

The difference in chemical trends in transition metal elements and compounds were summarized by Weber[11] and these differences were later confirmed using Mattheiss' self-consistent fields and energy bands to calculate the electron-phonon interaction. As an example, the trends in

T_c in Nb-Mo alloys[12] are deceptively simple. The values of $T_c(^\circ K)$, $N(E_F)$ in (states/eV atom) and $<|g|^2>_{\mathbf{k},\mathbf{k}'}$ in $(eV/\mathring{A})^2$, are 9.2, 0.9, 6.0 and 0.9, 0.3, 19, respectively for Nb and Mo. The T_c's differ by a factor of 10, but $N(E_F)<|g^2|>_{\mathbf{k},\mathbf{k}'}$ is nearly constant. This led McMillan to suppose (quite reasonably in 1968) that T_c was determined primarily by $[\omega^2]$ as defined by (1.5). With the further assumption that $\alpha^2(\omega)$ is constant, $[\omega^2]$ even reduces to $<\omega>/<\omega^{-1}>$. The latter can be obtained fairly accurately from neutron measurements of $G(\omega)$ which can be made on powder samples.

This simple and appealing picture dissolves after quantitative data become available on transition metal compounds. On the average it is probably still true that $N(E_F)<|g^2|>$ is nearly constant as is $\alpha^2(\omega)$. Similarly, most compounds have $T_c < 10^\circ K$ as the elements do. However, most compounds are prepared in the hope of finding high T_c's $> 10^\circ K$ and the ones that are carefully studied usually do have high T_c's. Thus we expect in advance that the high T_c materials are just the ones for which the McMillan approximations are most likely to fail. Similar remarks apply to attempts to estimate $<|g|^2>_{\mathbf{k},\mathbf{k}'}$ for compounds from elements. The soft modes will have anomalously large electron-phonon couplings in any high-T_c material, and the more complex the compound, the more possibilities arise for "accidental" conjunctions of large velocity differences, as in Eqn. (3.3), with large normal mode reciprocal effective masses M_α, as in Eqn. (1.3).

These general considerations have been implemented in detail for the A_3B compounds in the A-15 structure both by energy band calculations[13] and lattice-dynamical studies of the electron-phonon interaction.[14] These calculations show why the simple model in the preceding section of hydrostatic instabilities in the NbN or B-1 family cannot be extended to the A-15 tetragonal instability. (The latter is determined by a two-fold degenerate d-state which induces dimerization of the A chains, i.e., the macroscopic distortion is a by-product of an internal coordinate change.) Similarly, although there are important peaks in N(E) for E near E_F, these are again associated with the two-fold degenerate d states and so cannot be explained by three independent A chains, as in the Friedel model of the preceding section. Nevertheless, the calculations do confirm the importance of the A chains as the origin of the strong electron-phonon interactions responsible for high T_c's in the A-15 materials. All the soft modes with strong coupling involve either transverse or longitudinal modes in which nearest neighbor A atoms move oppositely to each other (optic or internal modes). These modes probably describe the "chain buckling" distortions as seen by radiation-damage induced large static Debye-Waller factors.[14] which accompany large reductions in T_c. The soft modes are generally not associated with any specific narrow feature of N(E), such as Fermi surface nesting or a peak at $E = E_F$, but more nearly reflect strong screening of electron-ion interactions made possible by the presence of many energy states within about 0.5 eV of $E = E_F$.

Point defects such as vacancies or impurities are not easily treated by energy-band theory, and so quantitative estimates of effects of such defects are difficult to obtain. The A15 compounds contain an amusing illustration which shows how such defects can enhance T_c by stabilizing a soft lattice and enhancing the electron-phonon interaction. By replacing B = Sn in Nb_3B with B = Ge one can raise T_c from 17° K to 22° K. This destabilizes the lattice, however, and it appears that Nb_3Ge with $T_c \sim 22°$ K is only a metastable phase. The phase can be prepared as a thin film by several epitaxial ruses,[15] one of which involves a graded concentration of O impurities. The presence of O impurities is used to expand the lattice constant (or decrease the density) by 2% (6%) of a very thin Nb_3GeO_x film, which is then used as a template to grow an expanded Nb_3Ge film epitaxially (nowadays called pseudomorphically, as in Ge on Si). This expanded film has $T_c = 22°$ K compared to $T_c = 17°$ K either for bulk Nb_3Ge or for $Nb_{3.2}Ge_{0.8}$. This suggests that the bulk Nb_3Ge has Nb atoms on Ge sites, and that such T_c-lowering antisite defects are suppressed in the expanded Nb_3Ge pseudomorphic film. Note that the expansion here is very large: density differences as small as 1% are often sufficient to produce new phases of intermetallic compounds at high pressures. The picture of T_c enhancement in metastable Nb_3Ge compared to Nb_3Sn is that in the expanded film the smaller Ge atom is "rattling around" in a large cavity, which corresponds to probably the first well-established example of what is described as T_c enhancement by fictive phonons in IV.13 and VI.3. As shown in Fig. 4 for the two cases, only the stoichiometric compound shows large phonon softening.[14] Matthias discussed various kinds of lattice instabilities and concluded that with binary compounds these instabilities would generally keep T_c below 20° K, so that ternary (or even quaternary) compounds were needed to increase T_c further.[16]

4. Solving the Gap Equation

BCS developed a self-consistent equation for the energy gap $E_g = 2\Delta$ which they solved in the limit of a small attractive electron-phonon interaction (weak-coupling limit) such that $E_g << \omega_D = \theta_D$. With the assumption that N(E) is slowly varying on a scale of E_g they obtained the gap equation

$$\frac{1}{N(E_F)V} = \int_0^{\omega_D} \frac{d\epsilon}{(\epsilon^2 + \Delta^2)^{1/2}} \tanh\left[\beta(\epsilon^2 + \Delta^2)^{1/2}/2\right] \quad (4.1)$$

with $\beta = (kT)^{-1}$. The condition $\Delta = 0$ determines T_c, which is (B.3.29)

$$T_c = 1.14\omega_D \exp[(N(E_F)V)^{-1}]. \quad (4.2)$$

Refinements of (4.2) have occurred in three stages. First the effects of electron-electron Coulomb repulsion[3] should be added; these replace $\lambda^{-1} = (N(E_F)V)^{-1}$ in (4.2) with $(\lambda - \mu)^{-1}$, where μ (sometimes denoted by μ^*) is of order 0.1 and can be calculated either directly or from tunneling data with an accuracy approaching 20%. For high-T_c materials $\mu << \lambda$ and it need not concern us greatly, although its presence is necessary to explain why all metals are not superconductors.

The second stage includes corrections for varying N(E), which occurs in Nb_3Sn (Fig. 3) and quite possibly in the new materials as well. It is interesting to examine these corrections in the weak-coupling limit where they can be evaluated analytically for a simple model, namely a Lorentzian peak of relative strength g and width a superimposed on a constant background.[17] If it is further assumed that $\Delta << a$ then

$$T_c = T_{co} \exp(g/4T_c) \quad (4.3)$$

where T_{co} is given by (4.2) and

$$\Delta = 2\omega_D \exp[-(N(E_F)V)^{-1}] \exp(g/2\Delta) \quad (4.4)$$

which gives the gap ratio $x = 2\Delta/kT_c$

$$x = 3.5 \exp[g(x^{-1} - 4^{-1})/T_c] \quad (4.5)$$

which almost reduces to I (2.1) when $g = 0$. With increasing g graphical analysis shows $3.5 < x < 4$. Note that this conclusion is valid only in the weak-coupling ($\lambda \lesssim 1$) limit.

The third state of refinement of the gap equation is the most interesting and the most important for high-T_c superconductivity. To analyze strong coupling ($\lambda \gtrsim 1$) the BCS treatment of Cooper pairs must be refined to include retardation and particle renormalization effects. This is done systematically in the Green's function theory described in Appendix B. Many experimentalists and theorists have found the discussion of these equations given by Allen and Dynes[18] both most balanced and most readable. They were the first to realize that even though the electron-phonon interaction may be attractive only over a range up to ω_D, T_c can become arbitrarily large and in fact as $\lambda \rightarrow \infty$,

$$T_c = 0.18\,\omega_D\,\lambda^{1/2}\,. \tag{4.6}$$

A recent compact discussion[19] of the Eliashberg equations gives the interpolation formula

$$T_c = 0.25\,\omega_D\left(e^{(2/\tilde{\lambda})} - 1\right)^{-1/2} \tag{4.7}$$

which reduces to (4.6) and in the strong-coupling limit $\lambda \gtrsim 1$ uses

$$\tilde{\lambda} = \lambda(1 + 2.6\,\mu)^{-1} \tag{4.8}$$

showing an increasing Coulomb effect for large λ, compared to the weak-coupling limit $\lambda \lesssim 1$, where $\tilde{\lambda} = \lambda - \mu$.

Further insight into the nature of strong-coupling effects, such as the relative importance of high-frequency phonons (where $F(\omega)$ is large) and low-frequency phonons (where $F(\omega)$ is smaller, but where indications of incipient lattice instabilities are often found) requires analysis of the dependence of the Eliashberg equations not only on the integrated quantity λ but also on different parts of $u(\omega) = \alpha^2(\omega)\ F(\omega)$. This analysis was undertaken by Bergmann and Rainer,[20] who calculated $\delta X/\delta u(\omega)$, that is, functional derivations of observables X, from measured tunneling characteristics. Results have since been derived for $X = T_c$, $\Delta(T=0)$, $x = \Delta/T_c$, $H_{c(1,2)}(0)$, $dH_{c(1,2)}(T_c)/dT$, $\Delta C_p(T_c)$, and other quantities as well, for most superconductors for which values of $\alpha^2(\omega)F(\omega)$ are available. This work leads to many qualitative conclusions, not all obvious. Some of these are summarized below.

Before proceeding to the results, we should recognize that for real materials one never changes only one part of $\alpha^2(\omega)\ F(\omega)$, even though lattice instabilities, for example, seem to be signaled by phonon softening for certain modes over restricted ranges of **k**-space. The reason for this lies in the readjustment or relaxation of the lattice, either in terms of density as in the TM examples (II.2), or in terms of other material properties (internal coordinate changes, site occupancies [including non-stoichiometry], static distortions, and so on). Such adjustments change $F(\omega)$ and especially $\alpha(\omega)$, at *all* values of ω, so that while mathematically one can imagine calculating ΔT_c from $\Delta u(\omega_o)$ for an isolated ω_o, physically this never occurs.[21] So far the strongest qualitative conclusions have been reached by comparing $\delta X/\delta u(\omega)$ for high and low frequencies for crystalline and amorphous materials.[22] The latter behave "anomalously" because they contain a high density of "low" frequencies $\omega_l \lesssim T_c$.

One of the most interesting questions which has been solved[23] is the maximum superconducting transition temperature T_c attainable in isotropic superconductors when

$$A = \int d\omega\alpha^2(\omega)F(\omega) \tag{4.9}$$

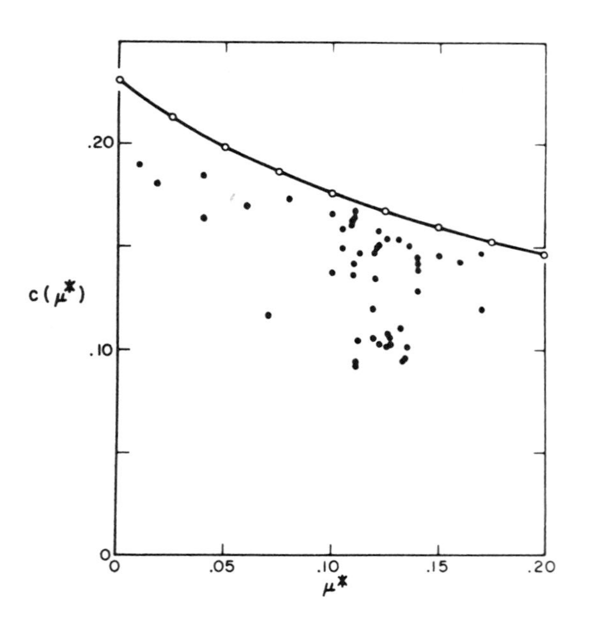

Fig. II.5 Comparison between experimental values of T_c/A and the theoretical least upper bound $c(\mu^*)$ from Eqn. (4.10) (ref. 23).

is kept fixed. The result (for $\mu = 0$) is $T_c \lesssim 0.23A$ and equality is obtained when $F(\omega) = \delta(\omega - \omega_E^*)$ with the Einstein frequency $\omega_E^* = 1.75A$, and $\lambda_E^* = 1.14$. Of course, the actual phonon frequency distribution is broad, so it is interesting to know how closely this least upper bound is approached in practice. In many cases the effects of phonon spectral broadening turn out to be small, as are the effects when μ (or μ^*) > 0. If we write

$$T_c \leq c(\mu^*)A \qquad (4.10)$$

with $c(0) = 0.23$, results for metallic and intermetallic superconductors for which α^2F is known from tunneling experiments are shown in Fig. 5. It is clear that the least upper bound is nearly obtained in practice, and that to achieve high T_c, a large value of A, i.e., a strong *overall* electron-phonon coupling for electrons at $E = E_F$ is necessary. Overall in the strong-coupling limit for high T_c's I would expect that (4.9) and (4.10) are at least as useful as (4.6) and (4.7).

Two other interesting quantities have been examined in the strong-coupling limit. The gap ratio $x = 2\Delta/kT_c$ can be increased substantially over the range of 3.5 - 4.0 associated with weak coupling and described by (4.5). In the limit $T_c >> \omega_D$ the maximum value[24] of x is near 11.5.

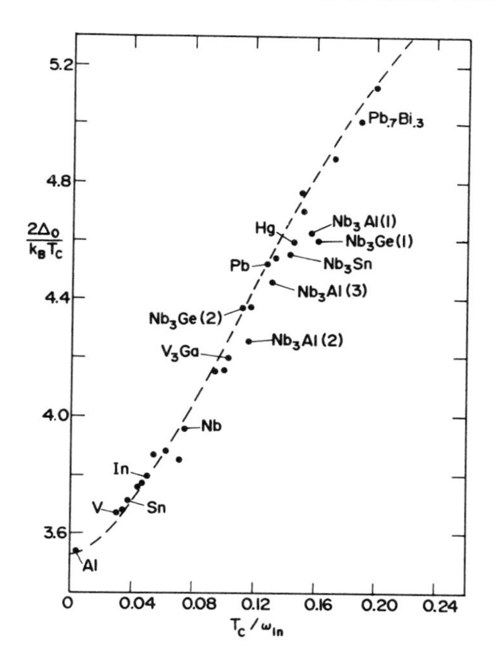

Fig. II.6. The gap ratio $x = 2\Delta/kT_c$ in the weak-to medium-coupling limit for elemental and transition-metal compound superconductors. The average ω_{ln} is $\ln\omega$ weighted by α^2F (ref. 25).

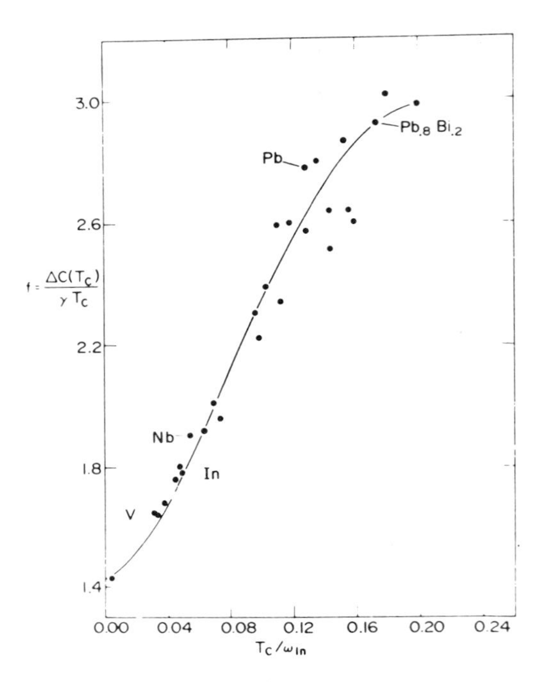

Fig. II.7. The jump in the specific heat $\Delta C(T_c)$ relative to the normal-state specific heat γT_c for weak-to medium-coupling (ref. 25).

In the weak-coupling limit the ratio $\Delta C_p/\gamma T_c$ of the jump in the specific heat to the electronic specific heat in the normal state is 1.4. For $T_c/\omega_{ln} \lesssim 0.2$ (medium coupling) good agreement between experiment and theory is obtained,[25] as shown in Figs. 6 and 7. The ratios $E_g/kT_c = x$ and $\Delta C/\gamma T_c$ are shown in the strong-coupling limit in Fig. 8. Note that whereas x continues to rise as T_c/ω_{ln} increases, $\Delta C/\gamma T_c$ reaches a peak near $T_c/\omega_{ln} = 0.2$ and then decreases rapidly.[25] This reflects the fact that for superstrong coupling the quasi-particles are mostly phonons (i.e., strong polarons) rather than merely renormalized electrons. Hence a large value of λ does not imply a large value of $\Delta C/\gamma T_c$, and even in the strong-coupling limit the

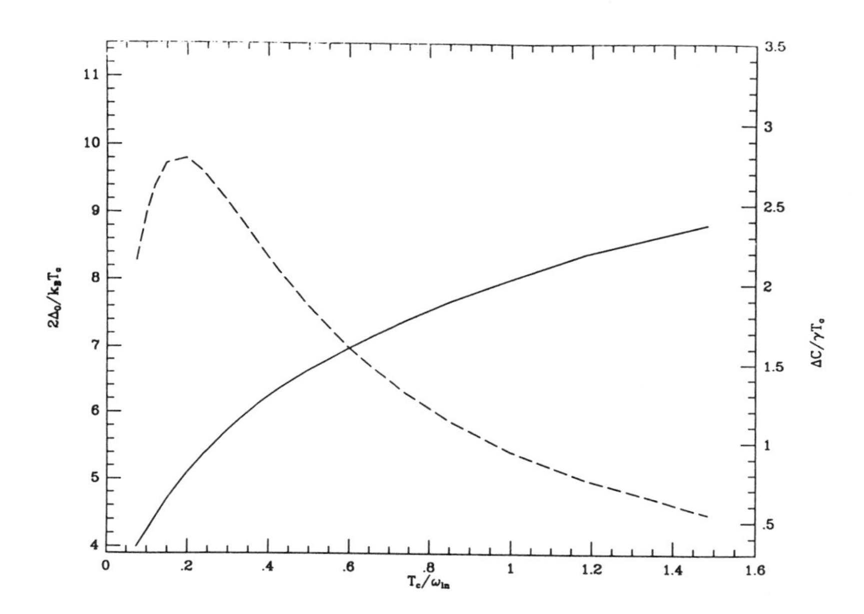

Fig. II.8. The gap/T_c ratio and $\Delta C(T_c)/\gamma T_c$ for weak-to-strong coupling obtained by solving the isotropic Eliashberg equations with Einstein phonons (J. P. Carbotte, unpublished).

measured values of $\Delta C/\gamma T_c$ will be similar to those expected for weak coupling.

5. Normal-State Properties

The electrical resistivity $\rho(T)$ is the most widely-measured and discussed normal-state property. It obviously depends on the electron-phonon interaction and for simple models[26] can be related directly to λ_{ep}. In all materials with strong e-p coupling (including those which condense into charge density wave states[27] as well as those which become high-T_c superconductors) $\rho(T)$ saturates at high T and can be described broadly by

$$\rho(T) = \rho_o + \rho_1 T + \rho_2 \exp(-T_o/T) \tag{5.1}$$

where the third term, rather than the second, is dominant at high T with $\lambda \gtrsim 1$. The best examples[28] of (5.1) are Nb_3Sn ($T_c = 18°$ K, $T_o = 90$K) and Nb_3Sb($T_c = 0.2$K, $T_o = 210$K), with $\rho(T)$ as shown in Fig. 9. At low temperatures the expansion

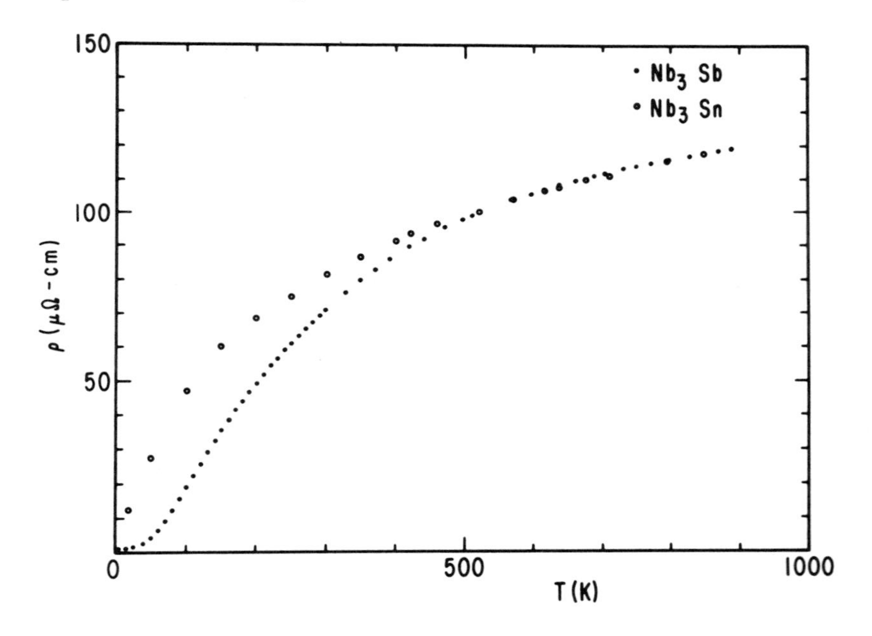

Fig. II.9. Comparison of the electrical resistivities $\rho(T)$ of Nb_3Sn and Nb_3Sb (ref. 28).

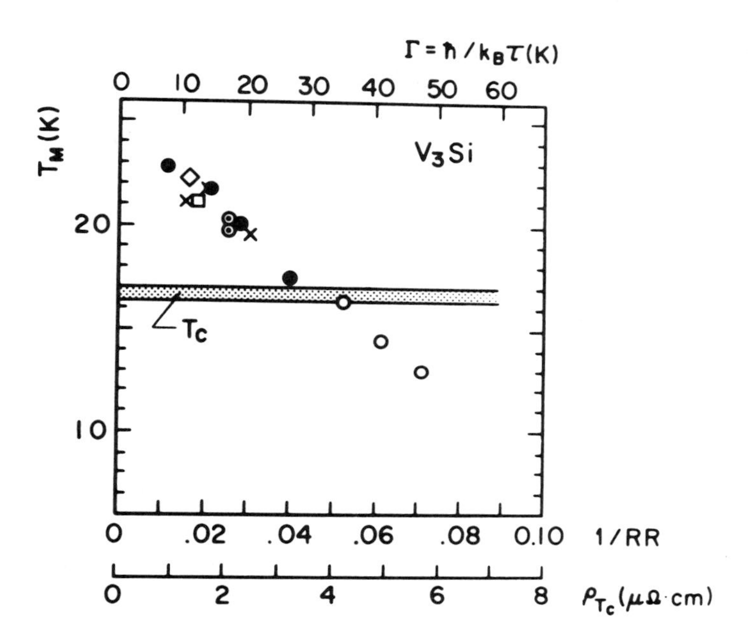

Fig. II.10. For V_3Si samples T_M and T_c are plotted against R^{-2}, where $R^{-2} = \rho(T)/\rho(300K)$ (ref. 30).

$$\rho(T) = \rho_o + aT^n \tag{5.2}$$

is often used.[29] Here for small λ one generally finds $n \gtrsim 3$, while for large λ one finds $n = 2$.

The central problem in analyzing $\rho(T)$ with either (5.1) or (5.2) is that ρ_o, the residual resistivity associated with defects, is *not* independent of the other parameters (such as T_o) and that this functional dependence is concealed in these equations which actually are only functional fits to a property which (especially for large λ) depends on e-p interactions in subtle ways. Thus atomic relaxation around defects is large when λ is large and this relaxation is itself strongly temperature-dependent, as is the cross-section for electron-defect scattering. In effect the third term in (5.1) is part of the first term ("static" scattering) and also part of the second term ("thermal" scattering). This non-separability has made it difficult to identify broad chemical trends in T_o or in the residual resistivity ratio (R^2 or R^3 in various notations) $= [\rho(T) - \rho_o]/\rho_o$ which can be related to T_c even for a group of materials with the same crystal structure (e.g., A15). Comparisons between materials with different crystal structures are almost certain to be misleading or wrong more often than right, especially when T_c is exceptionally high.

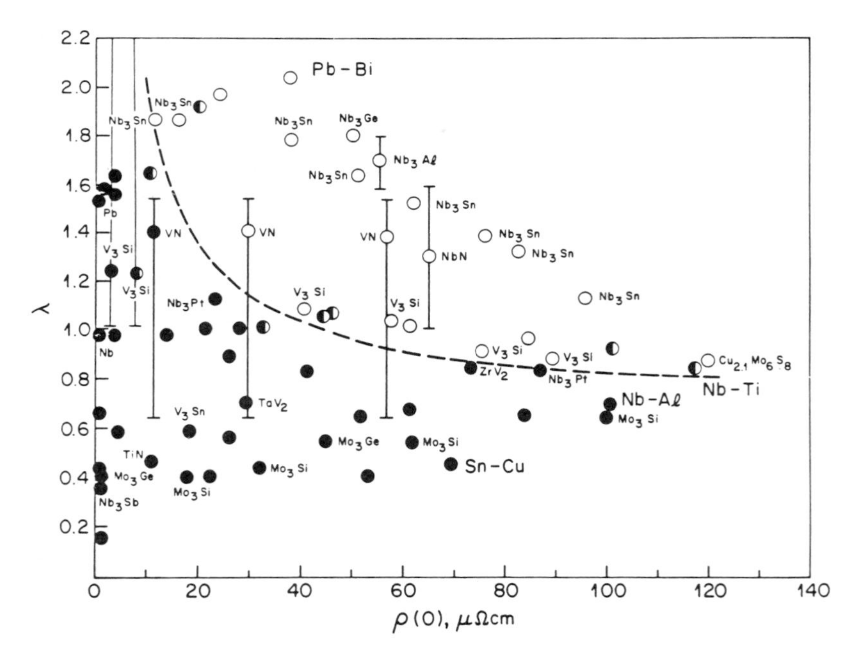

Fig. II.11. Electron-phonon coupling parameter λ estimated from tunneling data, vs. $\rho(o)$, normal-state resistivity at low T. Open circles, $n = 2$, filled circles, $n \geq 3$, half-filled, $2 < n < 3$ in Eqn. (5.2) (ref. 29).

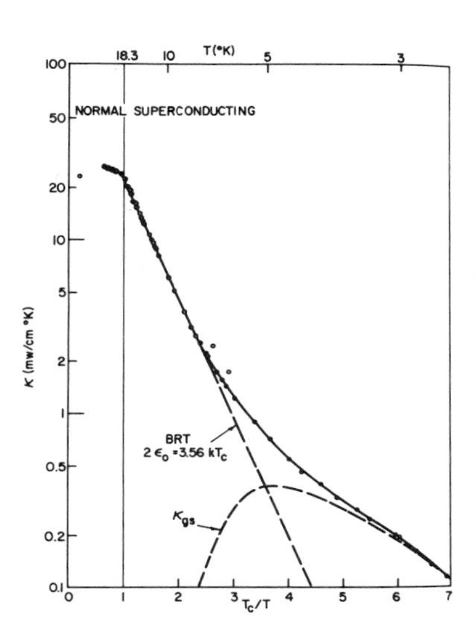

Fig. II.12. Thermal conductivity of a Nb_3Sn single crystal separated into electronic and lattice components. The dashed line is a theoretical fit to κ_e (ref. 32).

In what follows I mention some recent correlations which have been identified between T_c or T_M (a Martensitic phase transition temperature) and fits to $\rho(T)$ with (5.1) or (5.2). The reader is cautioned to note that the rôle played by defects varies greatly from one crystal structure to another.

Various samples of V_3Si exhibit a good correlation[30] between T_M (cubic/tetragonal Martensitic transition temperature) and $1/R^2$, as shown in Fig. 10, with only a small dependence of T_c on $1/R^2$. This shows that T_M is determined by defect pinning, but that for relatively low defect concentrations T_c is little effected by defects. For a large number of A15 compounds and for a few other metals with low defect densities Gurvitch has found[29] a correlation between $\rho(o)$, λ and n in (5.2) which is shown in Fig. 11. On the other hand, for highly defective samples one can have $\Delta\rho$ (thermal) = $\rho(T) - \rho(o) < 0$ as carriers are released from traps created by defect complexes. The same mechanism, with reduced resonant scattering by defects at high temperatures, can explain $\Delta\rho$ (thermal) > 0 in the case where the electron-defect interaction is weaker (lower defect densities). Sign reversal of $\Delta\rho$ occurs for resistivities near $\rho_o = 150\ \mu$ ohm cm in A15 and B1 superconductors (high defect densities).[31] For strongly disordered (both cation and anion vacancies, not a superconductor) TiO_x the sign reversal occurs[32] near $\rho_o = 300\ \mu$ ohm cm.

In addition to the electrical resistivity $\rho(T)$ the thermal conductivity $\kappa(T)$ changes drastically below $T = T_c$. The total thermal conductivity $\kappa(T)$ is the sum of the electronic and lattice contributions, $\kappa_e + \kappa_p$. In single crystals of good metals such as Nb_3Sn $\kappa_e(T)$ dominates in the normal state and in the superconducting state[33] for $T/T_c \gtrsim 1/3$. At very low temperatures κ_e decreases exponentially while κ_p follows a power law, so as $T/T_c \to O$ eventually κ_p dominates, as shown in Fig. 12. However, in very fine-grained powders or amorphous films such as the Pb and $Pb_{0.9}Cu_{0.1}$ films[34] shown in Fig. 13, the electronic mean free paths become so short that κ_p and κ_e are comparable even near T_c. For these materials κ_p is mainly used to probe point or extended nature of defects responsible for phonon scattering,[34] and relatively little information is gleaned concerning superconductive electron-phonon interactions.

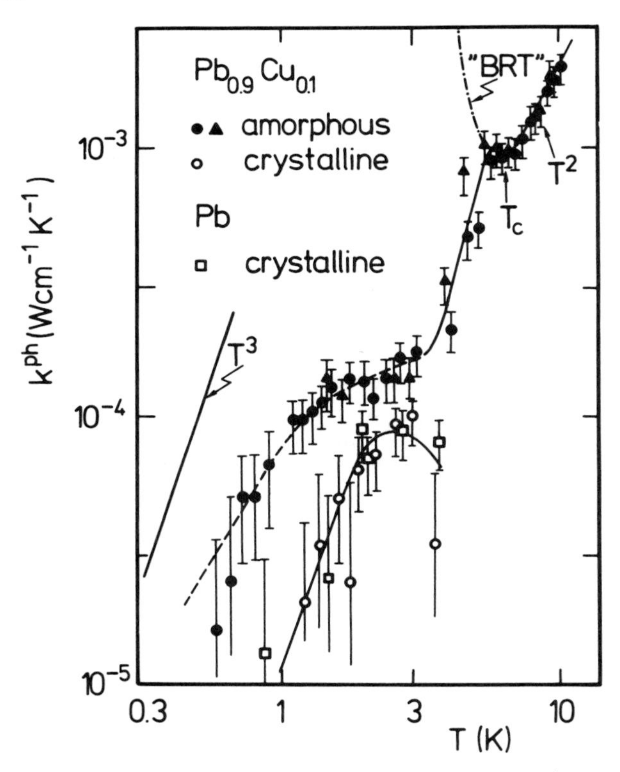

Fig. II.13. The phonon thermal conductivity κ^{ph} in Pb and $Pb_{0.9}Cu_{0.1}$ in polycrystalline and amorphous films (ref. 34).

Other normal-state properties can be obtained from specific heat and susceptibility measurements. On well-characterized single-crystal samples these measurements yield $N(E_F)$, which is related to the "bare" or band density of states $N_o(E_F)$ by $N(E_F)=(1+\lambda)N_o(E_F)$. This relation is well-satisfied for Nb and the Nb_3M compounds in the A15 structure shown in Fig. 3.

REFERENCES

1. J. Bardeen, Phys. Rev. *52*, 688 (1937).
2. M. H. Cohen and J. C. Phillips, Phys. Rev. *124*, 1818 (1961); M. L. Cohen, V. Heine and D. Weaire, in Solid State Physics *24* (Academic Press, N. Y., 1970).
3. P. B. Allen and M. L. Cohen, Phys. Rev. *187*, 525 (1969).
4. M. M. Dacorogna, M. L. Cohen and P. K. Lam, Phys. Rev. Lett. *55*, 837 (1985). For the student an excellent presentation of pseudopotential theory and electron-ion interactions is H. K. Leung, J. P. Carbotte, D. W. Taylor, and C. R. Leavens, Can. J. Phys. *54*, 1585 (1976).
5. D. Erskine, P. Y. Yu, K. J. Chang and M. L. Cohen, Phys. Rev. Lett. *57*, 2741 (1986).
6. J. C. Phillips, Phys. Rev. Lett. *26*, 543 (1971).
7. J. C. Phillips, in Douglass I, p. 339.
8. J. Labbe', S. Barisic and J. Friedel, Phys. Rev. Lett. *19*, 1039 (1967).
9. C. M. Varma and W. Weber, Phys. Rev. Lett. *39*, 1094 (1977).
10. W. L. McMillan, Phys. Rev. *167*, 331 (1968).
11. W. Weber, Phys. Rev. *B8*, 5093 (1973).
12. C. M. Varma, P. Vashista, W. Weber and E. I. Blount, Sol. St. Comm. *27*, 919 (1978).
13. L. F. Mattheiss and W. Weber, Phys. Rev. *B25*, 2248 (1982).
14. W. Weber, Physica *126B*, 217 (1984).
15. J. R. Gavaler, M. Askin, A. I. Braginski, and A. T. Santhanam, Appl. Phys. Lett. *33*, 359 (1978).
16. B. T. Matthias, Physica *69*, 54 (1973).
17. S. G. Lie and J. P. Carbotte, Sol. St. Comm. *34*, 599 (1980).
18. P. B. Allen and R. C. Dynes, Phys. Rev. *B12*, 905 (1975).

19. V. Z. Kresin, Phys. Lett. *A122*, 434 (1987); L. C. Bourne, A. Zettl, T. W. Barbee III, and M. L. Cohen, Phys. Rev. *B36*, 3990 (1987).

20. G. Bergmann and D. Rainer, Z. Phys. *263*, 59 (1973).

21. P. B. Allen, Sol. St. Comm. *14*, 937 (1974).

22. D. Rainer and G. Bergmann, J. Low Temp. Phys. *14*, 501 (1974).

23. C. R. Leavens, Sol. State Comm. *17*, 1499 (1975).

24. J. P. Carbotte, F. Marsiglio and B. Mitrović, Phys. Rev. *B33*, 6135 (1986).

25. F. Marsiglio and J. P. Carbotte, Phys. Rev. *B33*, 6141 (1986); F. Marsiglio, R. Akis and J. P. Carbotte, Phys. Rev. *B36*, 5245 (1987).

26. P. B. Allen in *Superconductivity in d- and f- Band Metals* (Ed. H. Suhl and M. B. Maple, Academic Press, N.Y. (1980), hereafter referred to as Douglass III), p. 291.

27. M. N. Regueiro, Sol. State Comm. *60*, 797 (1986).

28. Z. Fisk and G. W. Webb, Phys. Rev. Lett. *36*, 1084 (1976); S. J. Williamson and M. Milewitz, in Douglass II, p. 551.

29. M. Gurvitch, Physica, *B&C 135*, 276 (1985).

30. M. Kataoka and N. Toyota, Phase Transitions *8B*, 157 (1987).

31. N. Sauvides, J. Appl. Phys. *62*, 600 (1987); F. Rullieralbenque, L. Zuppiroli and F. Weiss, J. de Phys. *45*, 1689 (1984).

32. D. S. McLachlan, Phys. Rev. *B25*, 2285 (1982).

33. G. D. Cody and R. W. Cohen, Rev. Mod. Phys. *36*, 121 (1964).

34. H. V. Löhneysen and F. Steglich, Z. Phys. *B29*, 89 (1978).

III. New Materials

1. Chevrel Phases

More than 24,000 inorganic phases are known, approximately 16,000 binary and pseudobinary, and 8,000 ternary and pseudoternary. The transition metal-metalloids with A15 and related structures, as well as the carbides and nitrides with B1 and related structures, are all binaries and pseudobinaries. The discovery in 1971 by R. Chevrel and coworkers[1] of a family of ternary sulfides which proved to contain high-temperature superconductors thus represents a fundamental breakthrough in crystal chemistry.[2] Not only is the cluster structure of these materials apparently qualitatively different from that of the binaries, but they also exhibit many other properties which are different from those of binary intermetallic compounds. For practical purposes the Chevrel phases were the first new high-temperature superconductive materials, with T_c in $PbMo_6S_8$ being as high as 15K and H_{c2} as high as 60T.

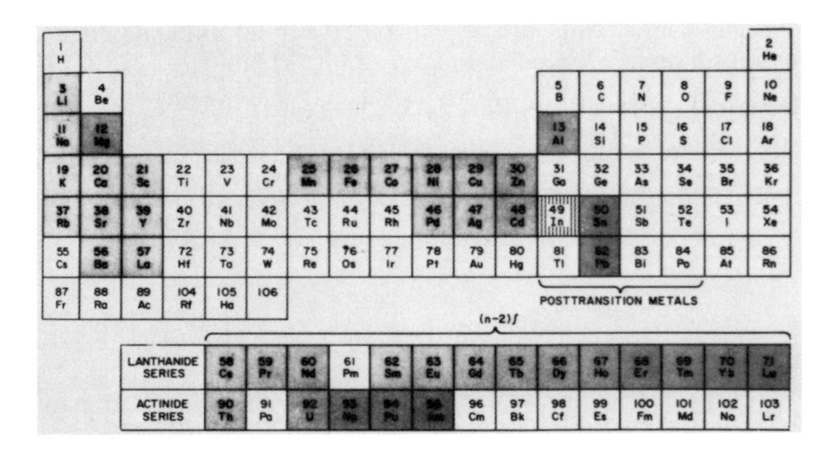

Fig. III.1. The M elements that have been found to form a ternary Chevrel phase M $Mo_6X_{(6,8)}$ are shaded on the periodic table (ref. 3).

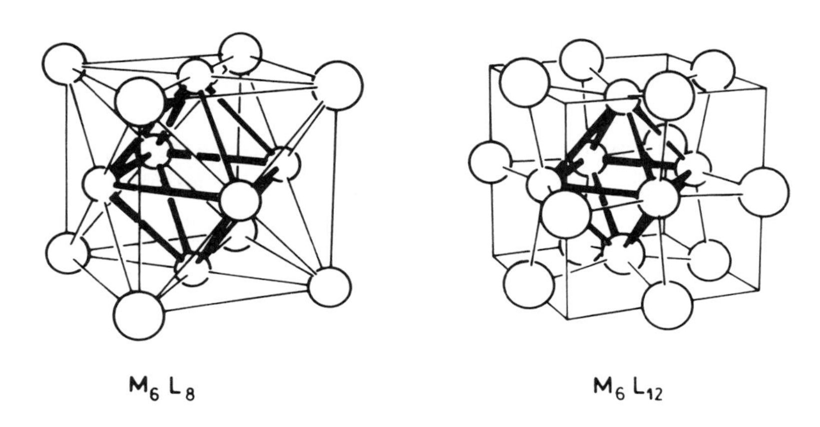

Fig. III.2. Environments of the Mo_6 cluster, in the Mo_6L_8 and Mo_6L_{12} units (ref. 4).

Chevrel phases are described by the chemical formula $M_xMo_6X_8$, where x is usually one, M is one of the more than 40 elements shown[3] in Fig. 1, and X is usually a chalcogenide (S, Se or Te) or occasionally a heavy strongly polarizable halide (Br or I). The novelty of the materials is due to the presence[4] of one of the two fundamental cubic structural units Mo_6L_8 or Mo_6L_{12} shown in Fig. 2. These building blocks contribute most of the valence electrons per formula unit (95% in the case of $PbMo_6S_8$, with $T_c = 15.2$ K). The Mo_6S_8 clusters in $PbMo_6S_8$ are combined with the Pb atoms to form a CsCl structure, with Pb replacing Cs and Mo_6S_8 replacing Cl and being slightly uniaxially distorted along its (111) axis.

From the foregoing we would expect most of the physical properties, including the electronic density of states at the Fermi energy, $N(E_F)$, the normal-state resistivity $\rho(T)$, and T_c and H_{c2} to depend mainly on X, with M playing a passive rôle, as suggested by several band calculations. However, these calculations predict band narrowing on going from X= S to Te, which would act to increase $N(E_F)$ and, all other things being equal, produce a gradual increase in T_c. The behavior which is actually observed, for instance in $LaMo_6(S, Se, Te)_8$ as shown in Fig. 3, is considerably more complex and has been attributed to random internal strains.[5]

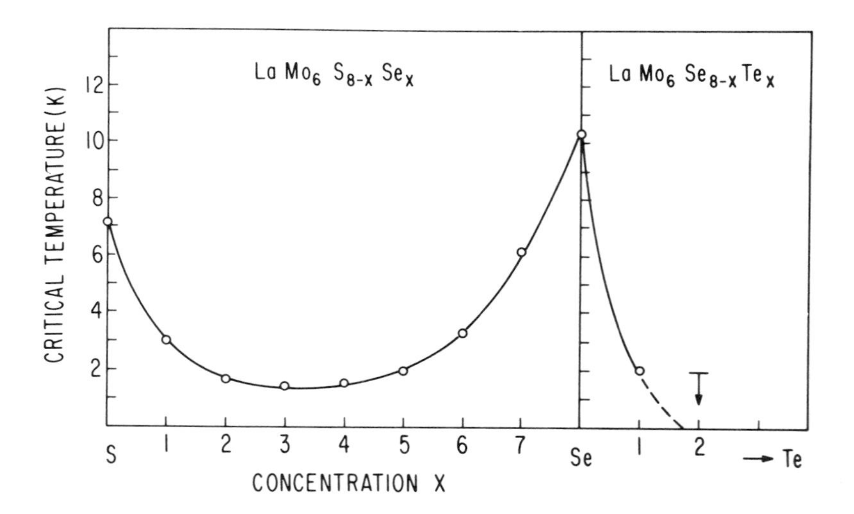

Fig. III.3. Critical temperature T_c vs. composition in the double alloy series $LaMo_6(S-Se-Te)_8$ (ref. 5).

The full complexity of superconductivity in the ternary Chevrel phases is brought out by examining the dependence of T_c on M. With M one of the nine Al, alkali, or alkaline earth elements shown in Fig. 1, and X = S, Se or Te, there are twenty seven possible compounds, but only one combination (M = Ba, X = Se) gives T_c (2.7K) greater than 1.1 K, although many form Chevrel phases. It seems that the M^{2+} element's remaining valence electrons near E_F (two in the cases of Sn and Pb, one in the cases of In, Tℓ and La) play an important rôle in producing an electrical bridge between the clusters. For the alkali and alkaline earth elements, this does not occur because the valence states lie too far above E_F, while Al^{2+} is too small to form a bridge. The internal bonding of the Mo_6L_8 or Mo_6L_{12} clusters is very strong because of valence electrons with E far below E_F (these are often called σ or framework electrons by organic chemists), so that the structural effects associated with the M^{2+} remaining valence electrons near E_F are small. (For example, in MMo_6S_8, the M-Mo spacing is 4.20 Å for M = Sn and Pb, even though Pauling's metallic radii for Sn and Pb differ by 0.08 Å, while his ionic radii differ by 0.13 Å.) It is wrong, however, to conclude from the passive structural rôle of M (which depends on *all* the valence electrons) that M's valence structure is irrelevant to T_c and to other properties depending on electrical paths associated with electrons with E near E_F. What is needed now is accurate theoretical calculations showing $N(E_F, \alpha)$ where α labels atomic sites. This kind of information is routinely available in quantum molecular calculations, but careful studies of the M dependence of $N(E_F, M)$ in Chevrel phases have apparently not been reported.

Another way to see the importance of the bridging element M is to notice that $T_c(Mo_6S_8) = 1.8$ K, so that direct cluster-cluster electron transfer and intracluster electron-phonon interactions alone produce a low T_c. The intercluster Mo — Mo spacing[6] in Mo_6S_8 increases by $\sim 6\%$ when M = Pb is added to form $PbMo_6S_8$ and T_c increases to 15.2 K. The bridge atom M serves two functions: it is an electrical bridge and a mechanical soft link, with a strong local electron — phonon interaction if $N(E_F, M)$ is large. The softness of the mechanical link would be reflected in the local vibrational density of states $G(\omega, M)$.

At present the band-structure calculations for Chevrel phases are too simple to justify an attempt to calculate $\omega_\alpha(\mathbf{k})$ and $G(\omega, M)$ as has been done for Nb_3Sn and NbN; the only available calculations are based on chemically unrealistic Lennard-Jones models.[7] However, the softening of $G(\omega)$ in $(Sn, Pb)Mo_6(S, Se)_8$ is evident in the polycrystalline vibrational spectra[7] shown in Fig. 4. This softening is very similar to the softening seen in Chaps. I and II for high T_c elements and A15 and B1 compounds, and of course it is characteristic of strong electron-phonon interactions.

Because Sn is a Mössbauer element the softening of the M = Sn site in $SnMo_6S_8$ can be measured very accurately in terms of the Sn Debye-Waller factor.[8] When this is fitted with a superposition of Debye and Einstein functions, the Einstein temperature $\theta_E(T)$ shows an abrupt (sample-dependent) break near T = 100 K, as shown in Fig. 5. The simple interpretation of this break is that it is associated with rhombohedral ferroelastic domain formation. This domain formation (which is analogous to the Martensitic transition in V_3Si) may not influence T_c directly, but it is another indication of strong M-centered electron-phonon interactions.

The Chevrel phases are interesting in another respect, as the first examples of an open structure in which some atoms (the M atoms) are so weakly bound that they can diffuse readily in and out of the sample, much as ions do in solid

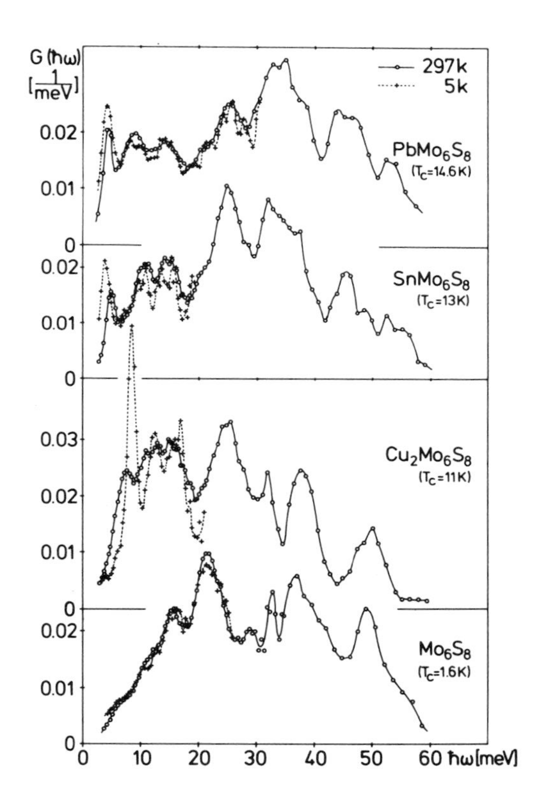

Fig. III.4 Experimental $G(\omega)$ for several high-T_c Chevrel phases, relative to $Mo_6(S,Se)_8$. Note the enhancement in the high-T_c compounds of $G(\omega)$ for $\hbar\omega \lesssim 15$ meV. It is important to remember that these enhancements are indications of stronger electron-phonon coupling primarily at the M atoms (ref. 7).

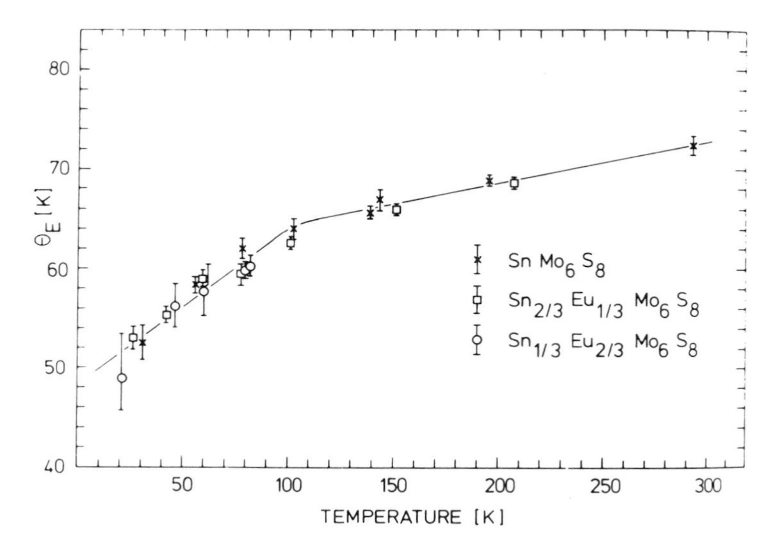

Fig. III.5. The softening of the effective Einstein contribution to the Sn Debye-Waller factor in $SnMo_6S_8$ as reflected by $\theta_E(T)$ (ref. 8).

electrolytes. This characteristic was used[3] to prepare high T_c $MMo_6(S, Se)_8$ compounds with M = Hg, In and $T\ell$ by low-temperature ($T \lesssim 500$ C) diffusion of M into $Mo_6(S, Se)_8$. The electrolytic chemistry here is quite interesting: $Hg,^{2+}$ In^{2+} and $T\ell^{2+}$ retain no more than one valence electron, not two, as in the case of Sn^{2+} and Pb^{2+}, and this reduces their σ electronic bonding to the Mo_6S_8 framework, facilitating the M^{2+} ion transport and compound synthesis. In general one can argue that (so long as $N(E_F, M)$ is large), strong Fermi-energy electron-phonon interactions at M facilitate ion transport, as these reduce the σ framework interactions.

2. Perovskite Superconductivity

The naturally occurring mineral perovskite (perewskite) or $CaTiO_3$ was first assigned the cubic ABO_3 structure (see

Fig. 6) in 1925. Technological interest in these materials has been high since the discovery of ferroelectricity in $BaTiO_3$ in 1943-6, followed by single-crystal growth in 1954, and by now hundreds of perovskite and pseudoperovskite ferroelectrics are known. An older partial listing[9] of these compounds contains more than 200 cubic or pseudocubic compounds, 50 tetragonal, 20 orthorhombic and 10 monoclinic materials. Another more recent listing[10] includes 11 Cu-containing perovskites A_2CuTO_6 and $A_3CuT_2O_9$, with A = Sr or Ba, and T = Nb, Ta or W, with none of the compounds in this 1971 list appearing in the 1960 list. Almost all of these old and new ABO_3 compounds are insulating, except when A and B are both heavy non-

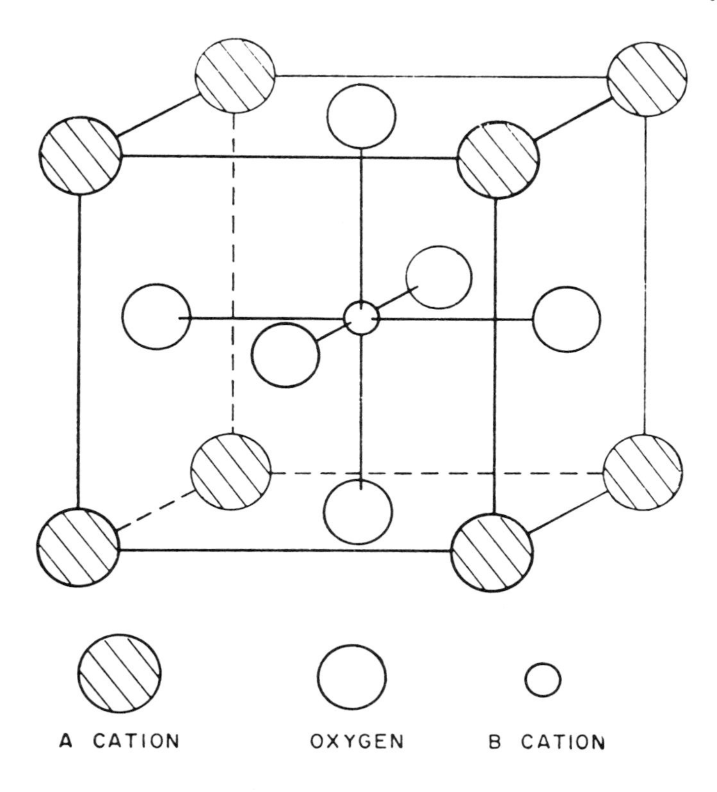

Fig. III.6. The ABO_3 perovskite structure.

transition metals, and more than 100 of them are ferroelectric.[10] The ferroelectric behavior results from off-center atomic displacements, so that the common occurrence of cubic distortions into tetragonal, orthorhombic, rhombohedral and monoclinic structures is not surprising.

Because only a few of these compounds are metallic, and then generally with low carrier densities (essentially semi-metallic), the discovery[11] by Sleight et al. of high-temperature superconductivity ($T_c \sim 13K$) in the pseudoternary alloy $BaPb_{1-x}Bi_xO_3$ was as surprising in its own way as that of Chevrel et al. in the sulfides. Their resistivity $\rho(T)$ is shown in Fig. 7 for $x = 0.2$. Note that in the normal state as T decreases towards $T_c = 11K$ the sample becomes less metallic and $d\rho/dT \rightarrow O$. For $x > 0.35$ these alloys are semiconductive (bronze color) and for $x < 0.35$ metallic (black).

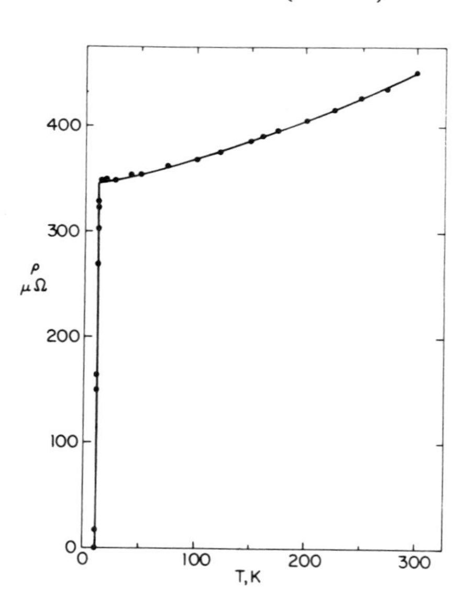

Fig. III.7. Resistivity $\rho(T)$ for $BaPb_{0.8}Bi_{0.2}O_3$ (ref. 11).

At first sight the semiconductive character of $BaBiO_3$ seems surprising, because the formula unit contains an odd number of valence electrons. The actual structure of powdered samples is not cubic, but is monoclinic ($\beta = 90.2(1)^\circ$) with two formula units and two inequivalent Bi sites per unit cell,[12] corresponding to Bi^{3+} and Bi^{5+} with the two $Bi(O_{1/2})_6$ octahedra relatively expanded/compressed by 7% and alternately tilted relative to the pseudocubic (110) axis by 10°. Thus $Bi^{3+}(O^-_{1/2})_6$ and $Bi^{5+}(O^-_{1/2})_6$ each separately contains an even number of valence electrons and need not be metallic. (The translation of this simple intuitive model into the language of one-electron wave functions is deferred to Chap. IV.) Remark here that single-crystal data have shown[13] that a complex phase transition (probably of the Cu_3Au order-disorder type) involving $2Bi^{4+} \leftrightarrow Bi^{3+} + Bi^{5+}$ occurs at high annealing quenching temperatures of order 800C; for this transition oxygen stoichiometry (vacancies) plays a secondary rôle. The phase transition to a static distortion is preceded by a transition to a dynamically distorted structure which is analogous to the fictive phonon model (IV.13).

As is the case for most oxides, most of the data reported for $BaPb_{1-x}Bi_xO_3$ alloys refer to powder samples. Great care and rigid control of factors such as composition of the melt, solution temperature, equilibration time, cooling rate and quench temperature were required to produce "homogeneous" single crystals, as defined by sharpness of high-angle diffraction reflections.[14] All these factors are the same ones that are familiar to scientists who prepare complex oxide glasses, or ferroelectric oxide crystals, but even when they are controlled carefully one still may not obtain "ideal" crystals. What is more likely is that tetragonal domain sizes will be somewhat increased and the volume fraction of extraneous phases (such as PbO)

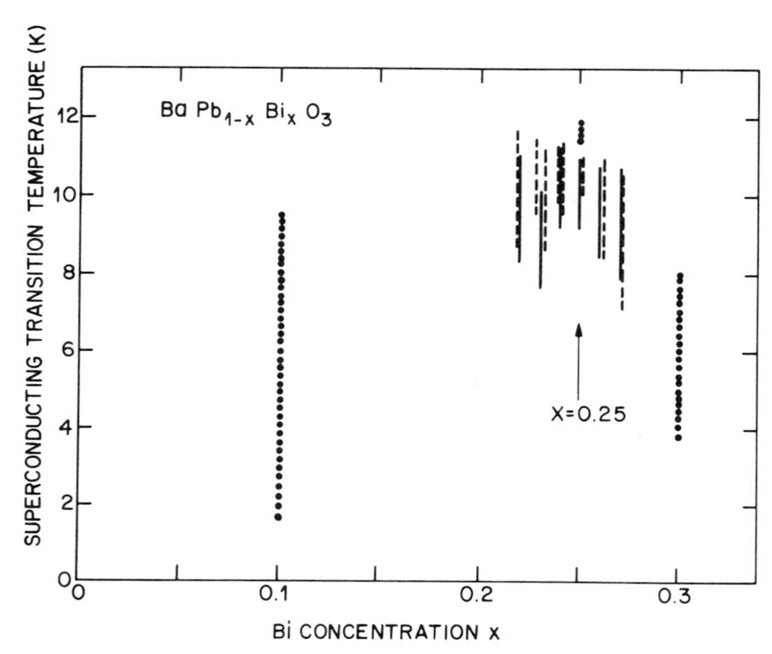

Fig. III.8. The variability and widths of the superconducting transition (measured by ac susceptibility) for $BaPb_{1-x}Bi_xO_{3-y}$ as a function of x. The different lines represent the results of four sets of crystals grown under slightly different conditions. The narrowest transitions are observed for x = 0.25 (ref. 14).

reduced. The resulting samples still exhibit great variability of $T_c(x)$, as shown in Fig. 8, with the smallest variability near $x = 0.25$, which suggests that the internal stress is minimized at this composition.

While there are many differences between the "old" metallic superconductors discussed in Chaps. I and II and the "new" sulfide and oxide superconductors, one factor which remains constant is the ubiquitous presence of lattice instabilities. The vibrational spectrum spans the range up to $\hbar\omega = 70$ meV, but the lower half of this range, which is shown[15] in Fig. 9, is the more interesting. Note the dramatically unstable optic modes at **k** = **M** and **R**, which a

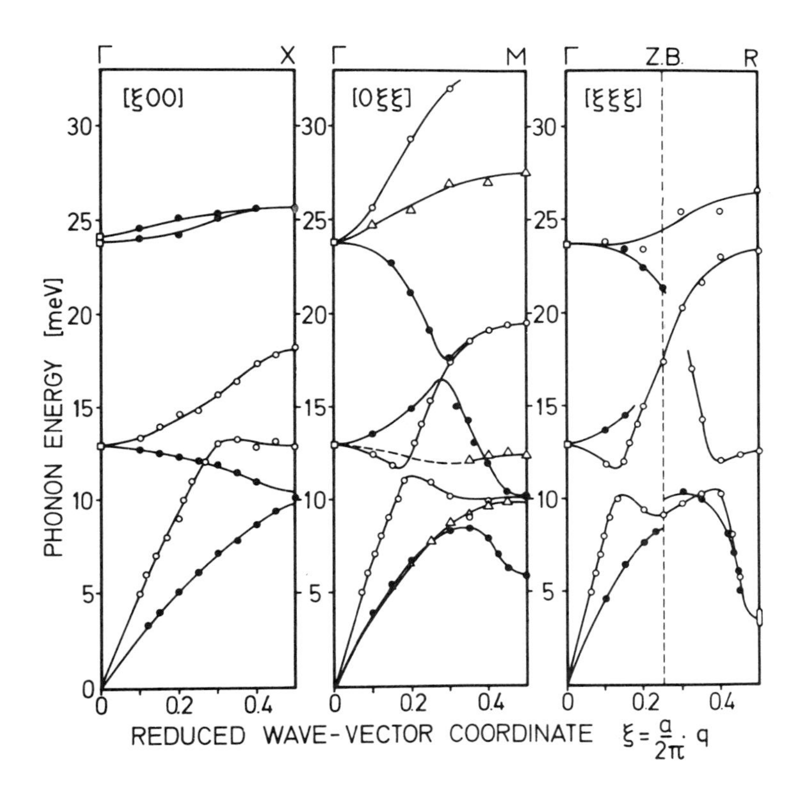

Fig. III.9. Phonon dispersion curves at 296K of $Ba(Pb_{0.75}Bi_{0.25})O_{3-y}$ (ref. 15).

force-constant model[15] assigns to rotation of MO_6 octahedra (M = Pb or Bi).

Phonon fine structure has been observed in $\alpha^2F(\omega)$ as measured by tunneling,[14] but as shown in Fig. 10 this correlates poorly with $G(\omega)$ as determined by neutron time-of-flight measurements.[15] This suggests that the sample surface (tunneling interface) may be oxygen deficient, leading to low-energy instabilities not representative of the bulk.

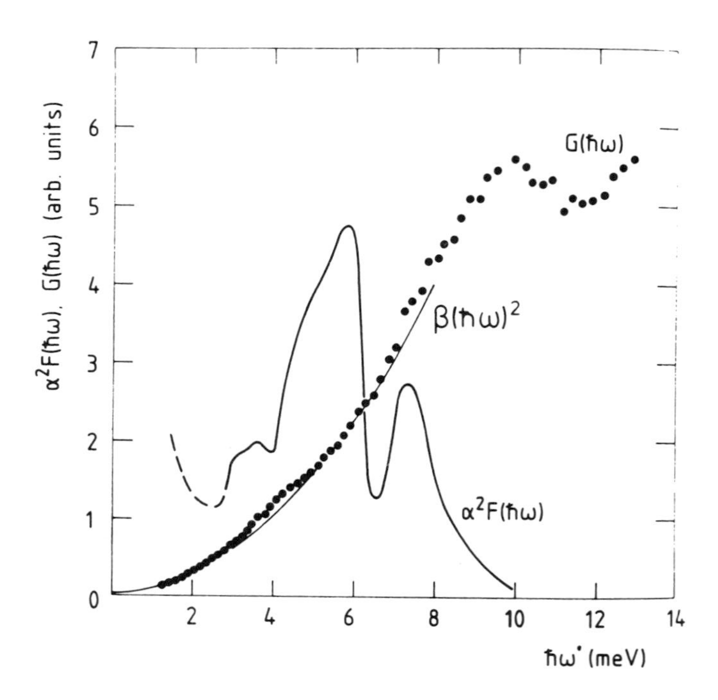

Fig. III.10. Low energy part of the generalized phonon density of states G($\hbar\omega$) and α^2F($\hbar\omega$) for Ba($Pb_{0.75}Bi_{0.25}$)O_3 (ref. 15).

It is possible that the maximum T_c(x) occurs at x = 0.25 because of chemical ordering of the Pb-Bi sublattice to form a Pb_3Bi superlattice. Two facets of the experimental data support this conjecture. The single-crystal data on variability of T_c shown in Fig. 8 shows narrowest widths and variabilities at x = 0.25. Also powder samples exhibit lattice *hardening* (not softening) at x = 0.25 (as measured by the change in shear modulus μ) which overall reflects T_c(x), as shown[16] in Fig. 11. These two correlations might both be explained by chemical ordering to form the $Ba_4(Pb_3Bi)O_{12}$ phase which is more acoustically rigid but has stronger electron-phonon interactions in the optic modes involving octahedral breathing and rotational modes.

The ideas that $BaPbO_3$ is a semimetal with small overlap between O(2p) and Pb(6p) bands near the Fermi energy,[12] while $BaBiO_3$ is a Peierls insulator, are consistent with the observation[17] that when Ba^{2+} is partially replaced by K^+, T_c is scarcely changed in $BaPb_{1-x}Bi_xO_3$ for $x = 0.25$, but that 30% of an $x = 1$ semiconductive sample becomes superconductive with $T_c \sim 30K$. Neither the $Ba(Pb, Bi)O_3$ nor the $(K, Ba)\,BiO_3$ alloy show any evidence of intrinsic magnetic character beyond that of a normal metal or semimetal, such as magnetic order, spin fluctuations, or large local moment paramagnetic susceptibility beyond that associated with trace impurities (such as unreacted K_2O_x). This was to be expected, because bismates and thallates are sometimes ferroelectric but are never magnetic except in the presence of obviously magnetic elements such as a suitable rare earth (Appendix C).

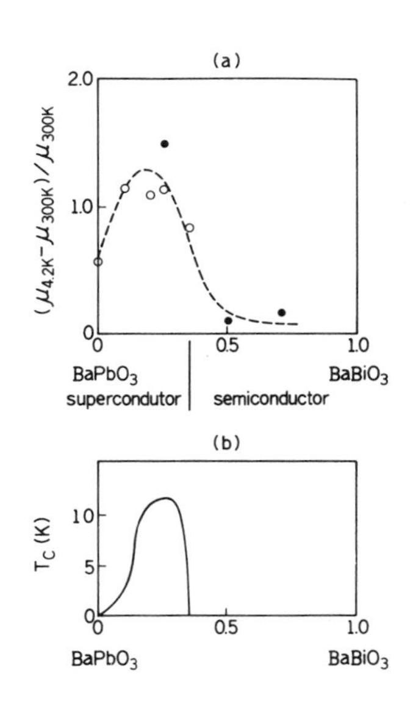

Fig. III.11. Composition dependence of (a) $\Delta\mu$ and (b) T_c in $BaPb_{1-x}Bi_xO_3$ alloys (ref. 16).

While the high $T_c \sim 30K$ attained in (K, Ba) BiO_3 alloys is virtually inexplicable in the context of various exotic magnetic theories (IV.12), there are several possible explanations based on lattice instabilities. First note that 30% or less of the sample exhibits a Meissner effect beginning near 30K. In general perovskites $A^{+m}B^{+n}O_3$ have (m, n) = (3,3) or (2,4), so that replacing Ba[2+] by K[1+] is likely to generate oxygen vacancies $O^{\square}$. In a small fraction of the sample volume (say 30%) these vacancies, together with $K_{[Ba]}$, may order to form domains with a typical dimension of order 100Å. The superlattices in these domains would be undetectable by diffraction methods, but on chemical grounds their most natural configuration would be tetragonal with xy planes consisting of $K_{[Ba]}O_{1-x}O^{\square}_{x[Ba]}K$ double layers, i.e., an xy plane of partial oxygen vacancies bounded by K-enriched sheets. Such a structure would be similar in some respects to the Y plane in $YBa_2Cu_3O_7$ discussed below, which is also a tetragonal perovskite defect structure or the Ca plane in $Ca_{0.86}Sr_{0.14}CuO_2$ discussed in App. C. The defect states associated with this layer could form a defect band with defect-enhanced electron-phonon interactions (IV.10).

3. Copper Oxides

The discovery of superconductivity in (La, Ba)CuO alloys with $T_c > 30$ K by Bednorz and Muller attracted wide attention to high-temperature superconductivity.[18] Subsequently it was shown[19] that the phase $La_{2-x}Ba_xCuO_4$ with $x \sim 0.15$ is responsible for bulk superconductivity with $T_c \sim 35$ K. We first discuss the crystal structure of this phase and then the novel methods of its preparation.

Several families of A_2BX_4 crystals are known, the most famous being the magnetic cubic spinels (ferrites). Here the A and B cations are small compared to the X anions, which form a close-packed lattice, with the A and B cations occupying one eight of the tetrahedral and one half of the octahedral interstices. In such an interstitial ionic compound the lattice constant depends primarily on the anion size and anion-anion contacts. The compound $LiTi_2O_4$($T_c = 13.7$ K) has the spinel structure.

The compound La_2CuO_4 forms in the tetragonal K_2MgF_4 or K_2NiF_4 structure. Here the B atoms lie in planes and are octahedrally coordinated by X atoms, while the A atoms are approximately nine-fold coordinated.[20] In this tetragonal ionic (halide) structure, the packing along the c axis is usually very tight, as shown by one abnormally short A-X distance in all of them. (The average K-F distance in K_2BF_4, B = Mg, Ni, Cu, is 2.61Å compared to 2.67 in KF, while La-O in La_2NiO_4 is 2.40Å, compared to 2.77Å in perovskites.) The shape of the octahedra around the B atoms is usually nearly regular, but in K_2CuF_4 the tetragonal (4+2) distortion of the octahedra decreases the z-axis Cu-F bond lengths by 6% relative to the planar bond

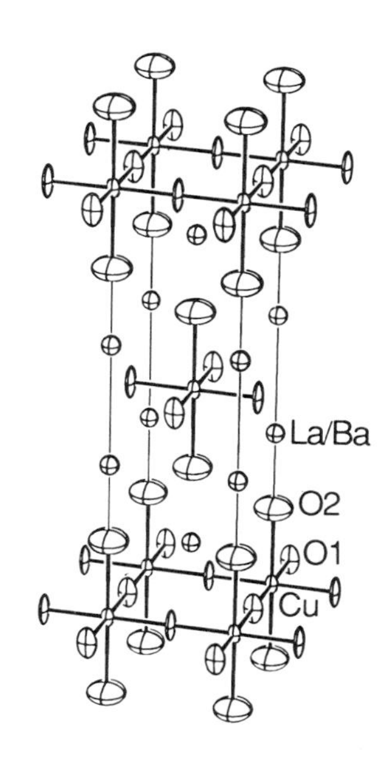

Fig. III.12. Tetragonal structure of $La_{1.85}Ba_{0.15}CuO_4$. All metal atoms are shown, but for clarity only the oxygens in the Cu octahedra are shown and labelled (ref. 23).

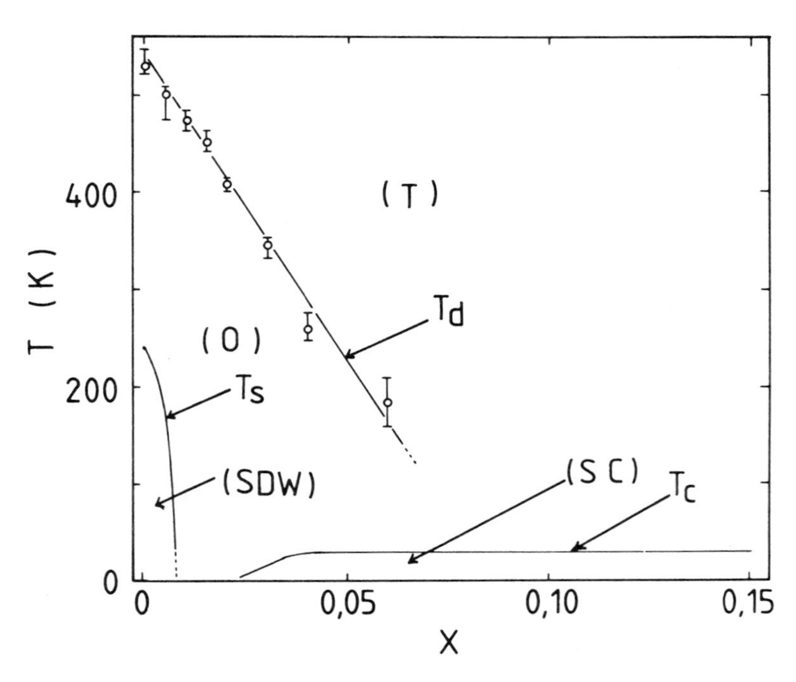

Fig. III.13. Phase diagram for $(La, Ba)_2CuO_4$ showing antiferromagnetic and lattice phase transition boundaries, as well as T_c (ref. 25).

lengths. In La_2NiO_4 the Ni-O z-axis bonds are relatively elongated by 11%, which increases greatly to $\sim 25\%$ in $(La, Sr)_2CuO_4$ and $(La, Ba)_2CuO_4$ crystals. Qualitatively one expects the greatest octahedral distortions for Cu^{2+} (d^9) because the odd d electron gives the largest Jahn-Teller effect. However, the distortion can have either sign for X = F or O, ionic or covalent bonding, and also occurs even for Ni^{2+} (d^8), so that a simple ionic crystal field model is inadequate, and covalency effects (obtainable in principle from the energy-band models discussed in Chap. IV) must be included.[20]

The structural chemistry of cupric compounds is exceptionally complex; "in one oxidation state this element shows a greater diversity in its sterochemical behavior than any other element," and even in sulfides it may be difficult to assign oxidation numbers to Cu atoms.[21] Generally Cu^{2+} is found in hydrated salts in square planar complexes augmented by one or two bonds normal to the plane, i.e., (4+1) and (4+2) complexes, but flattened tetrahedral and trigonal bipyramidal coordination are also observed. Square-planar oxocuprates may be linked in one-, two-, or three- dimensional arrays. Perhaps the most remarkable feature of all the oxocuprates is that in the entire lanthanide series $(RE)_2CuO_4$, only with RE = La is Cu weakly octahedrally coordinated. For all the other RE, Cu is planar coordinated, with the vertical O sites being replaced by $[RE_2O_2]^{2+}$ cationic groups.[22] None of the latter structures can be doped to be metallic or superconductive.

The tetragonal structure of $La_{2-x}Ba_xCuO_4$ with x = 0.15 is shown[23] in Fig. 12. Note the anisotropic thermal ellipsoids and in particular the large transverse amplitude of the vertical 02 atom, which is actually about 3% closer to La than to Cu, as well as the large vertical amplitude of the 01 atom. Here the onset temperature T_{co} is near 35K with $\rho = 0$ at $T_{coo} = 22$ K. Strong electron-phonon coupling presumably occurs at the 02 atoms, and by removing some of these T_{co} can be increased[24] to above 50K in islands while T_{coo} decreases. A careful analysis of atomic pair distribution functions has shown[23] substantial static displacements and additional structure which is composition-dependent. This has led to the conclusion that substantial local structural disorder may be present in these crystals, and that this disorder is related to unique superconducting behavior (especially doping dependence).

The pseudoternary alloys $La_{2-x}(Ba, Sr)_xCuO_{4-y}$ exhibit both magnetic and lattice instabilities, being antiferromagnetic for $T < T_s$ near $x-y=0$ and also showing an orthorhombic-tetragonal phase transition at $T=T_d$. Generally T_c reaches its maximum when x is such that $T_c = T_d$, while $T_c = 0$ when $T_s > 0$. Phase diagrams are shown[25,26] for M = Ba and M = Sr in Fig. 13 and Fig. 14. From these we conclude that antiferromagnetic correlations are of secondary importance to superconductivity, but that lattice softening (which probably reaches its maximum at T_d) is the underlying mechanism responsible for $T_c \sim 35$ K in these alloys, just as in V_3Si, for example (Fig. 10 of Chap. I). The displacive normal mode responsible for the tetragonal-orthorhombic distortion tilts the CuO_6 units about the (110) direction and this explains the anisotropy of the 01 and 02 thermal ellipsoids.[24]

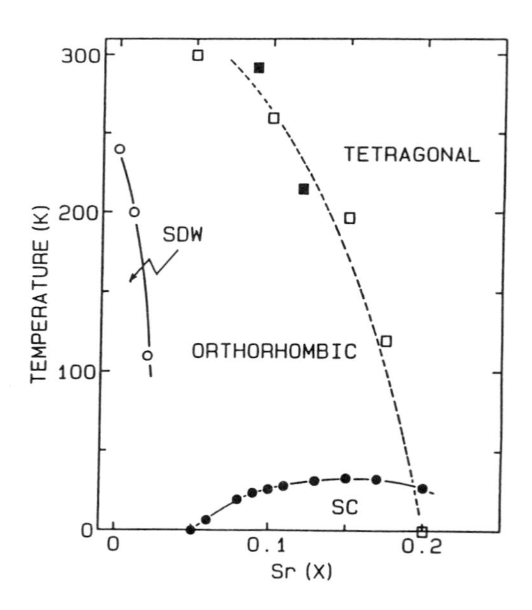

Fig. III.14. Phase diagram for $(La, Sr)_2CuO_4$ single crystals, showing how T_d intersects T_c near $x \sim 0.18$ (ref. 26).

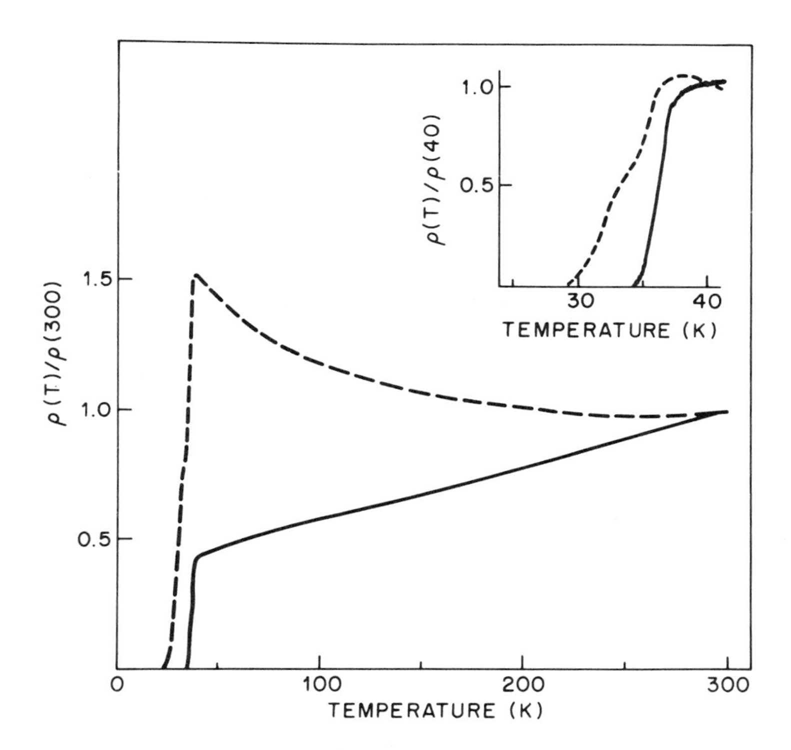

Fig. III.15. Resistivity $\rho(\mathrm{T})$ of $La_{1.8}Sr_{0.2}CuO_4$, the dashed (solid) curves referring to samples annealed in air (oxygen) (ref. 27).

The earliest report[19] of high temperature superconductivity in $La_{2-x}Ba_xCuO_4$ showed a volume Meissner effect of only 30% and an incomplete resistivity transition beginning at $T_{co} = 35$ K with a "foot" extending down to $T_{coo} \sim 22$ K, $\Delta T_c \sim 13$ K. Subsequently T_{co} was found[27] in increase to 38.5K and narrow to $\Delta T_c \sim 1.4$ K with a 60-70% volume Meissner effect with Ba replaced by Sr, while replacement of Ba by Ca gives only $T_{co} \sim 18$ K. Drastic differences in the normal-state resistivity $\rho(\mathrm{T})$ and T_c were found, depending on whether the sample was annealed at 1000-1100C in air or in oxygen, as shown in Fig. 15. The more highly oxidized sample shows more metallic normal-state $\rho(T)$ and a higher T_c.

We saw in the last section that the concentration of oxygen vacancies plays a minor role in perovskite superconductivity ($Ba(Pb, Bi)O_3$), but in $(La, [Ba, Sr, Ca])_2CuO_4$ the concentration of oxygen vacancies plays an important rôle. It seems likely that the activation energy for oxygen diffusion is larger in the orthorhombic phase than in the tetragonal phase, so that oxidation is more complete (and T_c is higher) in the latter, all other things being equal. This would help to explain the remarkable correlation[28] between T_d (the orthorhombic-tetragonal transition temperature) and the volume Meissner fraction shown in Fig. 16. The latter peaks at $x = 0.2$ in $La_{2-x}Sr_xCuO_{4-y}$, where $T_d \rightarrow 0$, and probably decreases for larger x because of a rapid increase in y. At the same time that the volume Meissner fraction is maximized, the

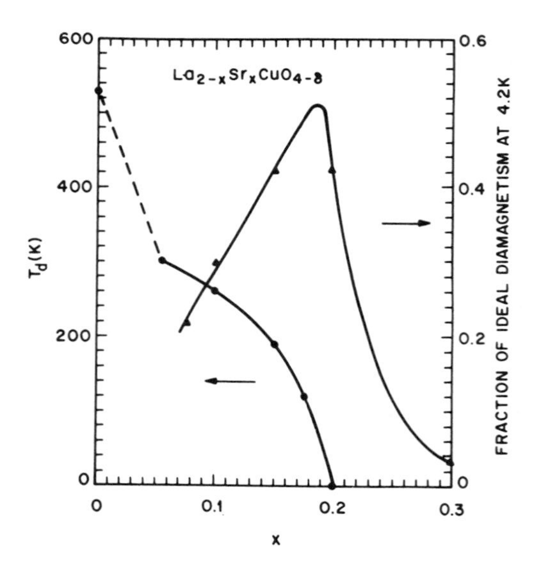

Fig. III.16. Tetragonal-orthorhombic $T_d(x)$ and volume Meissner fraction in $La_{2-x}Sr_xCuO_{4-y}$ (ref. 28).

normal-state resistivity $\rho(T_c+)$ is minimized,[28] which suggests that the nontransformed volume is semiconductive. This semiconductive volume can occur in grain cores and also at grain boundaries. Even more important for full oxidation may be the internal stress associated with differences in the A and B ionic sizes in $La_{2-x}M_xCuO_4$. The Pauling ionic radii for La^{3+}, Ca^{2+}, Sr^{2+} and Ba^{2+} are respectively (Å), 1.15, 0.99, 1.13 and 1.35. Thus La^{3+} and Sr^{2+} are very well matched, which correlates very well with the highest T_c and smallest ΔT_c for M= Sr.

One of the most remarkable experimental results on the compound La-Ba-Cu-O ($T_c \sim 36K$) is the anomalously large[29] value of $dT_c/dP = 0.63$ K/kbar, which is about 100 times that observed for most A15 superconductors and is by far the largest value ever observed for any superconductor. It seems likely that this gigantic value may be associated with the vicinal orthorhombic-tetragonal phase transition and the rocking of the $Cu(O_{1/2})_6$ octahedra. Magnetic susceptibility measurements[30] show that $N(E_F)$ changes relatively slowly with pressure, $d\ln N(E_F)/dP = (0 \pm 2) \cdot 10^{-3}$ compared to $d\ln T_c/dP = 8 \cdot 10^{-3}$/kbar. Thus the factor that is pressure-sensitive is the (electron-phonon) interaction parameter $\overline{V}$ in $\lambda = N(E_F)\overline{V}$ in Eqn. (I.2.1).

So far in discussing the special materials properties characteristic of high-temperature superconductors we have reviewed mainly the vibrational instabilities associated with strong electron-phonon interactions. These mechanical instabilities have as thermodynamic analogues the occurrence of metastable phases formed at high temperatures and often appearing as mixed phases in quenched materials, as shown[31] in Fig. 17. It seems quite likely that mechanical instabilities (and strong electron-phonon interactions) could be enhanced in metastable phases and perhaps especially so near interfaces between such phases. To prepare such phases the materials scientist

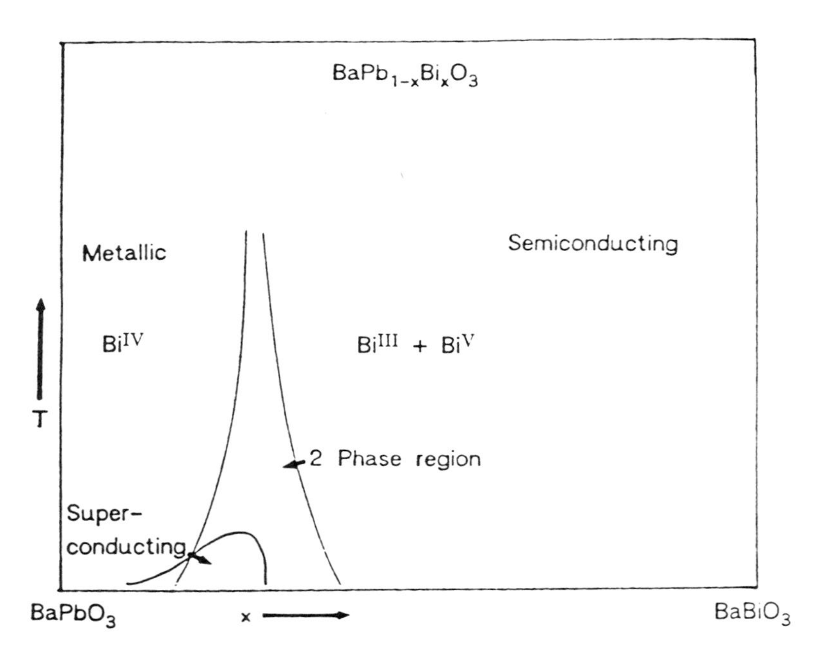

Fig. III.17. A sketch of metastable phases and the resulting electrical properties of $BiPb_{1-x}Bi_xO_3$ alloys (ref. 31).

must manage to "trick nature".[32]

Perhaps the first example of a high-T_c superconductor was the oxygen-stabilized metastable $Nb_3Ge(T_c = 25\,K)$ film discussed in Chap. II. Many metastable phases containing oxygen are known, especially in thin films, and it appears that the presence of oxygen tends to favor the formation of such phases both structurally and kinetically. Here we have seen in Fig. 15 how the oxidation conditions can effect T_c drastically in $(La, Sr)_2CuO_{4-y}$ alloys. Optimization of these conditions has been the primary focus of many efforts at synthesizing new high-T_c

superconductors.

These remarks are probably obvious to most chemists, but many physicists accustomed to idealizing solids as low-temperature perfect single crystals should remind themselves early and often of the complex atomic and molecular structures which invariably are associated with high-T_c materials. Oxidation by annealing at high temperatures below the melting point may easily produce a variety of metastable phases, and the K_2NiF_4 phase of La_2CuO_4 itself could easily be such a phase. It is also important to realize that because of the coherent nature of diffraction, structural studies (especially those based on powder patterns, but even including data from inhomogeneously oxidized single crystals) may preferentially emphasize regions which are more uniformly crystallized at the expense of more metastable regions which are defective or misoriented relative to the more ideal regions. For this reason we discuss not only the reported structures of a given material but also the structures of chemically related materials. In the case of La_2CuO_4 the drastic structural changes associated with small chemical differences (such as replacement of Cu by Ni or of La by any other RE) are already suggestive of metastable phases and strong electron-phonon interactions.

Another point that is implicitly accepted by chemists but is often overlooked by physicists is that the phrase "metastable phases" is generally an indication of the presence of a high density of complex defects which have characteristic atomic or molecular structure. For simple ionic crystals such as the alkali halides point defects, such as the F centers formed by halide vacancies, are well known. As the crystal structure becomes more complex and the bonding more covalent and more metallic, line and plane defects appear to reduce internal stress fields. In oxides with open structures especially, planar shear faulted regions (called Wadsley defects) have been extensively observed by electron microscopists.[33] These extensive defects are easily exploited for electrolytic applications,[34] but more generally one should assume that extensive defects are present in every open multicomponent oxide sample until proven otherwise. Thus annealing may reduce the concentration of point defects, but it may also increase the concentration of line and plane defects as the point defects migrate to the former.

The absence of superconductivity in La_2CuO_4 is easily understood in terms of its normal state resistivity $\rho(T)$ which increases rapidly, as in a semiconductor, as $T \rightarrow 0$. [With an odd number of valence electrons/formula unit the material might be expected to be metallic, but in complex oxides such simple rules are seldom valid. Thus $BaPb(Bi)O_3$ has an even (odd) number of valence electrons/formula unit, but is metallic (semiconductive). We return to this question for La_2CuO_4 in the next

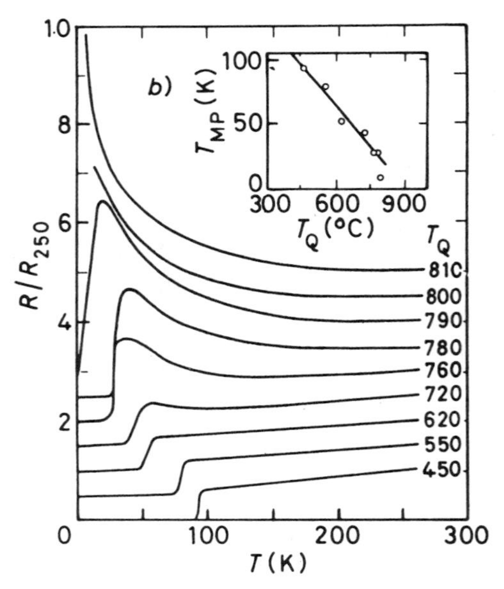

Fig. III.18. Resistivity vs T in sintered La_2CuO_4 at P = 0 and P = 10kbar. Inset: magnetoresistance at P = 0 (ref. 35).

chapter.] However, filamentary superconductivity in a fractional Meissner volume less than 10% has been observed (see Fig. 18) in an La_2CuO_4 sample[35] which must be associated with a metastable phase of unknown composition (possibly La deficient).[35]

Another mysterious La_2CuO_4-based alloy is $La_{2-x}Na_xCuO_{4-y}$, which exhibits[35a] resistive superconductivity for $0.11 \leq x \leq 0.41$ in the range $T_c \sim 16-18K$, but exhibits a substantial Meissner effect beginning at $T_{on} = 36K$ for $x = 0.41$. This Meissner effect may be associated with a metastable phase located on the surfaces of isolated voids. It is striking that T_{on} is so close to the maximum value of T_c (resistive) for $La_{2-x}(Ba, Sr)_xCuO_4$ alloys, but this may reflect only the same accident as occurs in the (La deficient) La_2CuO_4 samples discussed above.

4. Quaternary Copper Oxides

The discovery that $T_c[(La, Ba)_2CuO_4] > 30K$ by Bednorz and Müller immediately touched off a world-wide search for related oxides with higher T_c's. Most scientists assumed that oxides with higher T_c's should be sought in materials similar in composition to $(La, Ba)_2CuO_4$ with probably also the K_2NiF_4 structure. It was quickly found that replacing Ba by Sr increased T_c slightly and improved sample homogeneity. However, the most important observation was made by Chu and Wu after they had found that dT_c/dP was positive and very large in $(La, Ba)_2CuO_4$. This suggested to them the possibility of greatly increasing T_c

through an internal pressure generated by a large A-B size difference in an ABCuO compound, with possibly a new crystal structure. Their hopes were confirmed with the discovery[36] of $YBa_2Cu_3O_7$ ($T_c = 93K$) and $La(La_{2-x}Ba_x)Cu_3O_7$ ($T_c = 80K$) in the entirely new structure[37] shown in Fig. 19. This is apparently the first true quaternary metallic structure, all previous metallic compounds being binaries, ternaries, pseudo-binaries or pseudo-ternaries (see Appendix C). Quaternary insulators, many of them naturally occurring minerals, are well known, so that the metallic properties of these quaternary compounds alone are almost as remarkable as their superconductive properties.

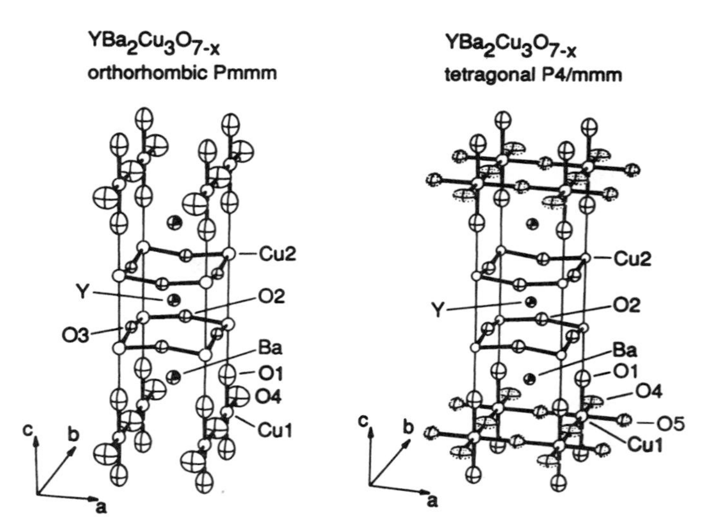

Fig. III.19. Structures of the orthorhombic ($x \sim o$) and tetragonal ($x \sim 0.7$) phases of $YBa_2Cu_3O_{7-x}$, as measured by neutron scattering from a powder sample, including thermal ellipsoids (ref. 36).

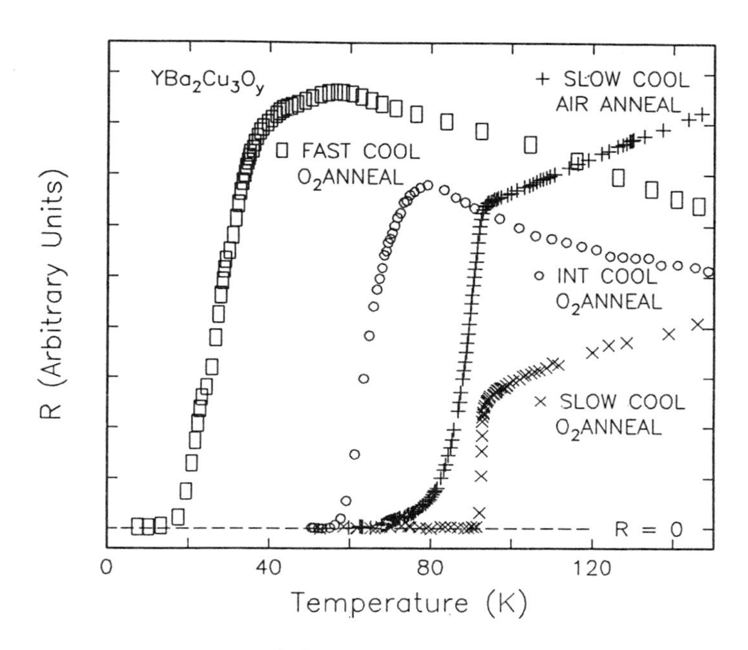

Fig. III.20. Resistivity $\rho(\mathrm{T})$ for $\mathrm{YBa_2Cu_3O_{7-x}}$ prepared in ways that reduce or increase x in the range $0 \lesssim x \lesssim 1$ (ref. 38).

The structure of $\mathrm{YBa_2Cu_3O_{7-x}}$ is similar to that of $\mathrm{(La, Sr)_2CuO_4}$ in several respects: both are open, layered structures, with some similarity to perovskite in the sense that Cu could be octahedrally coordinated, i.e., in a formally filled structure $\mathrm{YBa_2Cu_3O_9}$. The most obvious differences are the tripling of the unit cell of the latter along the c axis, with two inequivalent Cu planes, labelled Cu1 and Cu2, sandwiched between Ba planes for Cu1 and Ba and Y planes for Cu2, and the removal of oxygen from the Y plane and the Cu1 plane. If no oxygen were removed, the formal valence of $\mathrm{Cu^{+n}}$ would be $n = 3$; in $\mathrm{YBa_2Cu_3O_7}$, $n = 2.33$. No stable Cu compounds are known with $n = 3$, which suggests perhaps that $\mathrm{YBa_2Cu_3O_{7-x}}$ should not be stable at all! In practice, even more than with $(La, Sr)_2CuO_4$, annealing in oxygen is needed to reduce x to zero and produce high T_c's. An excellent overview of pseudoperovskite rare earth alkaline earth copper oxides shows how the K_2NiF_4 structure is related to the perovskite structure, and how this is indexed by the primary metal ion framework, with the superconductive properties being strongly influenced by the dimensionality of the secondary oxygen vacancy array.[38]

A very important difference between $(La, Sr)_2CuO_4$ and $YBa_2Cu_3O_7$ is the rôle played by the orthorhombic and tetragonal phases. In the former material the orthorhombic phase is semiconductive and can be made superconductive only by doping with Sr, Ba or Ca which eventually produces a tetragonal structure. Just the reverse is true for $YBa_2Cu_3O_{7-x}$: the tetragonal phase (obtained for $x \gtrsim 0.7$) is semiconductive, and T_c is maximized by reducing x to $\lesssim 0.1$ and obtaining a fully orthorhombic phase. This point is illustrated[38] in Fig. 20 which shows $\rho(T)$ for $YBa_2Cu_3O_{7-x}$ samples prepared in various ways, which alter x primarily.

In both the orthorhombic and tetragonal phases oxygen sites in the Y plane are vacant, leaving Cu2 five-fold coordinated. In the ideal orthorhombic phase in the Cu1 plane the 04 and 05 sites are occupied differently. As $x \longrightarrow 0$ the 04 sites are occupied along the b axis and similar 05 sites are vacant along the a axis, giving b-axis Cu1-04 chains. In the tetragonal phase $YBa_2Cu_3O_6$ both 04 and 05 sites are vacant. The Cu2 planes and Cu1 chains are dramatized in Fig. 21.

The ideal crystallographic structures of $(La, Sr)_2CuO_4$ and $YBa_2Cu_3O_7$ shown in Figs. 12, 19 and 21 do not necessarily describe the complete atomic structure of either superconductor, even in the volume of the sample which exhibits an ac Meissner effect with a small probe field ($\lesssim 1$ gauss). It is important to realize that diffraction describes only those parts of the sample which contain coherently scattering lattice planes. Most structural discussions focus on the crystallographic models, because these contain most of the available information. Regions of the samples which are less well ordered may contribute only a small background to measured diffraction spectra and yet still represent a substantial volume fraction of samples which are inhomogeneously oxidized and possible only partially ordered with respect to metallic layering as well. With three metallic components we have no reason to expect such perfect ordering, and at present the level of residual disorder which persists after repeated sintering, grinding and annealing is unknown. Some experiments which are not based on diffraction suggest substantial disorder. For example, X-ray absorption near-edge CuK spectra show

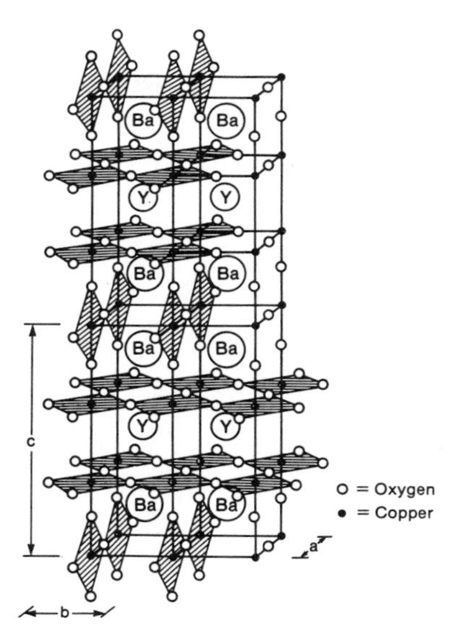

Fig. III.21. Idealized $YBa_2Cu_3O_7$ structure showing $Cu(O_{1/2})_4$ or $(CuO_2)_\infty$ planes and $(CuO)_\infty$ chains (ref. 38).

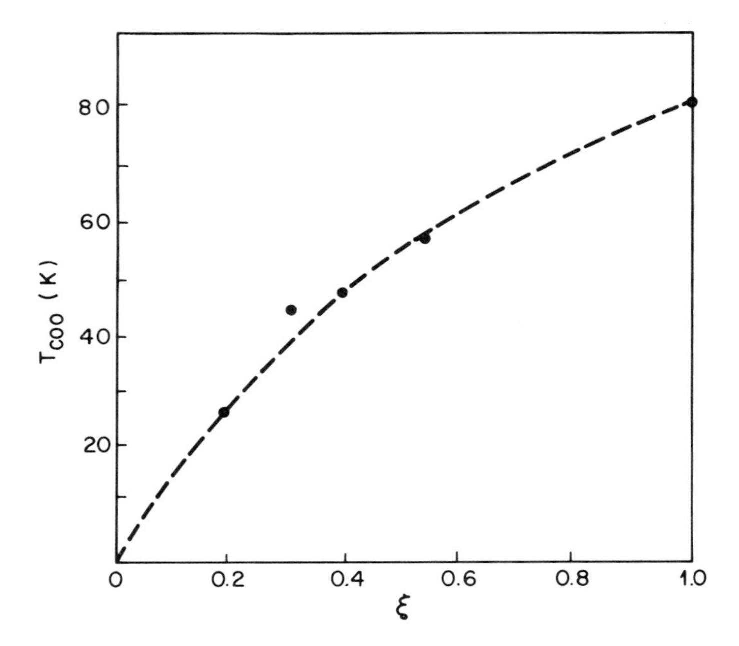

Fig. III.22. Zero resistivity temperature T_{coo} vs chain oxygen ordering parameter ξ in $La(Ba_{2-x}La_x)Cu_3O_{7+\delta}$ alloys (ref. 41).

structure which has been interpreted[39] as evidence that Cu is substituting into the Y/La sites as Cu_x where $x = 0.2-0.3$, but more generally one could say that there may be some reconstructed interfacial regions (which do not contribute to the diffraction spectra) which contain such Cu complexes, and that these interfacial regions represent 20-30% of the sample. Similarly it has been argued[40] that grains of $YBa_2Cu_3O_7$, even with a narrow transition at $T_c \sim 90K$, actually consist only of thin ($<1\ \mu m$) superconductive shells surrounding normal (and probably semiconductive) cores.

While the $YBa_2Cu_3O_7$ phase forms in multiphase samples, $La(Ba_{2-x}La_x)Cu_3O_{7+\delta}$ forms in the same orthorhombic structure only over a narrow composition range centered near $0 < x \lesssim 0.1$. Other values of x degrade

T_c rapidly, and this degradation has been found[41] to correlate well with $\xi = [n(04) - n(05)] / [n(04) + n(05)]$, which measures the integrity of the Cu1-04b chains. This correlation is shown in Fig. 22.

Apart from disordering the (04b, 05a) occupation probabilities one can also change T_c by adding or removing oxygen, and this is an additional effect in $YBa_2Cu_3O_{7-y}$. The results,[42] which are shown in Fig. 23, are spectacular and indicate a phase transition between $y = 0$ and $y = 0.4$ centered on $y = 0.24$. The orthorhombic structure of the $y = 0.4$ phase is similar to that of $y = 0$, but the additional oxygen vacancies in the Cu1 plane are ordered, as discussed in the next section. The samples shown in Fig. 23 were prepared by a special anneal to remove oxygen slowly and facilitate defect ordering. Less spectacular but similar results for $T_c(x)$ were obtained by quenching in vacuum with a range of quench temperatures.[43]

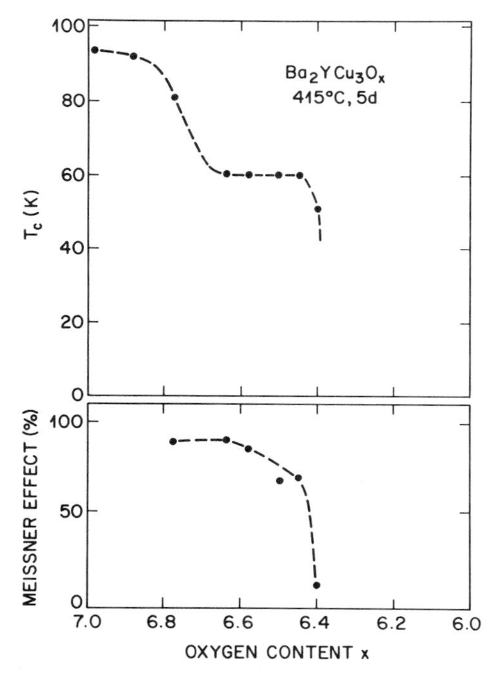

Fig. III.23. Plateaus in T_c suggest at least two different phases near $x = 7.0$ and 6.6 in $YBa_2Cu_3O_x$. The Meissner volume shows a high filling factor for $x > 6.4$ and no appreciable super-conductivity for $x < 6.4$ (ref. 42).

5. Oxygen Vacancy Ordering in $YBa_2Cu_3O_{7-x}$

Perhaps the most remarkable aspect of the Chu-Wu material is the facile diffusion of oxygen both within the sample and its evaporation (or reversible absorption) in equilibrium with O_2 gas, making this material a strong candidate for the best oxygen sensor known, especially in the temperature range 500-900C. Facile diffusion of cations is the basis for solid electrolytes, and the chemical trends in cation mobilities in many solid electrolytes correlate well with lattice softness, as reflected by low frequency optic vibrational modes, small coordination numbers, and lattice instabilities.[44] Thus while we are surprised by such facile anion diffusion, we should not be surprised to find that such facile diffusion is correlated with high-T_c superconductivity, especially if the diffusing O ions form part of the metallic and superconductive paths embedded in the complex structures of these materials. The possibility of "tricking nature"[32] by utilizing such soft lattice correlations was suspected long ago.[44]

Several features of the crystal structure change with variable x. The O atomic sites in the unit cell shown in Fig. 19 which are most strongly effected are the 04 and 05 sites in the Cu1 (chain) plane. Samples are normally oxidized near 900C, but the maximum degree of oxidation (usually taken as $x = 0$) occurs only when the sample is further annealed in O_2 near $T_a = 350C$. Thermogravimetric data[45] for heating and cooling in O_2 and in air are shown in Fig. 24. The oxygen gain (loss) near $T = 350C$ corresponds to $\Delta x = 0.5$ and it is reversible in O_2 but not in air (probably because of H_2O in the latter). While annealing near $T = 900C$ is necessary to form the ordered $YBa_2Cu_3O_{7-x}$ phase, the final value of x and the ordering of O onto 04 chains in the orthorhombic phase results from annealing near 350C. This temperature is unusually low (less than half the melting point), and it again reflects the

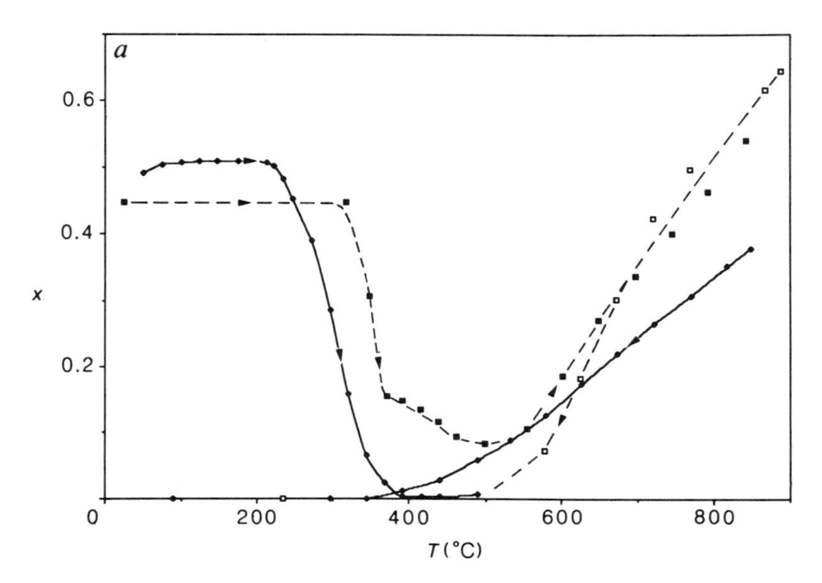

Fig. III.24. Thermogravimetric curves of $YBa_2Cu_3O_{7-x}$ in O_2 (continuous line) and in air (dashed line). Arrowheads indicate heating and cooling trajectories (ref. 45).

lattice softness. The vacant 05 sites provide obvious channels for O diffusion, and this presumably explains why x cannot be reduced to < 0. The maximum value of dx/dT_a is found near the orthorhombic/tetragonal phase transition composition, where domain walls probably increase oxygen diffusivity. The interaction energy between oxygen on nearest neighbor sites is estimated[46] to be 0.07-0.10 eV.

Selected area electron diffraction patterns (from single domains) have shown a wide variety of narrow diffuse streaks as a function of preparative treatment, macroscopic x values, quenching and annealing procedures. Very simple crystal structures have been proposed[47] for $x=0$ [04 occupied, 05 vacant, Fig. 19], for $x=1$ [both 04 and 05 vacant], and for $x=0.5$ [the $x=0$ and $x=1$ configurations alternate between b axis chains, with $N = 2 \cdot 13 - 1 = 25$ atoms cell, see Fig. 25]. Perhaps the most attractive superlattice pattern of all is that for $x = 0.12$, corresponding to $T_c \sim 93K$, as shown in Fig. 26. Here the oxygen vacancies occur at every fourth site on alternating chains, with $N = 8 \cdot 13 - 1 = 103$ atoms/cell, giving rise to alternating metallic chains and semiconductive chain segments. It seems likely that the plateaus in T_c shown in Fig. 23, with an abrupt transition near $x = 0.23$, can be explained in terms of percolation of domains which have structures similar to those shown in Figs. 25 and 26. Other combinations of metallic and semiconductive b-axis Cu1-04 chains may be obtained by somewhat different preparative methods. However, the separation of vacancies leaving short-range ordering into metallic and superconductive chain segments is expected to persist in samples which are nominally described as $YBa_2Cu_3O_7$, that is $x = 0$ is probably not an equilibrium phase. Instead one should

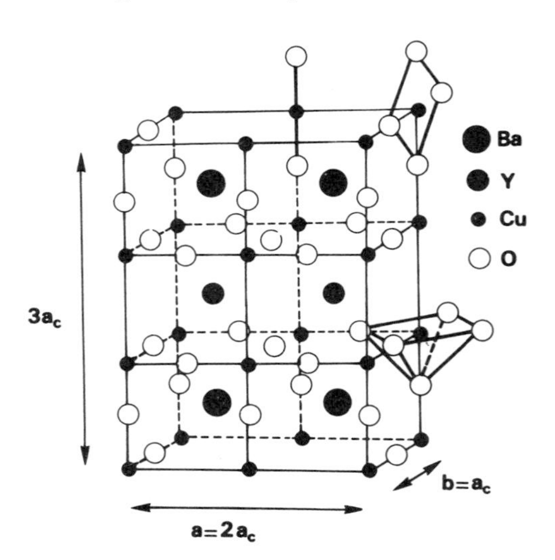

Fig. III.25. Proposed supercell for $YBa_2Cu_3O_{6.5}$, twice as large as the unit cell for $YBa_2Cu_3O_7$ (ref. 47).

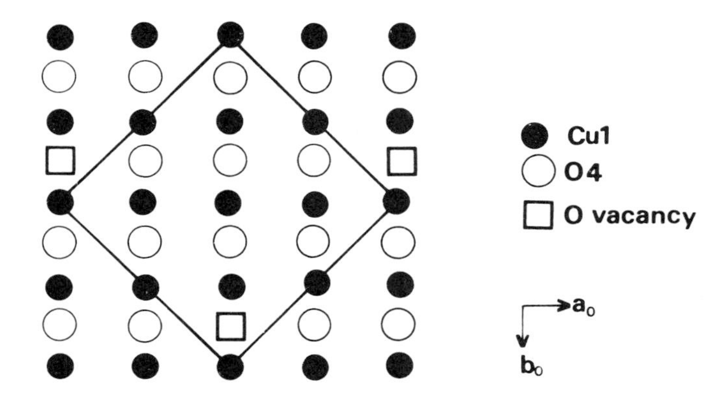

Fig. III.26. Proposed supercell (basal or Cu1 plane) for $YBa_2Cu_3O_{6.87}$, eight times as large as the $YBa_2Cu_3O_7$ unit cell (ref. 47).

visualize substantial short-range order even in samples which have been quenched too rapidly to exhibit the domain pattern shown in Fig. 25. Similarly, short-range metallic chain segments may also persist even in the absence of macroscopic orthorhombic symmetry or identifiable twinning.

The structure shown in Fig. 23 provides a natural basis for understanding the very high O diffusivity in $YBa_2Cu_3O_{7-x}$. The idea is that with an activation energy for O ion diffusion of order 0.1 eV (IX.3), we have a situation similar to that of Ag diffusion in the superionic conductors $\alpha-AgI$ and Ag_2S, where the activation energy is also about 0.1 eV. There tracer studies have shown that the ionic motion is highly correlated, and this has led to the proposal of the caterpillar mechanism.[48] Here this mechanism would involve correlated motion of O4 and O5 atoms, permitting the O4 vacancies to move down the Cu1-O4 chains (Fig. 19) by passing through the O5 sites. This motion is diffusive, not vibrational, and is the extreme limit of "anharmonic" motion. Its possible relation to strong electron-phonon coupling is discussed in IV.7. Anomalous low-frequency optic modes, which are the vibrational analogue of the caterpillar diffusive mode, have been found near 40 cm^{-1} ($\sim$ 60K) and are discussed in VI.2. Notice that these modes are in the classical limit (vibrational occupation number $n(T) \propto T$) for $T \gtrsim 100K$.

6. Copper and Other Cation Replacements

Many metals M are known to alloy substitutionally with Cu over a wide solubility range and to add or subtract electrons from the Cu conduction band according to simple valence considerations. Thus it is of interest to study the effect of such alloying on $La_{2-x}[Ca, Sr, Ba]_x Cu_{1-y} M_y O_{4-\delta}$ and $YBa_2Cu_{3-y}M_yO_{7-\delta}$ superconductive alloys. In such experiments considerable attention must be paid to correlations between y and δ, but bearing this in mind the results may be quite instructive.

The most obvious substitutions are M = Ni and Zn, which respectively subtract and add one conduction electron/atom, with the magnetic moment of Ni larger than Zn. In the case of $La_{2-x}Sr_xCuO_{4-\delta}$ it was found[49] that both Ni and Zn depress T_c rapidly, but that Zn depresses T_c more rapidly than Ni. This is probably because the electrons added by Zn compensate the holes added by Sr, and it is even possible that these impurities are spatially associated as well.

For $YBa_2Cu(1)Cu(2)_2O_{7-\delta}$ the question immediately arises whether M replaces Cu(1) or Cu(2). For $M = Co^{+3}$ and $A\ell^{+3}$ combined single-crystal X-ray diffraction, thermogravimetric and neutron powder diffraction studies have shown[50] that M replaces Cu(1). This is what one would expect on energetic grounds, because the chemical and strain energies associated with size and electronegativity misfits should be smaller for four-fold coordinated Cu(1) than for five-fold coordinated Cu(2). The thermogravimetric studies also suggest that Ni^{2+} and Zn^{2+} enter the Cu^{2+} sites without altering the local oxygen environment, which could imply substitution at Cu(2) sites, since these remain +2 even while oxygen is being removed from the Cu(1) chains and the latter are being converted to +1 at $\delta = 1$. On the other hand, substitution of Co or Aℓ at Cu(1) sites apparently tends to increase δ, or at least to hinder evaporation of oxygen from the O_4 sites. With this increasing stability these impurities may also segregate inhomogeneously.[51]

More accurate X-ray and neutron TOF diffraction experiments have shown[51] a more interesting and complex pattern of Co, Ni and Zn substitution for Cu in YBCO. The data show that the Ni atoms substitute *only* for Cu(2),

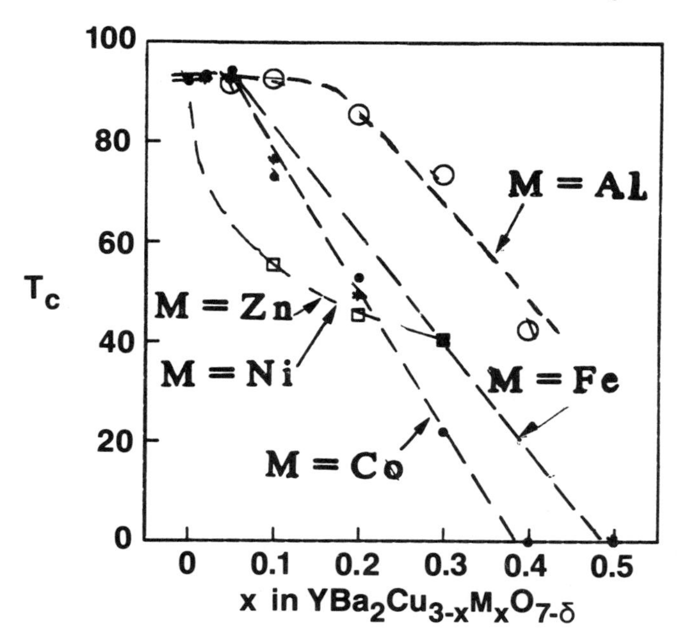

Fig. III.27. Variation of T_c with substitution of M for Cu (from ref. 51).

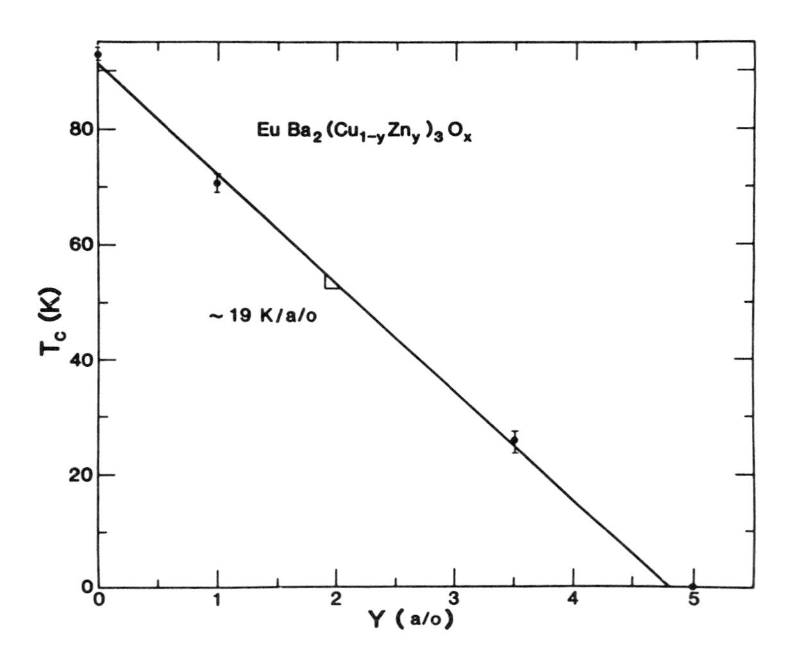

Fig. III.28. Variation of T_c with substitution of Eu for Y and partial substitution of Zn for Cu (from ref. 52).

whereas two-thirds of the Zn atoms replace Cu(1). Also both Ni and Zn impurities are accompanied by $O^{\square}$ oxygen vacancies in the ratios $Ni[O^{\square}(1)]_{0.3}[O^{\square}(2)]_{0.1}$ and $Zn[O^{\square}(1)]_{0.3}[O^{\square}(2)]_{0.2}$, where $O^{\square}(1)$ and $O^{\square}(2)$ refer to oxygen vacancies in the Cu(1)-O(1) chains and Cu(2) O_2 (2) planes, respectively. With these data it appears that the much greater reduction of T_c by Zn than by Ni (see below) may be explained by the larger concentration of $O^{\square}$ (2) which accompany Zn compared to Ni.

The depression of T_c for M= Zn and Ni is qualitatively different than for M= Aℓ, Co and Fe. As shown in Fig. 27, initially the latter have little effect, which also suggests inhomogeneities. The curvature seen in the Ni and Zn data is apparently also due to inhomogeneities. Studies of $EuBa_2(Cu_{1-y}Zn_y)_3\ O_{7-\delta}$ alloys have shown[51] that $T_c(y)$ is very nearly linear in y, as shown in Fig. 28. (The difference between the Y and Eu compound is that the latter can be formed at a higher temperature, producing a more homogeneous Zn distribution.) Interesting effects of annealing in O_2 on Co-doped samples have recently been reported.[51]

Perhaps the most striking differences between $La_{2-x}Sr_xCu_{1-y}M_yO_{4-\delta}$ and $YBa_2Cu_{3-y}M_yO_{7-\delta}$ alloys are the slopes d log $(T_c, T_m)/dy$ in the two materials. Typically the scale for the former alloy system is set by x, the Sr dopant level.[49] This suggests that the M = Ni or Zn impurities are compensating the Sr dopant either electrically (Zn) or magnetically (Ni). The situation is quite different for the latter alloys, where d log $(T_c, T_m)/dy \sim 10^2$, or about three times larger than for the former alloys. In particular, the normal-state metal-semiconductor transition temperature T_m (defined by Eqn. (7-1)) equals T_c for y about 0.04 in the former alloys, but in the latter alloys this condition is reached for $y < 0.01$. This is a very drastic effect on the the normal-state conductivity. It cannot be explained simply as the result of carrier trapping, because in the former alloys the formal valence of Cu is $[(2 + x - y) +]$, whereas in the latter alloys it is $[(2.33 - y) +]$ and with $x = 0.15$ one would expect the doping effect to be twice as small in the latter as in the former.

In addition to direct substitution of Cu(1), one can also alter the Cu(1) chains by removing 04 atoms, which tends to remove Cu(1) atoms, for example, in $YBa_2Cu_{3-\epsilon}O_{7-x}$, $La(Ba_{2-u}La_u)Cu_{3-\epsilon}O_{7-x}$ alloys, where values of ϵ of order x/4 are typical.[52] If the $Cu^{\square}$ were nearly equivalent to Zn substitution, then the depression of T_c in these alloys with increasing x could just as well be ascribed to increasing ϵ.

In any event, the effort [52] to unify T_c depression by relating it entirely to doping $(CuO_2)_\infty$ planes in both $(La, Sr)_2CuO_{4-y}$ and the 90-K alloys above is based on several erroneous ($\epsilon = 0$; $T_c(x = 0.4) = 90K$) and one dubious (only the planes count) assumptions and cannot be accurate.

One of the questions concerning cation substitution involves replacing Y (no magnetic moment) with another rare earth (RE) with a magnetic moment, which in many early papers was shown to have almost no effect on T_c. This is not surprising, because SCF energy band calculations show that $N_\alpha(E_F) = 0$ for $\alpha = Y$. A more interesting exchange is a magnetic RE for Ba, because although $N_{Ba}(E_F) << N_{Cu}(E_F)$, it is not negligible. The system $Nd \cdot (Ba_{2-x}Nd_x)Cu_3O_7$ has been studied carefully and the results compared to those of $La(Ba_{2-x}La_x)Cu_3O_7$ alloys.[52a] The general trends in terms of depression of T_c are similar, but there is a great broadening ΔT_c in the former alloys which is absent from the latter. It seems likely that this broadening reflects pair-breaking by the Nd_{Ba} spins, much as localized spins break pairs in intermetallic superconductors. This broadening may also reflect partial segregation of the additional Nd, corresponding to incipient phase separation. Even in the most careful experiments such macroscopic (rather than microscopic) sources of broadening are not easily excluded. We note that the atoms on the Ba site seem to be coupled anharmonically to the CuO_2 planes (V.4).

7. Normal-State Transport Properties

The normal-state resistivity is often observed to be closely correlated with T_c when both are varied as a function of composition. Typically we say the resistivity is metallic when $d\rho/dT > 0$ and semiconductive when $d\rho/dT < 0$. The semiconductive resistivity is probably associated with carrier freeze out, but in practice it has so far not proved useful to attempt to define an activation energy, for instance from an impurity level to a band edge, presumably because of sample inhomogeneities. These inhomogeneities also give the normal-state resistivity a percolative character when the material is metallic at high T and semiconductive at low T. In these mixed cases one can define a metal-semiconductor transition temperature T_m as

$$d\rho/dT_m = 0. \tag{1}$$

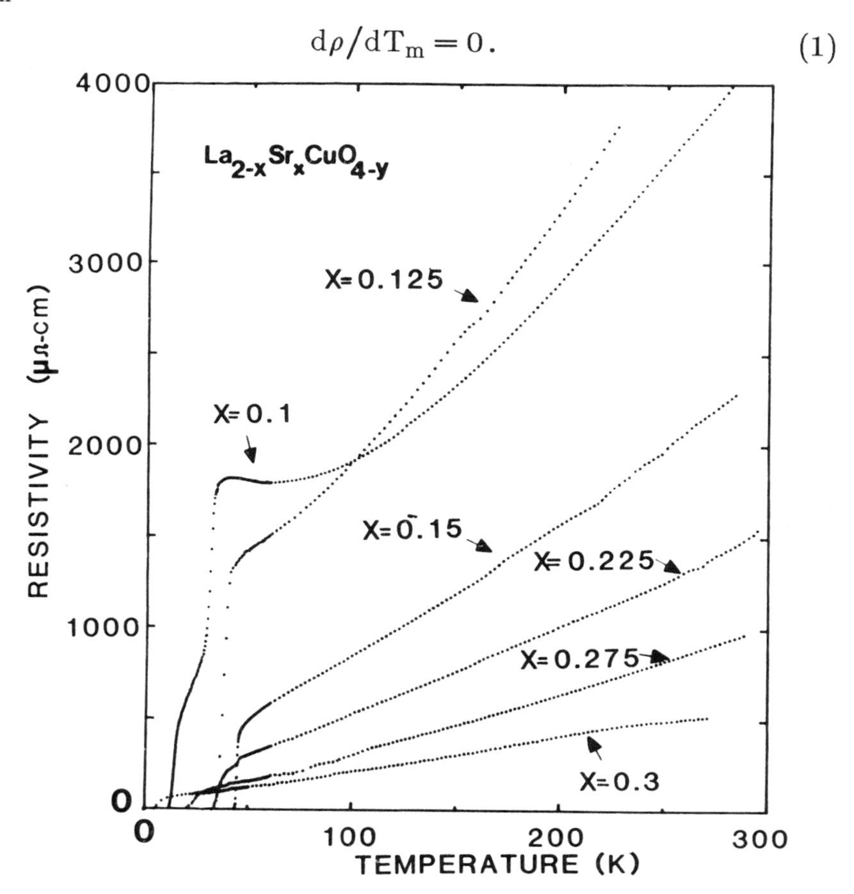

Fig. III.29. Resistivity $\rho(T, x)$ in $La_{2-x}Sr_xCuO_4$ sintered samples.

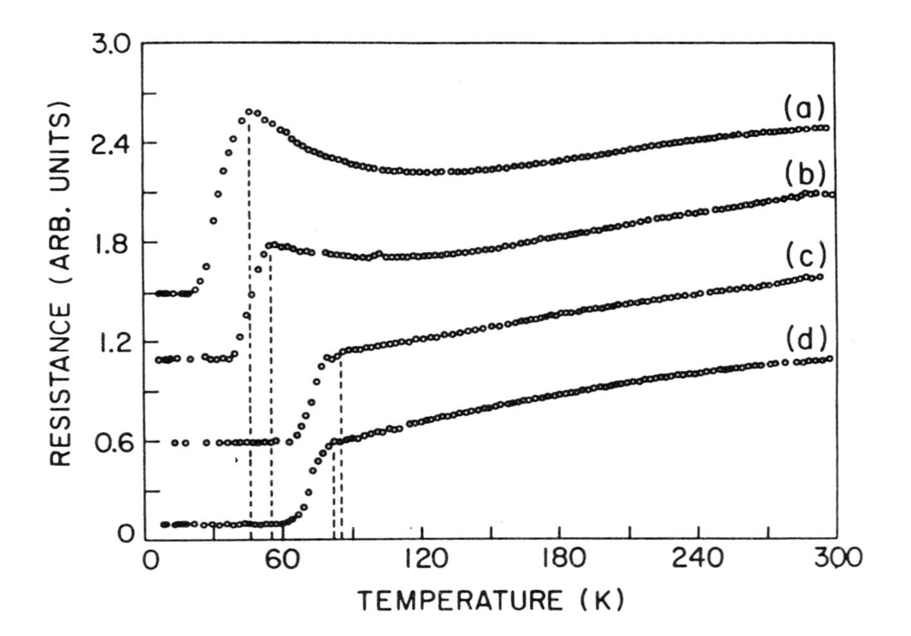

Fig. III.30. Resistivity $\rho(T, x)$ in $La(Ba_{2-x}La_x)Cu_3O_{7-\delta}$ sintered samples. Curves (a)-(d) refer to x = 0.12 annealed in various oxygen partial pressures increasing to 7 atm.

Higher values of T_m correspond to more semiconductive behavior, lower values to more metallic behavior. Generally decreasing values of T_m are correlated with increasing values of T_c.

The normal-state resistivity $\rho(T, x)$ for $La_{2-x}Sr_xCuO_4$ shown[53] in Fig. 29 exhibits the usual correlation between T_m and T_c, while the two-stage drops for x = 0.1 and 0.225 are suggestive of phase separation. Even when $T_c << T_m$ superconductive filaments occupying a small fraction of the sample can be present; the extreme case x = 0 was discussed with Fig. 18. More homogeneous behavior seems to be observed in the $La(Ba_{2-x}La_x)$ CuO_7 alloys shown[54] in

Fig. 30, where it appears that the maximum value of T_c is reached for $T_c = T_m$.

The most striking correlations between $\rho(T, x)$ and T_c are those obtained for $YBa_2Cu_3O_{7-x}$. We already saw in Fig. 23 that in samples specially prepared for homogeneity, $T_c(x)$ exhibits plateaus associated with separation of phases[42] with ordered oxygen vacancies at $x = 1/8$, $1/2$ and 1. Near the percolative boundary at $x = 0.22$ between the $x = 1/8$ and $x = 1/2$ phases the resistivity behaves anomalously in the sense that T_m reaches a maximum $> 300K$ (see Fig. 31) near $x = 0.22$. It is true that at this value T_c has dropped to $\sim 60K$, but at $x = 0.7$ (where $T_c \sim 30K$), T_m has returned to $\sim 100K$. Also, as shown in Fig. 31, $\rho(300K)$ is larger near $x = 0.22$ than it is near $x = 0.1$ or 0.4. These effects are presumably due to scattering at interfaces between $x = 1/8$ and $x = 1/2$ regions.

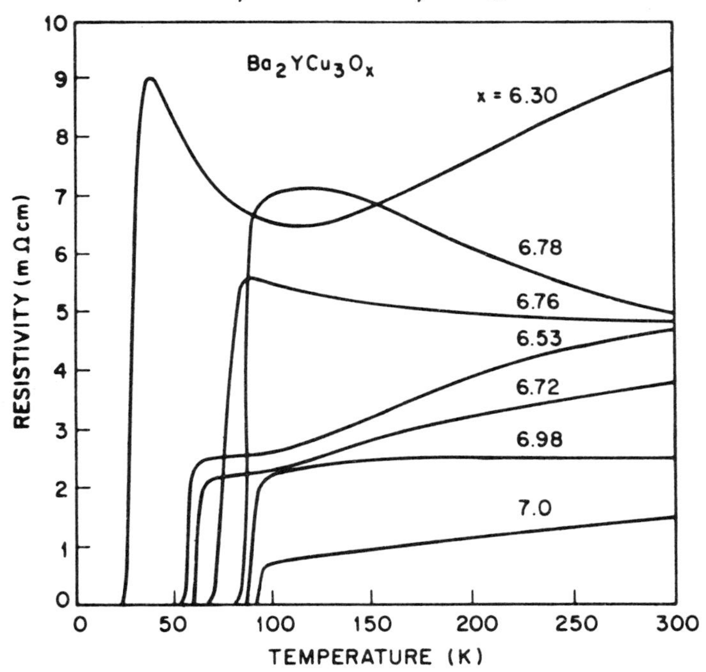

Fig. III.31. Representative resistivities $\rho(T, x)$ in $YBa_2Cu_3O_x$ sintered samples. Fully metallic behavior ($T_m \lesssim T_c$) is observed near $x = 6.53$ and 6.72, but not for $x = 6.76$ and 6.78.

At temperatures well above T_m generally $\rho(T)$ is approximately linear, see $x = 0.15$ in Fig. 29 for example. For $T > 300K$ oxygen vacancies may be formed, contributing extrinsically to the increase in $\rho(T)$. However, below 300K vacancies may disorder and this may contribute to $\rho(T)$. However, the electron-phonon interaction with soft phonons seems the most likely explanation for the linear increase in $\rho(T)$ which gives $d \ln \rho/dT \sim 1.5 \cdot 10^{-2}/K$ for $x = 0.15$. This value, although large, is considerably smaller than the value $5 \cdot 10^{-2}/K$ observed[55] for $ErRh_4B_4$, which is a normal intermetallic compound similar to the A15 materials (see Appendix C).

With layer crystals a large anisotropy of the resistivity tensor is expected, and this means that the Hall coefficient can be measured reliably only on single crystals. The Hall resistivity R_H is simply related to the carrier density n only in the case of a single, nondegenerate parabolic energy band, but in general it does provide qualitative information on the electron- or hole-like character if the transport properties are dominated by a single type of carrier. Data on a nominally single-crystal epitaxial thin film have been reported[56] for the in-plane resistivity and the Hall coefficient in $La_{1.94}Sr_{0.06}CuO_4$ ($T_c \sim 15K$). This composition is barely metallic and in general this doped material is less easily grown homogeneously than $YBa_2Cu_3O_{6.9}$. The data show hole-like conduction, and polycrystalline data on $La_{2-x}Sr_xCuO_{4-y}$ show this changes over to electron-like for $x > 0.2$. It seems likely that there is a high density of oxygen vacancies in these films, which increases rapidly for $x > 0.2$, but is substantial even for $x < 0.2$. These vacancies trap carriers and probably have a broad distribution of binding energies. This would explain the linear (a-bT) reduction in the positive Hall number n (defined as $1/eR_H$, as for a single parabolic energy band), as shown in Fig. 32. The apparent increase in n for $T < 100K$

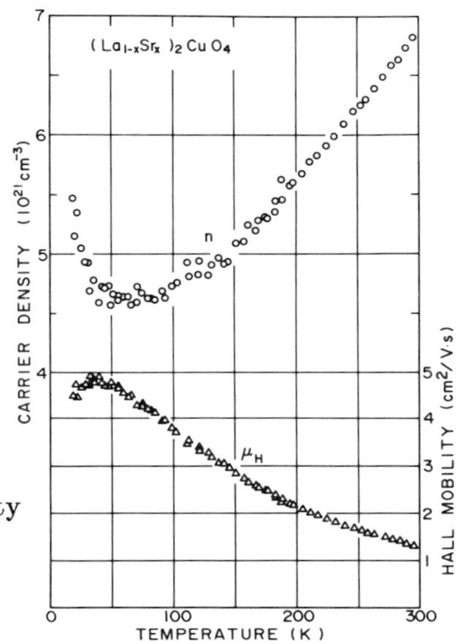

Fig. III.32. Nominal carrier density and Hall mobility in $(La_{1-x}Sr_x)_2CuO_4$.

is probably associated with inhomogeneities and percolative currents.

Twinned single-crystal data[57] for the anisotropic resistivity and Hall resistivity R_H (**H** parallel or perpendicular to the c axis) of $YBa_2Cu_3O_{6.9}$ exhibit surprising temperature dependences. The resistivity anisotropy, $\rho_{zz} = \rho_{\parallel} \sim 100\,\rho_{\perp}$, is not itself surprising, because the layered structure means that current flows along the c axis between planes probably along highly resistive semiconductive defects, leading to semiconductive $\rho_{\parallel}$ for

T < 200K, as shown in Fig. 33. The single-crystal in-plane resistivity is accurately linear in T. The in-plane ($\mathbf{H}\|_c$) Hall number $n_\|$ is positive and is given by a+bT, with a = 0, as shown in Fig. 34. Qualitatively this is similar to the behavior of $n_\|$ in epitaxial thin films of $La_{1.85}Sr_{0.15}CuO_4$ (Fig. 32), but the feature a = 0 is indeed unexpected. This can be explained, together with the linear resistivity $\sigma_\|$, by soft phonons (IV.9).

An ordered defect structure in epitaxial films of Y–Ba–Cu–O shows superconductive properties similar to those of $YBa_2Cu_3O_{7-x}$ but its normal-state properties differ drastically from those of all other high-T_c cuprates, including the recently discovered bismates and thallates (XI). The defect structure has the formula

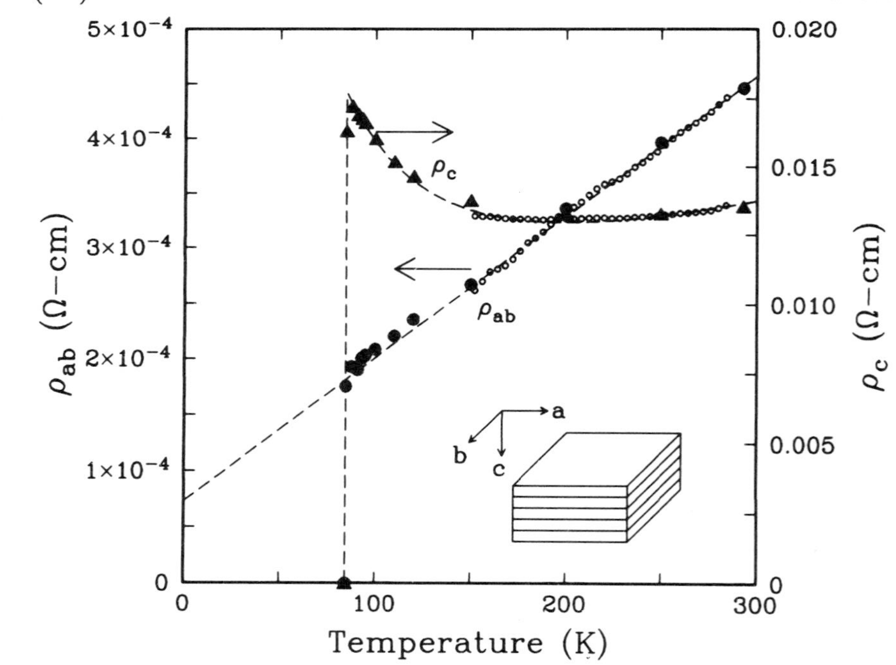

Fig. III.33. Temperature-dependent principal resistivity components of $YBa_2CuO_{6.9}$ measured on a single crystal.

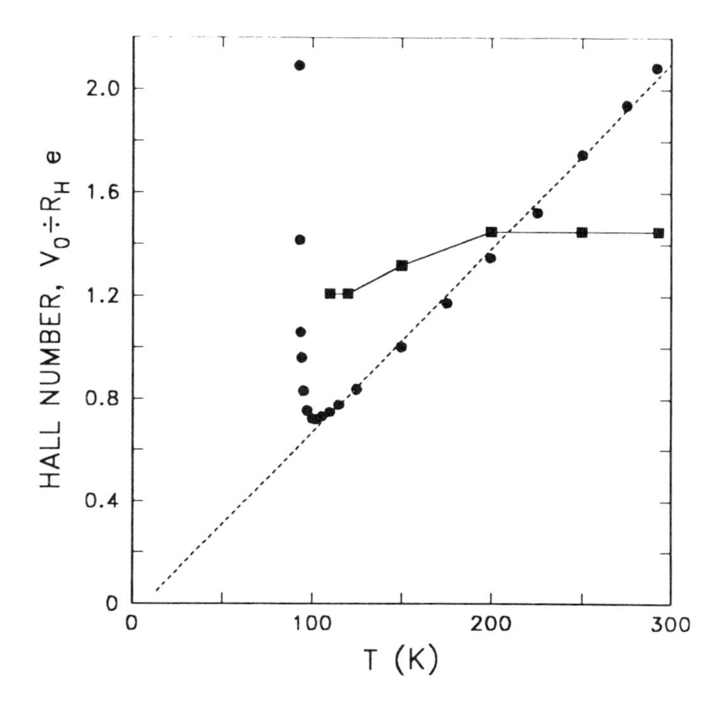

Fig. III.34. The Hall number $n = V_o/R_H e$ for **H**$\parallel$c (circles) and **H**$\perp$c (squares). The dashed line shows that $n_\parallel$ is nearly proportional to T.

$Y_2Ba_4Cu_8O_{20-x} = 2(YBa_2Cu_3O_7) + 2CuO_3$ and it has a structure similar to $YBa_2Cu_3O_7$ except that every CuO chain layer becomes a chain bilayer. This presence of this chain bilayer has little effect on T_c, except that at low T it weakens the coupling between CuO_2 planes, which reduce T_c by about 10K.

The chain bilayers drastically modify the normal-state properties, doubling the normal-state conductivity and nominal Hall carrier density n_H at high temperatures $T \gtrsim 300K$. This suggests that the bilayer chains are highly conductive at high T while the single chains in $YBa_2Cu_3O_7$ are not. Near $T = 125K$ n_H reaches a maximum and with decreasing T it extrapolates to zero near $T = 55K$. Whereas in other cuprate high-T_c materials $\rho_{\parallel}(T)$ extrapolates to zero near $T = 0$, with bilayer chains $\rho_{\parallel}(T)$ extrapolates to zero near $T = 55K$, just as $n_H(T)$ does.

The most natural explanation for these results is that a Jahn-Teller transformation of the chain bilayers begins near 125K which causes the bilayer carriers to freeze out. There is no evidence in the data that this freeze-out has much effect on the CuO_2 planar carriers or the superconductivity associated with them. Thus the peculiar behavior of the bilayer chains is chiefly a reminder of the multiplicity of internal order parameters which can exist in cuprates with large unit cells and several distinct cuprate structural subunits.

8. Pressure Effects

One of the advantages of pressure experiments is that to the extent that they are reversible one might optimistically assume that percolative effects due to inhomogeneities are less important to dT_c/dp than to T_c, and doping effects also are relatively constant under pressure. The general trend is that at high pressure resistive transitions are broadened, especially near $\rho = 0$. Pressure effects are expected to be especially large near lattice or electronic phase transitions.

Most of the experimental data have been reported for dT_{co}/dp, where T_{co} is the resistively measured "onset" temperature. The general result[59] is that dT_{co}/dp is small in $YBa_2Cu_3O_7$ (the resistively measured values scatter around 0), but is very large and positive in $La_{2-x}(Ba, Sr)_xCuO_4$. For x near 0.15 observed values range between 0.1 and 0.3, whereas for A15 compounds typically $dT_{co}/dp \lesssim 10^{-2}$K/kbar. Inductive (ac susceptibility) measurements[59] at 90% value (near T_{co}) give +0.30K/kbar for $La_{1.8}Sr_{0.2}CuO_4$ in reasonable agreement with the resistive results, but +0.16 for $YBa_2Cu_3O_7$, about twice as large as any of the resistive values. This suggests that inhomogeneous percolation effects may still be a problem in determining dT_{co}/dp resistively. A noteworthy difference between the inductive and resistive techniques is that the former show the transition *narrowing* under pressure, apparently as a result of two-phase behavior which may be more easily resolved inductively.

We saw in Sec. 6 and Fig. 28 that replacement of Cu by Zn apparently occurs more homogeneously in $XBa_2Cu_3O_7$ for X= Eu than for X= Y. The pressure dependence of ρ for X= Eu for 1% Zn doping[52] is shown in Fig. 35. We see in the resistivity a small "foot" near $\rho = 0$ for P = 0 which narrows with increasing P, analogous to the inductive

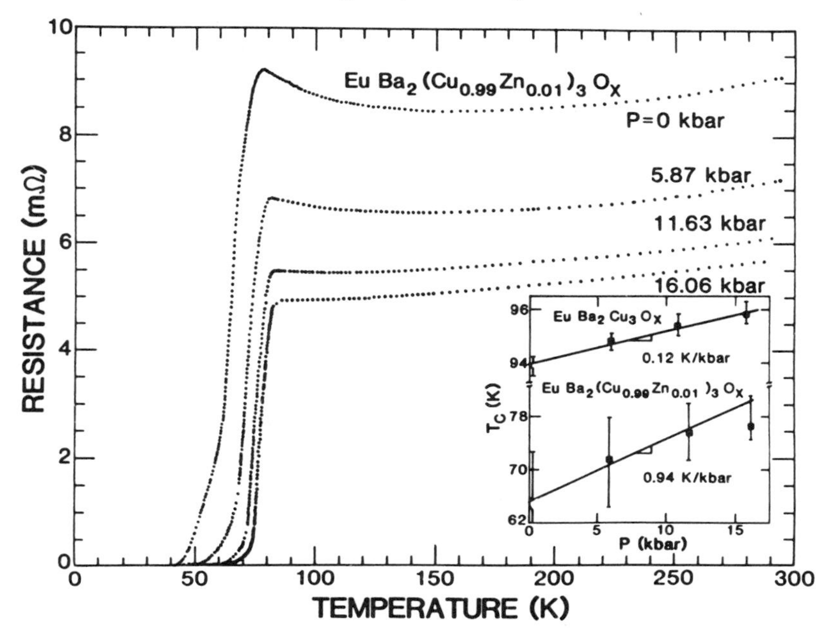

Fig. III.35. Resistance $\rho(T,P)$ of $EuBa_2(Cu_{0.99}Zn_{0.01})_3O_7$ at four pressures. The inset compares the pressure dependence of the midpoint transition temperature T_{cm} for doped and undoped samples. Note the change in ordinate scales in the inset.

results[59] for X = Y. Also for $x = 0$ the value of $dT_{cm}/dP = +0.12K/kbar$ is close to the inductive value (+0.16 K/kbar), as shown in the inset. But most remarkable is the value $dT_{cm}(x = 0.01)/dP = 0.94 K/kbar$. This is by far the largest pressure dependence of T_c yet observed, about three times larger than $(La, Sr)_2CuO_4$ and about 100 times larger than most A15 intermetallic compounds.

Under pressure the orthorhombic unit cell of $YBa_2Cu_3O_7$ contracts anisotropically.[60] In a powder measurement it has so far not proved possible to decide whether or not the orthorhombic distortion is maintained or reduced at pressures of order 100kbar, but even at these pressures the distortion is at most reduced by about a factor of two.

The smaller value of dT_c/dp in $YBa_2Cu_3O_{7-x}$ may be understood as the result of oxygen vacancies (for $x > 0$) or even as the result of missing oxygen atoms relative to the "filled" pseudoperovskite formula $YBa_2Cu_3O_9$. The evidence for this is the composition dependence[61] of $dT_c(x)/dp$ in $La_{2-x}Sr_xCuO_{4-y}$ which for $x < 0.15$ increases with T_c but which for $x > 0.15$ decreases much faster than T_c, as shown in Fig. 36. This rapid decrease is probably correlated with y increasing from near zero[62] starting near $x = 0.15$. The internal relaxation around the oxygen vacancies may compensate to a considerable extent the macroscopic strain generated by the external pressure.

9. Metallic (But Not Superconducting) Copper Oxide Pseudoperovskites

In the preceding two sections we have seen many examples of high-T_c superconductors where T_c correlates well with the metallic conductivity in the normal state and the metal-semiconductor transition temperature T_m. From

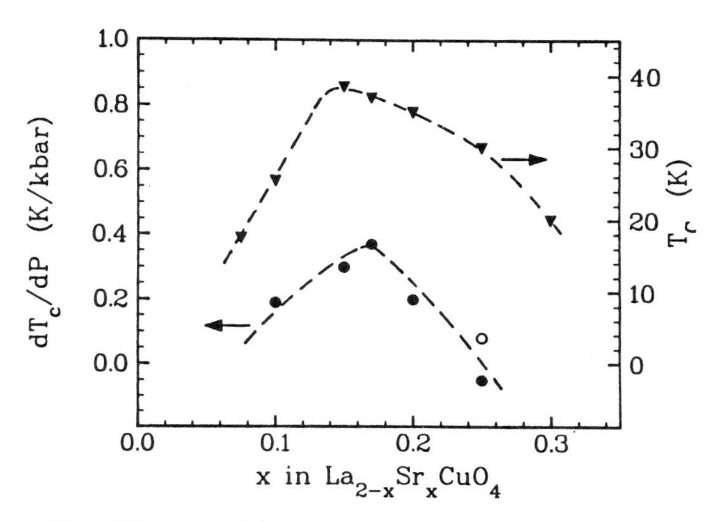

Fig. III.36. $T_c(x)$ and dT_c/dp in $La_{2-x}Sr_xCuO_{4-y}$.

these examples one might conclude that whatever mechanism is responsible for high-T_c superconductivity becomes effective in these materials so long as they are not semiconductive, that is, so long as they are metallic in the normal state. In this section we discuss counterexamples which show that this conjecture is *not* correct. Metallic normal-state behavior is a necessary condition for high-T_c superconductivity (a conclusion which may be näively obvious, but which is not reached by several exotic theories), but it is not sufficient.

Apart from the two families of superconductors $La_{2-x}(Ca, Sr, Ba)_xCuO_4$ and $YBa_2Cu_3O_{7-x}$ which we have discussed extensively, there are many pseudo-perovskite quaternary compounds containing copper octahedra (sometimes reduced to capped or uncapped copper squares) which have been prepared by Michel, Raveau and coworkers.[62,63] In the reduced state most of these compounds are semiconductive, but some become metallic

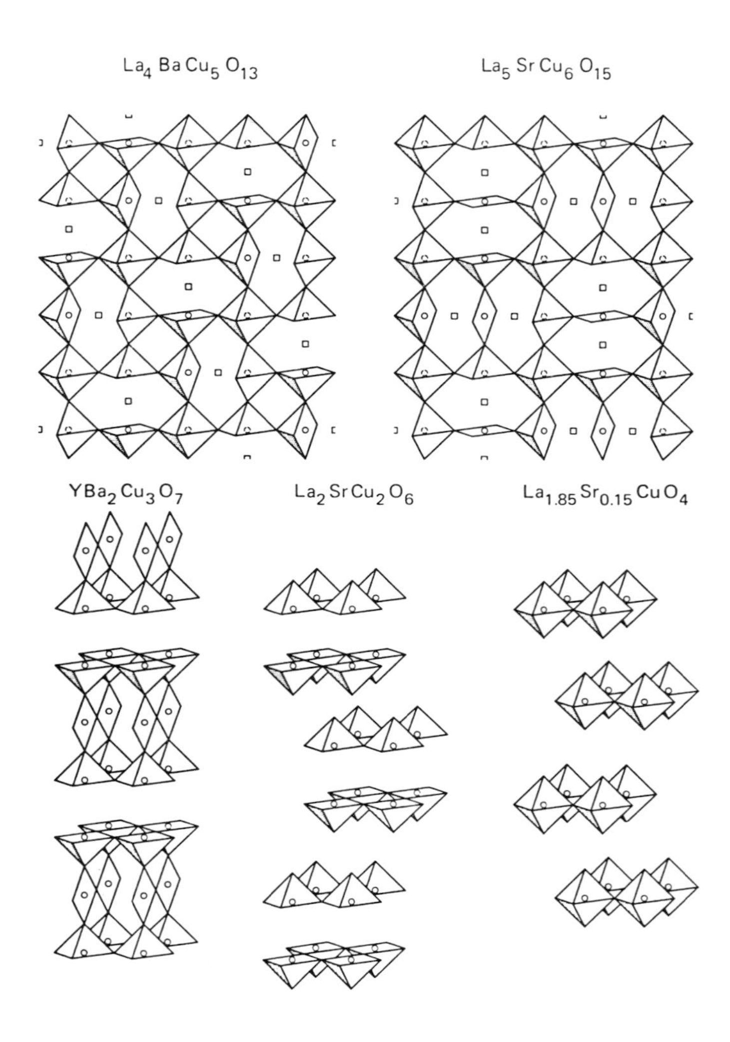

Fig. III.37. A comparative sketch of the structures of metallic copper oxides, some superconductive and some not.

upon being oxidized at temperatures near to or somewhat higher than 1000° C.

Three metallic pseudoperovskite quaternary Ca oxides have been prepared, found to be metallic, but not superconductive,[38,64] down to 5 K. These compounds are $La_4BaCu_5O_{13}$, $La_5SrCu_6O_{15}$, and $La_2SrCu_2O_{6.2}$. The structures of these compounds are complex (not surprisingly) and are illustrated in Fig. 37. Data on the magnetic susceptibility $\chi(T)$ are featureless and similar in magnitude to those for the high-T_c materials $(La, Sr)_2CuO_4$ and $YBa_2Cu_3O_7$, so that the magnetic susceptibility appears to be an irrelevant property. Thus the materials are fully metallic, like $YBa_2Cu_3O_7$, and the absence of superconductivity must arise from structural differences. This absence occurs when there are no CuO_2 planes or CuO chains, but only chains of corner-sharing octahedra or square pyramids. We will see in (VI.3, 4) that the key feature of the CuO_2 planes is the strong coupling of transverse O atom vibrations normal to the plane to superconductive electrons.

REFERENCES

1. R. Chevrel, M. Sergent and J. Prigent, J. Solid State Chem. *3*, 515 (1971).

2. B. T. Matthias, M. Marezio, E. Corenzwit, A. S. Cooper and H. E. Barz, Science *175*, 1465 (1972).

3. J. M. Tarascon, F. J. DiSalvo, D. W. Murphy, G. Hull, and J. V. Waszczak, Phys. Rev. *B29*, 172 (1984).

4. R. Chevrel, M. Hirrien and M. Sergent, Polyhedron *5*, 87 (1986).

5. O. Fischer et al., in Douglass III, p. 485.

6. R. Chevrel and M. Sergent, in *Superconductivity in Ternary Compounds I* (Top. Cur. Phys., Springer, N. Y., 1982, Ed. O. Fischer and B. Maple), p. 25.

7. S. D. Bader, S. K. Sinha, B. P. Scheweiss and B. Renker, ibid., p. 223; D. J. Holmgren, R. T. Demers, M. V. Klein and D. M. Ginsberg, Phys. Rev. *B36*, 5572 (1987).

8. J. Bolz, J. Hauck and F. Pobell, Z. Phys. *B25*, 2351 (1976).

9. R. W. G. Wyckoff, *Crystal Structures* Vol. 2 (Interscience, N.Y., 1960) p. 359-77. See also F. S. Galasso, *Structure, Properties and Preparation of Perovskite-Type Compounds* (Pergamon, Oxford, 1969) for a list of some 400 additional pseudo-ternary perovskite-type compounds $A(B, B')O_3$.

10. Y. N. Venevtsev, Mat. Res. Bull. *6*, 1085 (1971).

11. A. W. Sleight, J. L. Gillson, and P. E. Bierstedt, Sol. State Comm. *17*, 27 (1975).

12. D. E. Cox and A. W. Sleight, Sol. State Comm. *19*, 969 (1976); Y. J. Uemura et al., Nature *335*, 151 (1988).

13. C. Chaillout et al., Sol. State Comm. *65*, 1363 (1988).

14. B. Batlogg, Physica *126B*, 275 (1984).

15. W. Reichardt, B. Batlogg, and J. P. Remeika, Physica *135B*, 501 (1985).

16. T. Hikata et al., Phys. Rev. *B36*, 5578 (1987).

17. L. F. Mattheiss, E. M. Gyorgy, and D. W. Johnson, Jr., Phys. Rev. *B37*, 3745 (1988); R. J. Cava et al., Nature *332*, 814 (1988).

18. J. G. Bednorz and K. A. Muller, Z. Phys. *B64*, 189 (1986).

19. H. Takagi, S. Uchida, K. Kitazawa and S. Tanaka, Jap. J. Appl. Phys. Lett. *26*, L123 (1987).

20. K. Knox, J. Chem. Phys. *30*, 991 (1959).

21. A. F. Wells, *Structural Inorganic Chemistry* (Fifth Edition, Oxford, 1984), p. 1116 ff.

22. H. Müller-Buschbaum, Angew. Chem. Int. Ed. Engl. *16*, 674 (1977).

23. J. D. Jorgensen et al., Phys. Rev. Lett. *58*, 1024 (1987); P. Day et al., J. Phys. *C20*, L429 (1987); T. Egami et al., Int. Conf.

Supercond., Drexel, 1987 (World Scien. Pub.).

24. N. Terada et al., Jap. J. Appl. Phys. *26* L510 (1987).

25. T. Fujita et al., Jap. J. Appl. Phys. *26*, L368 (1987).

26. D. Jérome, W. Kang, and S. S. P. Parkin, Proc. Mag. and Mag. Mat., Chicago, 1987.

27. R. J. Cava, R. B. van Dover, B. Batlogg and E. A. Rietman, Phys. Rev. Lett. *58*, 408 (1987).

28. R. M. Fleming, B. Batlogg, R. J. Cava and E. A. Rietman, Phys. Rev. *B35*, 7191 (1987).

29. C. W. Chu et al., Phys. Rev. Lett. *58*, 405 (1987).

30. A. Allgeier et al., Sol. State Comm. *64*, 227 (1987).

31. A. W. Sleight in *Chemistry of High-Temperature Superconductors* (Ed. D. L. Nelson, M. S. Whittingham, and T. F. George, ACS Symp. Ser. 351, ACS, Washington, 1987), p. 1.

32. B. T. Matthias, Physica *69*, 54 (1973).

33. J. G. Allpress, J. V. Sanders and A. D. Wadsley, Acta Cryst. *B25*, 1156 (1969).

34. R. J. Cava, D. W. Murphy and S. M. Zahurak, J. Electrochem. Soc. *130*, 2345 (1983).

35. P. M. Grant et al., Phys. Rev. Lett. *58*, 2482 (1987); S. A. Shaheen, Phys. Rev. *B36*, 7214 (1987).

35a. J. T. Markert, C. L. Seaman, H. Zhou, and M. B. Maple, Sol. State Comm.*66*, 387 (1988).

36. M. K. Wu et al., Phys. Rev. Lett. *58*, 908 (1987).

37. J. D. Jorgensen et al., Phys. Rev. *B36*, 3608 (1987); R. J. Cava et al., Phys. Rev. Lett. *58*, 1676 (1987).

38. J. B. Torrance, Y. Tokura, and A. Nazzai, Chemtronics *2*, 120 (1987); E. M. Engler, Chemtech, *17*, 542 (1987).

39. F. W. Lytle, R. B. Greegor and A. J. Panson, Phys. Rev. *B37*, 1550 (1988).

40. D. S. Ginley et al., Phys Rev. *B36*, 829 (1987).

41. C. U. Segre et al., Nature *329*, 227 (1987); A. Maeda, T. Yabe, K. Uchinokura and S. Tanaka, Jap. J. Appl Phys. *26*, L1368 (1987).

42. R. J. Cava et al., Phys. Rev. *B36*, 5719 (1987); A. Davidson et al., Appl. Phys. Lett. *52*, 157 (1988).

43. M. Tokumoto et al., Jap. J. Appl. Phys. *26*, L1565 (1987).

44. J. C. Phillips, J. Electrochem. Soc. *123*, 934 (1976); J. C. Phillips, in Douglass II, p. 413.

45. P. Strobel, J. J. Capponi, C. Chaillout, M. Marezio and J. L. Tholence, Nature *327*, 306 (1987).

46. P. Strobel, J. J. Capponi, M. Marezio, and P. Monod, Sol. State Comm. *64*, 513 (1987).

47. M. A. Alario-Franco, J. J. Capponi, C. Chaillout, J. Chenevas and M. Marezio, MRS Symp. Proc. *99* (1987); C. Chaillout et al., Phys. Rev. *B36*, 7118 (1987); Solid State Comm. *64*, 283 (1988); B. Barbara et al., Phys. Lett. *A127*, 366 (1988); D. J. Werder, C. H. Chen, R. J. Cava, and B. Batlogg, Phys. Rev. *B38*, 5130 (1988).

48. I. Yokota, J. Phys. Soc. Jap. *21*, 420 (1966).

49. J. M. Tarascon et al., Phys. Rev. *B36*, 8393 (1987).

50. T. Siegrist et al., MRS Symp. Proc. *99* (1987).

51. J. M. Tarascon, Phys. Rev. *B37* (1988); H. A. Borges et al., Physica *148B*, (1987); Y. Shimakawa, Y. Kubo, K. Utsumi, Y. Takeda, and M. Takano, Jap. J. Appl. Phys. *27*, L1071 (1988); T. Kajitani, K. Kusaba, M. Kikuchi, Y. Syono and M. Hirabayashi, Jap. J. Appl Phys. *27*, L354 (1988); S. Horn et al., Phys. Rev. *B38*, 2930 (1988).

52. F. Izumi et al., Jap. J. Appl. Phys. *26*, L1616 (1987); K. Takita, Jap. J. Appl. Phys. *27*, L67 (1988).

52a. H. Nozaki, S. Takekawa and Y. Ishizawa, Jap. J. Appl Phys. *27*, L31 (1988).

53. J. M. Tarascon, L. H. Greene, W. R. McKinnon, G. W. Hulle and T. H. Geballe, Science *235*, 1373 (1987).

54. D. Mitzi et al., unpublished.

55. L. D. Woolf, D. C. Johnston, H. B. McKay, R. W. McCallum and M. B. Maple, J. Low Temp. Phys. *35*, 651 (1979).

56. M. Suzuki and T. Murakami, Jpn. J. Appl. Phys. *26*, L524 (1987).

57. S. W. Tozer, A. W. Kleinsasser, T. Penney, D. Kaiser and F. Holtzberg, Phys. Rev. Lett. *59*, 1768 (1987); T. Penney, S. von

Molnař, D. Kaiser, F. Holtzberg, and A. W. Kleinsasser, Phys. Rev. *B38*, 2918 (1988).

58. K. Char et al., Phys. Rev. *B38*, 834 (1988).

59. R. Griessen (unpublished); M. R. Dietrich, W. H. Fietz, J. Ecke and C. Politis, Jap. J. Appl. Phys. *26*, Suppl. 26-3 (1987).

60. W. H. Fietz, M. R. Deitrich and J. Ecke, Z. Phys. *B69*, 17 (1987).

61. J. E. Schirber, E. L. Venturini, J. F. Kwak, D. S. Ginley and B. Morosin, Phys. Rev. *B35*, 8709 (1987).

62. C. Michel and B. Raveau, Rev. Chim. Miner. *21*, 407 (1984).

63. C. Michel, L. Er-Rakho, M. Hervieu, J. Pannetier, and B. Raveau, J. Sol. State Chem. *68*, 143 (1987).

64. J. B. Torrance, Y. Tokura, A. Nazzai and S. S. P. Parkin, Phys. Rev. Lett. *60*, 542 (1988).

IV. New Theory

1. Aims and Scope

It is tempting to suppose that the remarkable properties of high-T_c superconducting materials are an indication of an entirely new physical phenomenon unlike any previously known. However, almost all the superconductive phenomena observed to date in the oxides are qualitatively similar to those previously found in the old metallic materials. We have reviewed experiment and theory for these old materials in Chaps. I and II for two reasons. First we want to place the new materials in perspective against the many decades of effort on the old materials. Second for those whose memories do not reach back to the discovery of the phenomenon, or even the more recent development of the fundamental BCS theory, we hope to learn from the old materials. We assume that the physics of high-T_c superconductivity in the new materials has much in common with that of the old materials. Both the similarities and the differences can help us to understand the origin of the remarkably high T_c's of the oxides.

The most complex yet well studied old materials are probably those found in island A (App. C), such as the A_3B compounds, where A is an early transition metal (like Nb) and B is a metalloid (like Ge or Sn). Such compounds contain $N = 4$ atoms/cell and structural subunits (the A atom chains) which are relatively unusual for metals. Starting from the discovery of high T_c superconductivity in V_3Si in 1950 the full exploration of these materials experimentally (good tunneling data) and theoretically (good vibrational spectra, reasonably good calculations of λ) has taken more than 30 years. Over these years we have seen many technical advances in both experiment and theory. On the other hand, the new materials have $N \gtrsim 10$

and exhibit many properties, such as high anionic mobility and truly exotic crystal structures, which did not occur in intermetallic high-T_c compounds. It may be that the increased complexity of the oxide materials is just balanced by our increased technical sophistication. If this is the case, understanding the new materials and utilizing their technological potential may also require decades.

At this stage it would be rash to attempt to predict the future course of either experiment or theory. The aim of this chapter is to make it easier for the author (and hopefully the reader) to point in the right general direction. The subject has many obvious problems. I have tried to discuss only the most plausible solution(s) of these problems, especially in the context provided by crystal chemistry and materials physics. As discussed in Sec. IV.12, many alternative theories have been proposed based on other contexts (such as quark confinement, superfluidity in $He,^3$ or metallic superconductors with $T_c < 0.1K$) which seem to me to be too remote from the basic physics of oxides to be relevant. The complexity of these materials is such that one can ill afford the luxury of pursuing unproved "interesting" alternative theories if one is to succeed in testing the BCS theory itself; as we have seen, this project alone may require decades of research at the most advanced levels.

Throughout this chapter I will make suggestions regarding various features of the atomic and electronic structures which, if present, would tend to favor higher values of T_c. These suggestions are the theoretical analogues of the methods (essentially trial-and-error) which materials scientists have used to discover high-T_c materials. To achieve high T_c's several factors must simultaneously be optimized and theory, to the extent that it errs in estimating these factors, should predict lower T_c's than those observed, providing it is using the right general mechanism and errs on detailed technical factors. These factors are the scientific analogue of what logicians call proximate causes. Different approaches, based on new general mechanisms, or ultimate causes, may predict very high T_c's. I view these approaches with great skepticism, if only because such new general mechanisms, if correct, would generate far too many high-temperature superconductors, even among simple binary oxides. Some specific weaknesses of these novel or exotic general mechanisms are discussed near the end of this chapter.

2. Electronic Structure of $BaPb_{1-x}Bi_xO_3$

The material $BaPb_{1-x}Bi_xO_3$ is of special significance because it is the first oxide perovskite superconductor with $T_c > 10$ K. In Sec. III.2 we discussed the phase diagram and material properties of these alloys (Figs. III-8 and 11)

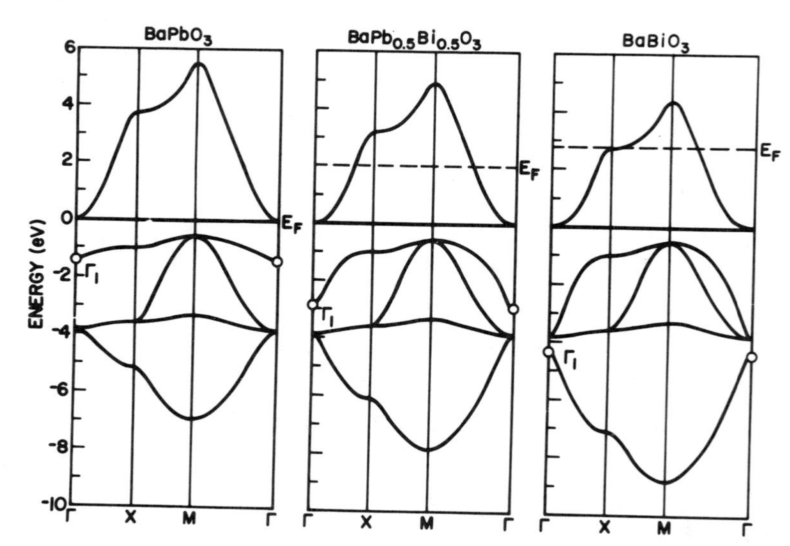

Fig. IV.1. A tight-binding fit to the energy bands of $BaPb_{1-x}Bi_xO_3$. In this fit the O-O second neighbor interactions are neglected, causing the flat nonbonding O bands at $E = E_F$ for $x = 0$ (ref. 2).

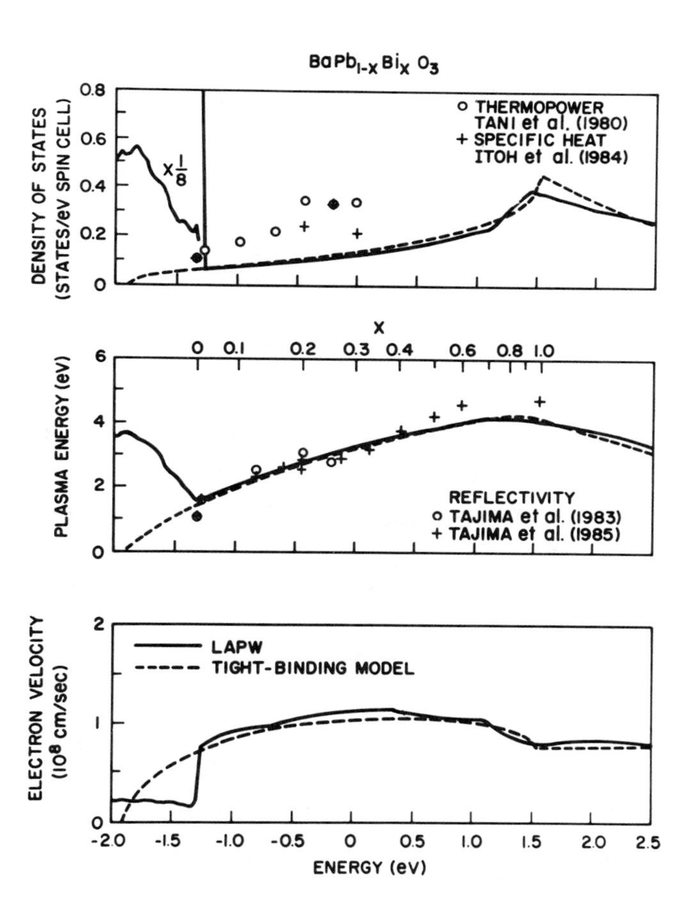

Fig. IV.2. Comparison of the predictions of the one-electron band model with $N(E_F)$ measured by specific heat and thermopower, and the band-filling measured optically (ref. 2).

and saw that T_c reached its maximum value near $x = 0.25$, and was metallic/semiconductive for $x \lessgtr 0.35$. We also saw that recent single-crystal data agree with earlier powder data as to the general features of the rotations of the (BiO_6)

octahedra which characterize the monoclinic semiconductor $BaBiO_3$.

The one-electron band structure of these alloys has been analyzed in depth.[1] With N = 5 atoms/cell (cubic) or N = 10 (tetragonal or monoclinic) and more than ten doubly occupied electronic orbitals in the cubic phase, the electronic structure is quite complex and requires expert analysis.[1] A simplified band structure[2] which contains the essential features of the complete calculations[1] is shown in Fig. 1.

The essential feature of these energy bands is the broad, nearly free electron band for $E > 0$ which consists of antibonding (Pb, Bi)6s−O2pσ hybrids. At $x = 1$ this band is half full and in the absence of superlattice ordering it generates a nearly spherical Fermi surface in the metallic region $0 \leq x \lesssim 0.3$. The general features of this band are consistent with the general alloy trends for the plasma energy (as measured by the Drude contribution to the reflectivity) as shown in the lower panel of Fig. 2.

Optical measurements in the infrared are in general a less sensitive probe of electronic structure near E_F than $N(E_F)$ as determined from electronic specific heat measurements of the Sommerfeld coefficient γ. These are shown in the upper panel of Fig. 2, which compares $N(E) = (1+\lambda)N_o(E)$ with the band structure calculations for $N_o(E)$. The experimental $N(E)$ is enhanced over the calculated $N_o(E)$, and the enhancement factor shows considerable scatter between the direct specific heat value and the indirect value measured in transport (thermopower) studies. It is interesting that the two measurements agree at $x = 0.25$. This could be a result of much better sample homogeneity, and this is consistent with the narrowed superconductive transition temperature at $x = 0.25$ shown in Fig. III-8, as well as the suggestion that at this composition the Pb_3Bi sublattice orders chemically to form a Cu_3Au superlattice.

Most interesting and difficult is not the superconductive behavior near $x = 0.25$ but rather the semiconductive behavior for x near 1. For $x = 1$ the Bi6s—02pσ antibonding band is half-full and it seems possible that the observed semiconductive gap of order 0.2 eV might be a Hubbard gap resulting from antiferromagnetic ordering (see the following section). In general, especially in materials containing no transition or rare earth metals, we are reluctant to abandon the usual one-electron self-consistent field (SCF) approach with which almost all the quantitative results on electronic structure and properties have been obtained. Indeed I know of no failures of this method which require Hubbard-like many body corrections in non-molecular materials in the absence of transition or rare earth elements.

The thorough analysis[1] of $BaBiO_3$ leads to the conclusion that a direct (indirect) energy gap of order 1 eV (0.2 eV) is expected near E_F in the monoclinic crystal structure. (The simplified bands in Fig. 1 are shown for the cubic structure; the full monoclinic bands are shown in Fig. 8 of ref. 1.) The actual bands as calculated are semimetallic (band overlap of order 0.1 eV), but this is easily explained either as a result of numerical limitations for the $N = 10$ atom unit cell, or as a result of local approximations to exchange and correlation. (These same local approximations make Ge a semimetal as well, and in general they underestimate the semiconductive energy gap by 0.5 eV in covalent materials with large dielectric constants.[3])

Also of interest is the charge exchange between Bi_I and Bi_{II}, formally $Bi^{(4-\delta)+}$ and $Bi^{(4+\delta)+}$ in the monoclinic unit cell. Chemically one can explain the semiconductive behavior by setting $\delta = 1$, but this extremely ionic model is shown to be unnecessary by the band calculation. In fact the calculation shows that the observed gap can be explained by $\delta \lesssim 0.01$, and this very small value of δ is consistent with measurements of core energy shifts.[2]

The band-structure model of these alloys predicts that if the carrier concentration could be brought closer to the half-filled band condition without producing either charge density waves or chemical ordering waves, T_c would be substantially increased. It turns out that this can be done better by valence substitution on the Ba site, i.e., by replacement of Ba by K, because this substitution has almost no effect on the band states near E_F, although it does shift E_F closer to the half-filled condition, where the electron-phonon interaction is a maximum.[4] So far this approach has increased T_c from 13K in $BaPb_{0.75}Bi_{0.25}O_3$ to $\sim$30K in $Ba_{0.6}K_{0.4}BiO_3$, with no evidence for either magnetic or alloy superlattice instabilities.[4]

In summary: as yet, no untoward many-body effects are needed to explain the structure and properties of $BaPb_{1-x}Bi_xO_3$ alloys. Moreover, the band model has successfully predicted the chemical path which increases T_c from 13K to $\sim$ 30K in the (Ba, K) BiO_3 alloys. (This is a local field effect, whether it arises purely electronically, as suggested by the band model, or whether it involves defects and superlattice ordering, a possibility mentioned at the end of III.2. By contrast, in exotic models based on anything but the electron-phonon interaction, it is customary to place all local-field effects in adjustable parameters, which are supposed to be determined by someone else.) But now, as we proceed to the pseudoperovskite copper oxides, this situation changes dramatically.

3. Formal Valence in the Cuprates

Given the complexities of $(La, Sr)_2CuO_4$ and $YBa_2Cu_3O_{7-x}$, both chemically and structurally, the theoretical discussion should begin at the simplest point. We can denote the formal ionic valence of a metal M by M[n+], for instance La[3+], Y[3+], Sr[2+], Ba[2+]. The one-electron self-consistent field (SCF) calculations discussed in the next section show that in these cases the actual charge distribution is approximately consistent with the formal valence picture. On the other hand, as we saw in the preceding section, the description of monoclinic $BaBiO_3$ as $Ba_2Bi[+3]Bi[+5]O_6$ is quite misleading: the actual charge transfer $\pm\delta e$ is better described with $\delta \sim 0.01$ according to the SCF calculations.

The formal description of La_2CuO_4 is quite simple: La[3+], Cu[2+], O[2−]. For the alloy $La_{2-x}Sr_xCuO_{4-y}$ this becomes La[3+], Sr[2+], O[2−] and Cu[(2+x−2y)+], if one arbitrarily assigns the formal valence discrepancy entirely to Cu. This could then be written as $Cu[2+]_{1-x+2y}$ and $Cu[3+]_{x-2y}$, and descriptions of this sort have appeared, often with $y=0$. These descriptions do not agree with the SCF charge densities, as discussed in the next section. Here we wish to review some experimental data bearing on the formal valence.

The normal valences for cuprates are +1 (as in Cu_2O) and +2 (as in CuO). Even a small fraction of Cu[+3] is surprising. One can test whether Cu[+3] is present by comparing core-level positions by photoemission spectroscopy. So far little evidence has been found[5] for Cu[3+]. It seems more reasonable to suppose that O is in the O[(2−δ)−] state than to describe Cu as Cu[3+].

When the doped alloy $La_{2-x}Sr_xCuO_{4-y}$ is prepared, Sr is an acceptor which contributes one fewer electron (or one more hole) to the Cu-O conduction band than the La it replaces. Similarly an $O^{\square}$ oxygen vacancy replaces O[2−], binds two fewer electrons and acts as a double donor. It is favorable energetically to form the defect complex $(Sr[1+])_2O^{\square}[2-]$ compared to distributing the Sr holes on O as O[(2−δ)−]. Apparently for $x \lesssim 0.15$ the oxygen vacancy concentration $y \leq 0.01$, but for $x \gtrsim 0.15$ there is a rapid increase in y according to measurements of $O^{\square}$ activation energies, positron line shapes, and direct thermogravimetric data. This dependence on x is roughly what one would expect if the Sr impurities form $O^{\square}$ whenever they are on nearest neighbor sites of the La sublattice.[6]

For $YBa_2Cu_3O_{7-x}$ the situation is more complex. With $x = 0$ the formal valence of Cu is [2.33+] if O is [2−]. This suggests that actually O is [(2−δ)−]. There are two inequivalent Cu sites, and for $x = 0$ the material is metallic and the coordination numbers are $N_c[Cu(1)] = 4$ and $N_c[Cu(2)] = 5$. For $x = 1$ we have $N_c[Cu(1)] = 2$ while the environment of Cu(2) is unchanged and the material is semiconductive. For $x = 1$ we might suppose that Cu(1) has formal valence [+1] while Cu(2) has formal valence [+2] while oxygen is O[2−]. For $x < 1$ oxygen vacancies are created on the Cu(1) chains, and the presence of these vacancies creates localized energy levels which may act as acceptors to dope the Cu(2) planes. We return to this point at the end of IV.5.

4. One-Electron Structures of La_2CuO_4 and $YBa_2Cu_3O_{7-x}$

In La_2CuO_4 with the valence states La[3+] and O[2−], one has Cu[2+] or a d^9 valence configuration. The Cu 3d orbitals hybridize with the O2p orbitals to form bonding and antibonding hybrids. By far the most significant aspect[7] of these energy bands is that when they are fitted with tight-binding or atomic orbital parameters $E_{3d}(Cu)$

and $E_{2p}(O)$ agree to within 0.2 eV. This agreement, the *Mattheiss relation*, is crucial to understanding the special chemical properties of cuprates, which include not only high-T_c superconductivity, but also the special square-planar coordination and the rich structural diversity of Cu compounds (III.3). The resulting energy bands near $E = E_F$ in the tetragonal phase are similar to those of $BaBiO_3$ (Fig. 1, IV.2) in the sense that the Fermi energy intersects a nondegenerate band with band width 4 (3.5) eV in $BaBiO_3$ (La_2CuO_4), and this band is half-full.[7] However, the tetragonal distortion in La_2CuO_4 of the Cu-centered oxygen octahedron is large and as a result the width of the Cu3d − O2pσ hybrid band for k in the z direction is small (about 0.2 eV). This gives the energy bands a strongly two-dimensional character. The energy bands in the two-dimensional limit have saddle points at the centers of the

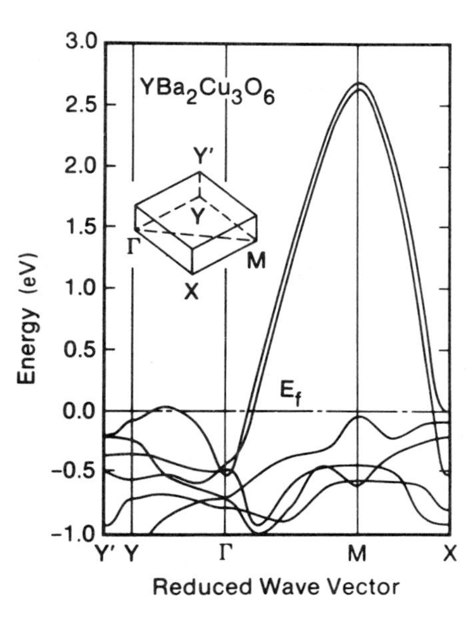

Fig. IV.3. Energy-band structure of $YBa_2Cu_3O_6$. Only the six highest Cu 3d − O2p bands are shown (ref. 10).

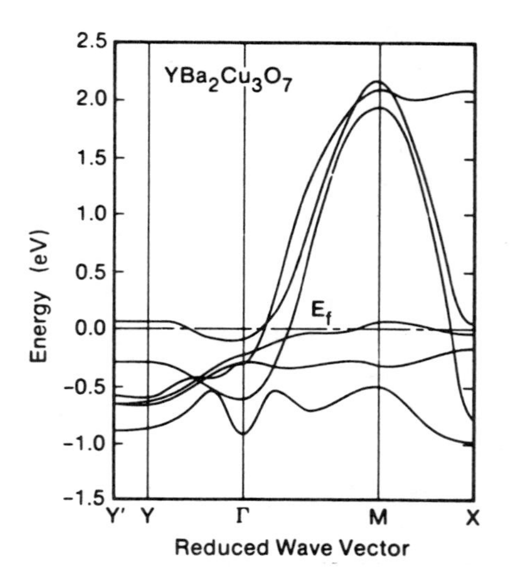

Fig. IV.4. Energy-band structure of $YBa_2Cu_3O_7$ (ref. 10).

zone edges which produce a logarithmic peak in the electronic density of states N(E) and E_F is centered on this peak. The effect of an orthorhombic distortion is to lift the degeneracy of the x- and y- axis peaks, so that E_F now falls symmetrically in a valley between the two peaks. However, this orthorhombic distortion by itself does not open an energy gap at centers of the zone edges[8,9] and so cannot explain the semiconductive character of La_2CuO_4 as shown in Fig. III. 29.

In $YBa_2Cu_3O_{7-x}$ there are three Cu planes, Cu(1) and two Cu(2), and as a result we must look for three antibonding $Cu3d-O2p\sigma$ antibonding energy bands near

E_F. Calculations have been reported [10] for x = 0 and 1. For x = 1 there are no O atoms in the x-y plane of the Cu(1) atom, which is connected to the Cu(2) planes only by Cu(2) – O(1) – Cu(1) – O(1) – Cu(2) chains, as in Fig. III. 19. As shown in Fig. 3, only two Cu d bands lie partly above E_F, so Cu(1) is essentially in a $3d^{10}$ state, or Cu(1) has formal valence +1. The coupling of the Cu(2) planes via the O(1)2p and Cu(1)4s states, as measured by the k_z band width, is relatively large (0.6 eV) and the overall band width is similar (3eV) to that in La_2CuO_4. As we saw in the preceding section, this leaves Cu(2) in a +2 state, with the two half-filled Cu(2) bands semiconductive by whatever mechanism explains the semiconductivity of La_2CuO_4.

For $YBa_2Cu_3O_7$ the Cu(2) bands are slightly changed and a third band associated with the Cu(1) – O(4) chains crosses E_F, as shown in Fig. 4. This band is not half full, and the metallic character is explained with Cu(1) in a

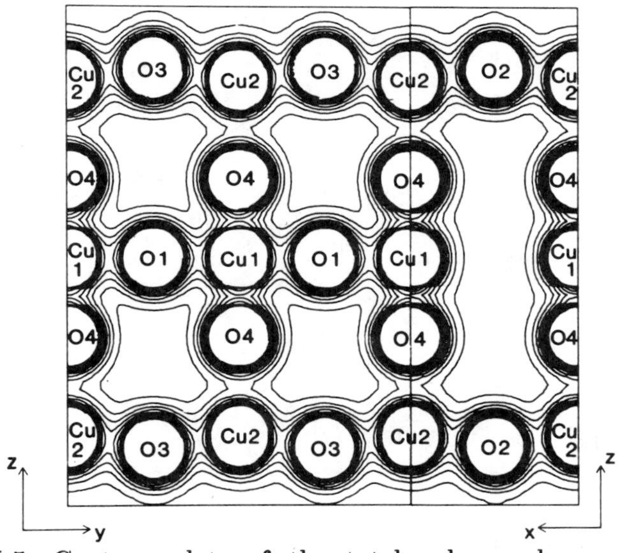

Fig. IV.5. Contour plots of the total valence charge density in $YBa_2Cu_3O_7$ with the Cu-O framework labelled and the Y and Ba atoms omitted (ref. **11**).

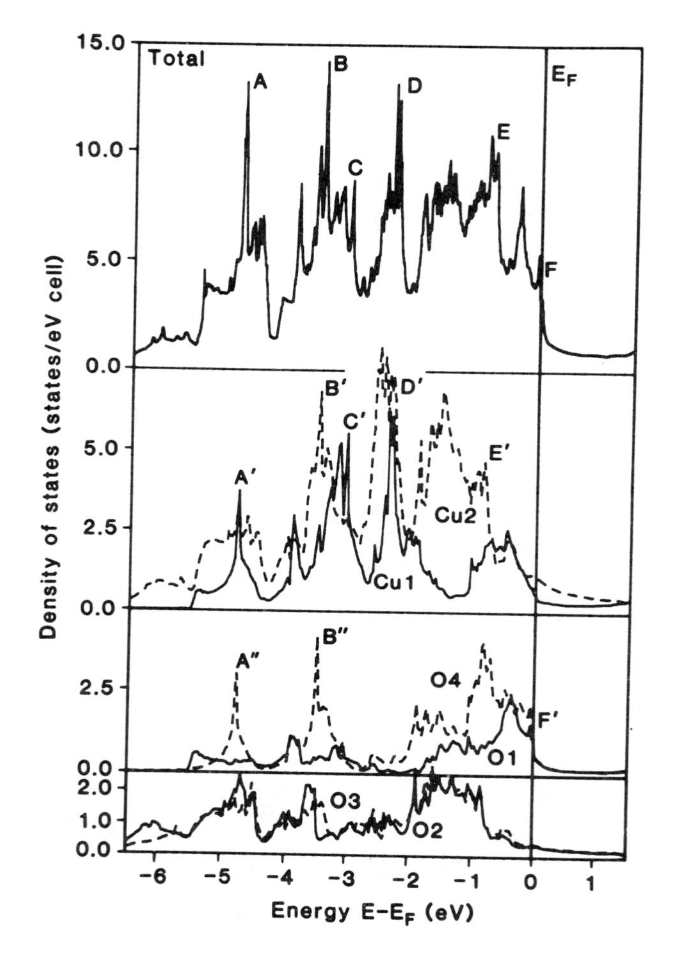

Fig. IV.6. Total density of states N(E) and projected densities of states $N_\alpha(E)$ in $YBa_2Cu_3O_7$ (ref. 11).

valence state closer to +2, the k_z valence band width is much reduced (to about 0.1eV), a somewhat surprising result of the SCF calculations.

The most detailed, instructive and complete calculations[11] of the electronic structure show explicitly that in addition to the σ antibonding bands the π states of the Cu1-O1 chains lie within 1eV of E_F. The contours of the total valence charge density are shown in Fig. 5. These have a tubular character which reflects the weakness of second neighbor electronic overlap. Because O4 is much closer to Cu1 (1.80Å) than to Cu2 (2.30Å), the buckled Cu2 — O2 - O3 plane is partially decoupled from the Cu1 — O1 chains with side O4s. This decoupling is reflected in the $N_\alpha(E)$ partial densities of states, which show O2, O3 peaks where there are Cu2 peaks, and O1, O4 peaks where there are Cu1 peaks (Fig. 6). In this figure it is clear that the planar Cu2 — O2 antibonding π band is centered near $E_F - 1.5$eV and has an upper edge near $E_F - 0.8$ eV. The Cu1 — O1 chain (with side O4's) is centered near $E_F - 0.8$ eV and has an upper edge just at E_F.

From Figs. 4 and 6 it is clear that current at $E = E_F$ is still carried by the broad Cud-Opσ antibonding bands in both the Cu2 O2 O3 planes and the Cu1 O1 O4 chains, with high-density π Cu1 O1 peaks which nearly overlap E_F, but that the πCu2 peaks merely represent highly polarizable quantum well walls placed between the chains and the Y non-polarizable walls. It is important to realize that the details of the electronic states within ± 0.5 eV of E_F shown in Fig. 4 and 6 cannot be relied on, both because of convergence problems in the one-electron calculation, and because dielectrically anisotropic nonlocal exchange and correlation energies are omitted entirely in these calculations at present. These corrections will be discussed later, but here we mention that their inclusion could raise the π chain bands, especially the O4 peak in Fig. 6, and increase $N(E_F)$ substantially. Such an upward shift of the O4 peak would also couple the planes and chains more effectively, which would favor higher T_c's.

5. Semiconductivity and Antiferromagnetism in La_2CuO_4 and $YBa_2Cu_3O_6$

The generic features common to La_2CuO_4 and $YBa_2Cu_3O_6$ are that both contain CuO_2 planes with Cu in a [2+] state when O is [2−] and as a result the highest antibonding $3d - 2p\sigma$ energy band is half-filled. In this case if we construct an antiferromagnetic state with 3d holes of opposite spin localized on alternating Cu [2+] $3d^9$ ions then the Coulomb repulsive energy is lowered as the overlap between electronic wave functions is reduced to a minimum. This antiferromagnetic state is a special case of a spin density wave (SDW) state with maximal amplitude and shortest wave length. It appeared in the early literature as part of a discussion between Mott and Slater[12] on the origin of the insulating properties of NiO.

Within the usual approximations of SCF techniques (which treat Coulomb exchange and correlation energies by the local density approximation, which evaluates these energies according to their estimated values in a free electron gas) these magnetic energies are generally underestimated, especially for narrow d and even narrower f bands. Hubbard therefore suggested[13] an idealized model Hamiltonian to describe the magnetic energies in the narrow-band or nearly atomic limit. Hubbard's Hamiltonian appears to be very simple, as it includes only two parameters, U which describes the repulsion between two electrons of opposite spin in the same nondegenerate 1s atomic orbital, and t to describe the overlap or transfer energy (the one-electron band width), in conventional notation $1s1s\sigma$.

For many years the Hubbard model appeared to be mainly of academic interest to field theorists. Hubbard originally designed the model to describe magnetism in elemental transition metals, but there the 3d orbital degeneracy both reduces the effects of Coulomb repulsion (due to intra-atomic electron-electron correlations) and greatly complicates (for practical purposes, renders impossible) the solution of his equations as suitably generalized. Also it turned out that in most cases the orbitally degenerate energy bands of transition metals were actually not narrow enough to make expansions in t/U reasonably convergent. Gradually interest therefore turned to rare earth metals where the Anderson model[14] (originally designed to discuss d-s interactions in transition metals) can be applied to f-d interactions, and where the f band width is small enough to treat t perturbatively.

Another way to generate examples with narrow bands is to reduce the effective dimensionality of the system, as with chains and planes of atoms (for example, the Friedel model

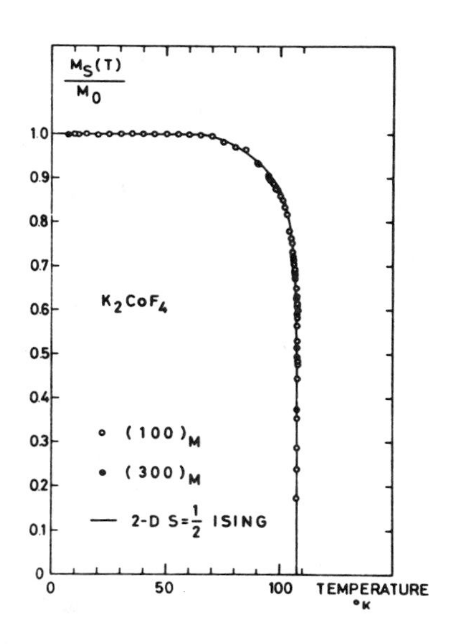

Fig. IV.7. Sublattice magnetization of a model two-dimensional Ising system, as monitored by two magnetic neutron Bragg scattering intensities (ref. 16).

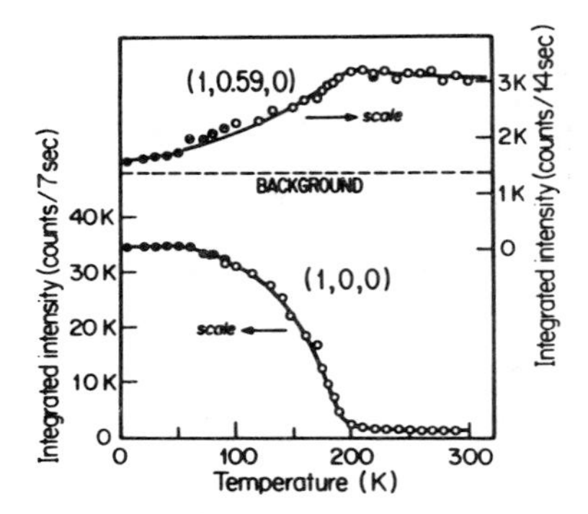

Fig. IV.8. Antiferromagnetic elastic Bragg peak (100) intensity and basal planar static (inelastic) rod intensity as a function of T. The latter measures the planar domain size, which below T_N (here about 200K) is transformed into three-dimensional long range order (ref. 18).

discussed in II.2). Then if N(E) contains a logarithmic singularity at the center of a two-dimensional band, as it does for 1s states in a square lattice, the half-filled band has an antiferromagnetic insulating ground state[15] in the Hubbard model for all values of U/t. By a happy coincidence this simplified, idealized model appears to be well suited to describing the electronic states near E_F of the CuO_2 planes in the chemically and structurally complex materials La_2CuO_4 and $YBa_2Cu_3O_6$.

Antiferromagnetism in two-dimensional magnetic systems is realized experimentally in fluorides[16] with the K_2NiF_4 structure, where the ratio of interplanar to intraplanar exchange can be less than 10^{-3}. The sublattice magnetization M(T) for K_2CoF_4 is shown in Fig. 7, and it is very close to the theoretical form (exponent 1/8, almost a step function) expected for a two-dimensional Ising system. The ordering magnetic moments may be as large as 0.5 Bohr magnetons/Cu atom.

We might expect that because La_2CuO_4 also has a similar layered structure (the small orthorhombic distortion is not significant), that it would be antiferromagnetic with two-dimensional Ising character. Indeed antiferromagnetic ordering is observed by neutron scattering from single crystals of La_2CuO_{4-y}, but there is little evidence for two-dimensional behavior in M(T). One reason for this could be that it is much more difficult to reduce the defect concentration in oxides than in fluorides, and usually the defects are distributed inhomogeneously.[17] The best single-crystal data,[18] from a sample grown in a copper oxide flux, are shown in the lower part of Fig. 8. We see that M(T) shows power-law behavior of a three-dimensional type. This means that it is unlikely that spin fluctuations will show substantial intrinsically two-dimensional features due to dynamical interactions alone.

The layered character of magnetic ordering in La_2CuO_4 is brought out by studying static two-dimensional magnetic domains, which produce not Bragg peaks but rods of inelastic scattering,[18] as shown in the upper half of Fig. 6. The planar domain size d is found to be larger than 200Å. It seems likely that these domains are spatially layered by inhomogeneities (such as oxygen vacancies, which would establish domain edges), but that these same inhomogeneities generate interlayer coupling which produces three-dimensional spin dynamics.

Antiferromagnetic ordering has also been observed[19] in the $Cu(2)O_2$ planes of semiconductive $YBa_2Cu_3O_6$, with again about $0.5\,\mu_B$ spins per Cu(2) atom. The three dimensional interlayer coupling is even stronger than in La_2CuO_4, which may be caused by a greater concentration of defects or by the three-fold increase in the $3d - 2p\sigma$ k_z band widths in $YBa_2Cu_3O_6$ compared to La_2CuO_4 as discussed in the preceding section. This leads to an increase in the Neél temperature from $< 300K$ in La_2CuO_4

to $> 500K$ in $YBa_2Cu_3O_6$.

The neutron scattering data thus qualitatively confirm the Hubbard half-filled band model for CuO_2 planes. In this model an energy gap (which optical data discussed in Chap. VII show to be about 2 eV) is opened at E_F in the electronic excitation spectrum, as a result of electron localization, and the material is not metallic and hence not superconductive. As x increases in $La_{2-x}Sr_xCuO_4$ or $YBa_2Cu_3O_{6+x}$ alloys, metallic behavior is recovered, T_N goes to zero and the materials become superconductive. The next problem is to what extent the anti-ferromagnetic and semiconductive localization effects persist in the metallic state and how they effect the interactions responsible for superconductivity.

Several theories, discussed briefly later in this chapter, have assumed that significant fluctuations of a magnetic and semiconductive type persist from the semiconductive $YBa_2Cu_3O_6$ phase into the metallic $YBa_2Cu_3O_{6.5}$ and $YBa_2Cu_3O_{6.9}$ phases. These theories were tested[20] on the high flux reactor at Grenoble using *32 detectors for 21 beam days* to look for diffuse magnetic scattering, which was not observed in the latter metallic phase ($T_c = 92K$). These experiments place an *upper limit* on the percent of metallic Cu atoms with spin 1/2 interacting on an energy scale of 300K of less than 2.5%. Even at this upper limit it is difficult to see how such spin fluctuations could produce a $T_c \sim 100K$, no matter how strongly they are coupled to orbital currents. When these data are combined with other data[20] on oxygen-deficient samples, the composition dependence of the mean-square magnetic moment shown in Fig. 9 is obtained. Also shown is $T_c(x)$ for these samples; a strong *anticorrelation* is observed. This anticorrelation appears to be sufficient to disprove suggestions[44] that magnon (rather than phonon) exchange is responsible for Cooper pair formation in the cuprates.

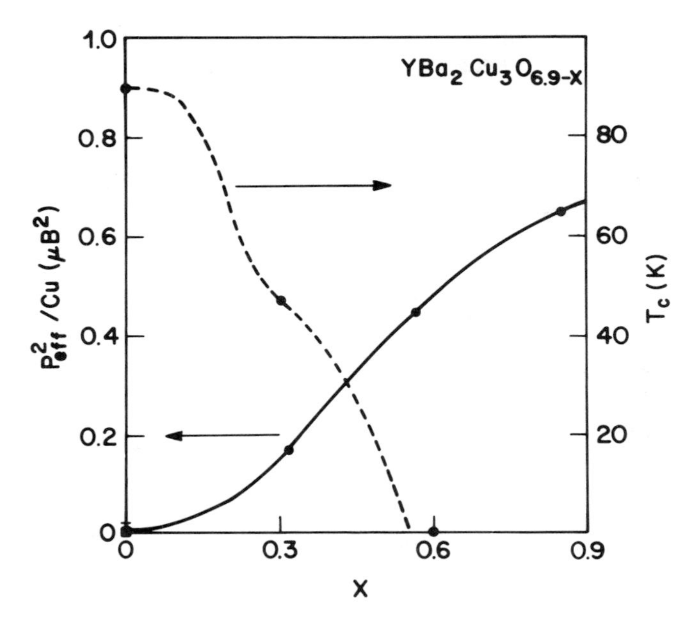

Fig. IV.9. Data from ref. 20 on spin fluctuations in $YBa_2Cu_3O_{6.9-x}$. The mean-square spins p^2_{eff}/Cu atom (in Bohr magnetons) are measured by diffuse scattering of polarized neutrons for several compositions x. For comparison $T_c(x)$ is shown for these samples.

A more subtle problem is why the presence of spin fluctuations in the CuO or CuO_2 planes, as implied, for example, in $YBa_2Cu_3O_{6.5}$, does not destroy superconductivity altogether, as the presence of local impurity magnetic moments (for instance of Fe impurities) does in the intermetallic compounds described in I and II. Because there are no such spin fluctuations on the materials with $T_c \gtrsim 90K$, this question seems to be of secondary importance, but it is not entirely moot. I believe that the most probable answer to this question is that in each plane the magnetic and superconductive regions are spatially disjoint, perhaps on a scale of 100Å. As the sample fraction associated with the former increases, the fraction of the sample which is superconductive and percolates decreases and in this decreasing fraction T_c decreases as well. If this is so, then ingenious theoretical models[46] designed simultaneously to decouple spins and charge and also to provide a high-T_c mechanism without phonons (but with adjustable parameters) are unnecessary.

A constructive approach to the effect of oxygen vacancies $O^{\square}$ and chain fragments on the electrical properties of $YBa_2Cu_3O_7O_x^{\square}$ is to analyze the acceptor levels introduced by $O^{\square}$ and their doping effects on the carrier concentration n using a simplified band structure model.[6] This model shows that for widely spaced $O^{\square}$ there are no acceptors, but as $O^{\square}$ spacing (or length of chain fragments) decreases, a critical length is reached where acceptor levels appear and there is, in effect, an electronic phase transition due to acceptor pinning of the Fermi level. This model, which is shown in Fig. 10, can explain both phases transitions shown in Fig. III-24, if it is assumed (as described there), that the evenly-spaced vacancies first occupy alternating chains for $x < 1/2$. To fit the data the critical chain length for $x > 1/2$ is four (Cu_4O_3 segments), (corresponding to T_c going from 0 to 60K), while that for $x < 1/2$ is two. Unfortunately there is at present no independent evidence for these parameters, and the observed phase transitions could result just as well from quantum percolation effects of a more complex nature (IV. 8). Nevertheless this model provides a simple and valuable illustration of such effects in a specific context.

6. Electron-Phonon Interactions in $(La, M)_2CuO_4$ and $YBa_2Cu_3O_{7-x}$ Alloys

In I.2, II.1 and II.3 we discussed what is known about electron-phonon interactions in simple and transition metal compounds. Special emphasis was placed on lattice instabilities, because even at short wave lengths these are readily observed by neutron scattering, and these instabilities frequently correlate with T_c. It is important to realize that unlike total energies, electron-phonon interactions are not subject to a variational principle and are not directly measurable. Almost everything that is known about electron-phonon interactions is based on chemical trends in lumped quantities (such as T_c or ρ) which contain other factors as well (such as vibrational frequencies). Moreover, the electron-phonon interaction is a derivative of a scattering potential, and such derivatives can be much more sensitive to computational and physical approximations.

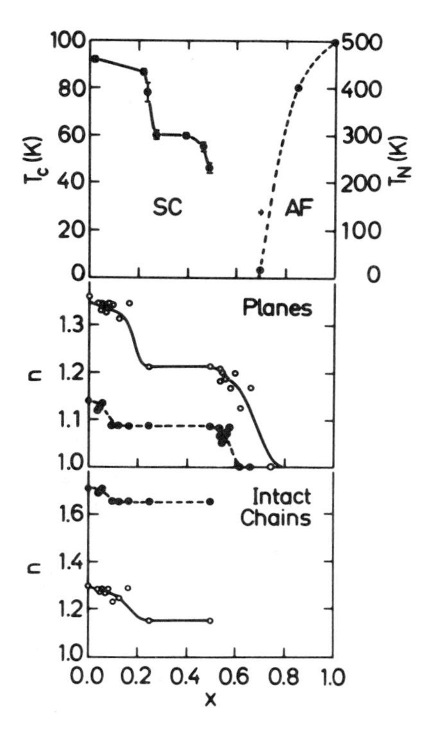

Fig. IV.10. Electronic model of chain-fragment doping and the phase diagram of $YBa_2Cu_3O_7O_x^{\square}$. The phase diagram is shown in the upper part of the figure (see also III-24), while several models for the carrier densities of planes and intact chains are shown below (ref. 6).

In summary, what we know about electron-phonon interactions is based mainly on our experience with metals, not oxides, and that experience itself contains many chemical averages. The approximations made, both computational and physical, are those suited to closely packed metals. When we apply this formalism to metallic yet much more open oxides we can expect some surprises.

Two kinds of calculations have been reported for electron-phonon interactions in metallic oxides. These are direct calculations based on the RMTA or rigid muffin-tin approximation[21] to the perturbed potential, and semi-empirical calculations based on tight-binding (TB) fits to the unperturbed energy bands.[22] The RMTA is in some ways more complex than the TB method, and it does not utilize information from measured lattice vibration curves. Moreover, as the coordination number decreases the fraction of the crystal volume filled by the muffin tins (inscribed spheres centered on each atom in the unit cell) decreases as well, and this is an especially serious problem for the oxygen ion spheres, which have only two contacts with this construction. This problem applies both to the energy bands calculated in equilibrium and to the perturbed ones, but it may be more serious for the latter.

There is a second problem associated with the MTA (muffin-tin approximation). This is specifically associated with bond charges and the bond charging region. It has been known for a long time that bond charges are small and that their overall affect on the electronic structure is small, as measured (for example) by the overall band width. An upper limit on bond charges, which is approached or attained only with purely covalent bonds, is $Ze/n\epsilon_0$, where Z is the atom valence, n its coordination number, and ϵ_0 is the static electronic dielectric constant (about 4.5 in these materials). Thus the O bond charge should be less than 2e/3, which is small but not negligible. In particular, the bond charge may be important in comparing bonding σ bands with non-bonding (or weakly bonding) π bands, and in calculating electron-phonon interactions in transverse acoustic modes.

In the MTA the bond charge region is reduced to a point - the contact point of inscribed spheres. In the simplest version of the MTA the electric field in the interstitial region is put equal to zero, which means that all the electronic charge is confined to the inscribed spheres. This makes the material much too ionic, virtually eliminates π band widths, and reduces the bond charge almost to zero. In nearly close-packed metals, such as Cu, where the inscribed spheres fill nearly 2/3 of the crystal volume, and ionic effects are small or absent by symmetry, the problems associated with the MTA may be quite small. In more open structures with low coordination numbers $\lesssim$ six, the volume fraction filled by inscribed spheres is typically 1/3 or less, and large potential discontinuities readily occur for reasonable choices of ionic sphere radii. Technically, when N (the number of atoms/unit cell) $\ll 10$, this problem has been solved by perturbation theory on the interstitial regions. The perturbative approach converges slowly, however, especially near the contact point. In the present materials, because $N \gtrsim 10$, the simple MTA converges extremely slowly. An idea of how slow the convergence is can be gained quickly enough. Three different calculations have been reported for $YBa_2Cu_3O_7$, one using 650 basis functions with the planar orthorhombic axes inadvertently interchanged,[23] one using 750 basis functions[23] and one using 850 basis functions with the correct crystal structure,[11] which means an increase in the Cu(1)-O chain bond length of 4%. The difference between using 750 and 850 basis functions is an increase of valence band width of ~ 1 eV, which may mean that the bonding states have converged for the latter. However, higher energy antibonding states converge more slowly than lower energy bonding states. At present it is estimated that the differences between the three calculations, which are quite significant and are discussed below in detail, come as much or more from differences in convergence as from differences in bond lengths. As we shall see, this means that the relative weights of σ and π contributions to electron-phonon interactions on the Cu(1) chains are at present indeterminate, and even the π contribution to $N(E_F)$ is uncertain (IV.4).

As for the lattice vibration problem, we still are far from completely disentangling the normal modes of these compounds, and so most calculations rely on analogies with the (much simpler) perovskite normal modes in compounds such as $BaPbO_3$. Thus both computational approaches to both atomic and electronic normal modes are still in their infancy.

In spite of these caveats, the results which have been obtained so far are extremely instructive. To examine the effect of doping, energy bands and electron-phonon couplings were calculated for both La_2CuO_4 and $LaBaCuO_4$, and both $N(E_F)$, the total density of states, and $N_\alpha(E_F)$, as projected on atom α, were obtained.[21] The effect of replacing La[3+] with Ba[2+] is to increase $N(E_F)$ by a factor of 3. As shown in Fig. 11, this large increase (which is reduced in rigid-band models) comes about in two ways. A second-neighbor bond forms between Ba and Cu, which redistributes $N_{Cu}(E)$, mostly to lower E, but also increases $N_{Cu}(E_F)$. The O_z atoms coplanar with Ba also now have a much larger $N(E_F)$, meaning that they should not be regarded as O[2−]. The second effect (which is contained in the rigid-band model) is a lowering of E_F because of the reduction by one electron of the number of valence electrons/unit cell. Overall for $La_{2-x}M_xCuO_4$ it appears that with $x \lesssim 0.2$ these calculations still agree with

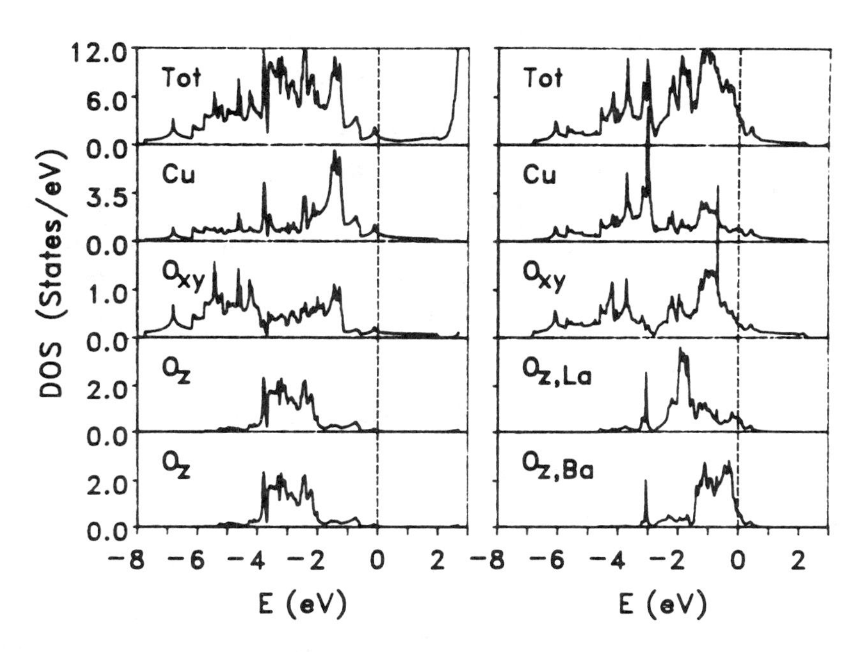

Fig. IV.11. Projected densities of states $N_\alpha(E)$ for inequivalent atoms in La_2CuO_4 and $LaBaCuO_4$. Note the peak in N(E) about 0.1 eV below E_F in La_2CuO_4. In $La_{2-x}Sr_xCuO_4$ one finds E_F centered on this peak at $x = 0.14$ (ref. 21).

the rigid band model, providing only the *average* effect of doping with M alkaline earths is considered. As we saw in III.4 and IV.3, for $x > 0.2$, increasing x creates $O^\square$, oxygen vacancies, which may be associated with $(Sr[1+])_2\ O^\square[2-]$ complexes. The increase in $N_\alpha(E_F)$ at $\alpha = O_{z,Ba}$ shown in Fig. 7 is the band-structure indication of this structural instability.

In the RMTA because only the inscribed spheres are displaced all electron-phonon interactions are forced into an extremely localized limit. If we define η_α as $N(E_F) < I_\alpha^2 >$ then from (II, 1.4)

$$\lambda = \Sigma\lambda_\alpha = \Sigma\eta_\alpha / M_\alpha < \omega^2 > . \tag{1}$$

While there are large changes in $N(E_F)$ between La_2CuO_4 and $LaBaCuO_4$, the changes in η_α are much smaller. In fact, $\Sigma\eta_O$ in the two compounds is almost the same, and the main difference is in η_{Cu}. From the calculated values[21] of η_α the derived results for λ and T_c (Allen-Dynes) are shown in Fig. 12. In $La_{2-x}Ba_xCuO_4$ with $x = 0.14$ and $\Omega_o = 225K$ (see VI.1) one obtains $T_c \simeq 35K$, which is better than one might expect. One concludes that band theory is able to explain T_c in $(La, M)_2CuO_{4-y}$ alloys reasonably well.[21,22]

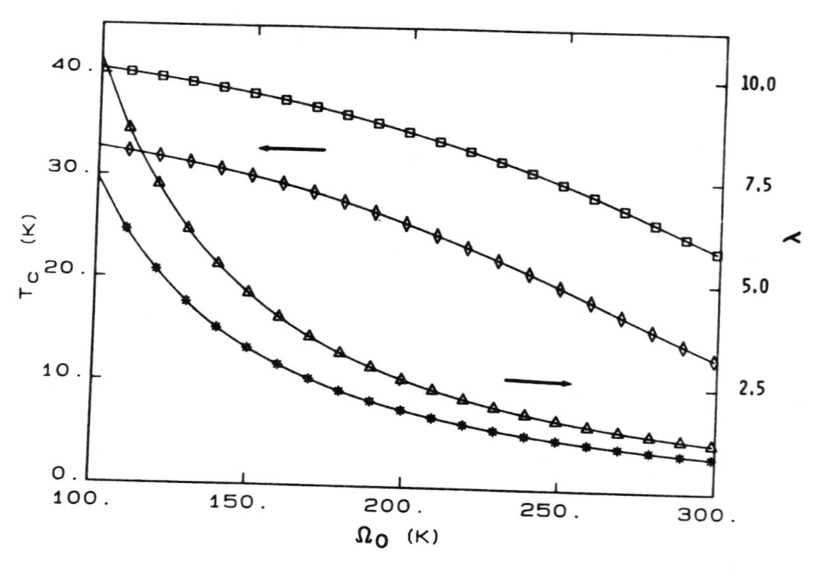

Fig. IV.12. Calculated values of λ and T_c in La_2CuO_4 and $La_{2-x}Ba_xCuO_4$ for $x = 0.14$ as a function of Einstein vibrational frequency Ω_o (ref. 21).

A more pedestrian but still important test of band theory is the normal-state anisotropic resistivity including the positive Hall constant R_H. This has been calculated[24] for $La_{2-x}M_xCuO_4$ and $YBa_2Cu_3O_7$ and the results are in reasonable agreement with experiment up to $x = 0.2$ (where $O^\square$ appear to trap holes). The absence of saturation (II.5) in $\rho_{xx}(T)$ at high T is explained by the absence of other bands within $\hbar/\tau \sim 0.6$ eV of the dpσ single band at E_F.

When we turn to $YBa_2Cu_3O_{7-x}$, the presence of both chains and planes greatly complicates the calculation of T_c. The planar Cu3dO2pσ bands are similar to those in $(La, Sr)_2CuO_4$, but now for x near 0 we find E_F is lower. Relative to the peak in $N_\sigma(E)$, near the center of the σ component of the density of states, the lowering of E_F reduces $N_\sigma(E_F)$ and the planar contribution to λ. This suggests, as one can easily guess, an important rôle for the chains.

Both conventional band theory[24] and the semiempirical TB method[25] have been used to estimate λ for $YBa_2Cu_3O_7$. As we would expect, because $N_\sigma(E_F)$ is reduced for this composition, if only the σ bands contribute, a small value of $\lambda \lesssim 1$ is obtained. However, if the Cu1 chain O1π and even O4 π bands (see IV.4) were to contribute, λ would increase accordingly. In the example studied by Weber,[25] only the O1 π band increase was included, and it is directly proportional to the increase in $N(E_F)$ due to $N_\pi(E_F)$, so that the couplings V_σ and V_π in his model are substantially the same.

The detailed analysis of the σ and π contributions to the phonon density of states $F(\omega)$ and the contributions to λ by $\alpha^2F(\omega)$ according to Eqn. I(2.2) is shown in Fig. 13. By comparing the σ and $\sigma + \pi$ panels we note the large density of vibrational states in $F(\omega)$ associated with Cu1-O1 chain buckling π modes at low frequencies. These give a very

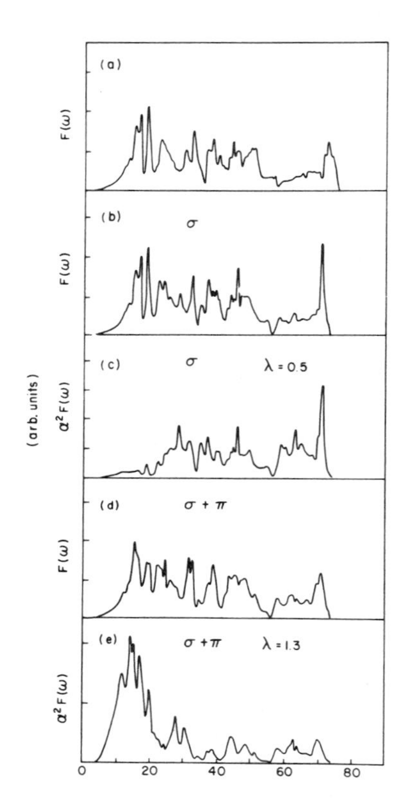

Fig. IV.13. Comparison of σ and $\sigma + \pi$ models for $F(\omega)$ and $\alpha^2 F(\omega)$ in $YBa_2Cu_3O_7$. In (a) the vibrational spectrum associated with short-range forces only and no valence electron screening is shown. In subsequent panels σ and π valence screening effects are added (ref. 25).

large contribution to λ because in I(2.2) there is a weighting factor for $\alpha^2 F(\omega)$ of ω^{-1}. However, the net result is still an increase of λ only to 1.3.

There are several factors that are omitted from Weber's calculation which could easily increase λ significantly. In the A15 systems comparison of theoretical calculations based on the local density approximation with the best tunneling data has consistently shown[26] a general tendency to underestimate transverse V_π compared to longitudinal V_σ couplings. In cubic metals this error is not large (at worst about 50%), but it could be much larger in oxides where dielectric anisotropy is accompanied by electron confinement to Cu layers by insulating or semiconductive Y-O or Ba-O layers. As a quantitative analogue we may note that confinement of excitons to two-dimensional layers (rather than three-dimensional volumes) increases their binding energy by a factor of four. Such confinement has not been observed[27] in transition metal dichalcogenides, which is chemically instructive. It suggests that confinement in the present case may be an important factor, and this confinement arises as a result of two structural changes: replacement of the chaleogenide by oxygen, and alternation of semimetallic CuO and insulating or semiconductive YO and BaO layers. The confinement effects would show up in a full band calculation[3] of nonlocal exchange and correlation effects using anisotropic dielectric screening, but so far such calculations have been completed only for simple systems such as Si. Direct evidence for confinement has been obtained by nuclear quadrupole resonance relaxation experiments (IX.2).

An equally important factor omitted from Weber's model is the possible overlap of the Cu1-O4 π band with E_F. If this were to occur, not only would it enhance N(E), but, taken together with the other factors mentioned above, it also could increase λ to values in the range $\lambda \sim 4$.

7. Defect Electronic Structure

Several preliminary calculations of the electronic structure of $(La, Sr)_2CuO_{4-y}$ with $y = 1/4$ have been reported.[28,29] These calculations used quadruple supercells, i.e., the basic structural units were $[La_2CuO_4]_4O^{\square}$, which means that the results are far from convergent or self-consistent, especially near $O^{\square}$, either in terms of electronic or relaxed atomic structures. Nevertheless, several general features emerged which are likely to persist in more accurate calculations. These are:

(1) After allowance has been made for projection of bands in the extended Brillouin zone back into the reduced zone by the supercell geometry, the vacancy pushes one or two additional π-like bands into the π gap near E_F where in the absence of the vacancy there are only σ bands. These π bands are presumably expelled by the repulsive $O^{\square}$ potential from the O1 and O4 π bands 1-2 eV below E_F, as shown in Fig. IV-6.

(2) In these calculations vacancy-vacancy interactions give these defect bands a large band width of order 0.5 eV because at this density the vacancies are closely spaced. The vacancy bands are pinned at or very near E_F by self-consistency of their charge state, which is likely to be close to $O^{\square}(Z-)$ relative to nearest neighbor Cu [M+], with $2M - Z = 2$. For the widely-spaced vacancies on Cu1 chains in $YBa_2Cu_3O_{6.87}$, as shown in Fig. III-26, this band width would be reduced to 0.1 eV or less. Note that the $CuO^{\square}Cu$ complex binds two holes, which are probably in a singlet (non-magnetic) state because they are extended over three atomic volumes.

On chemical grounds we expect that the Mattheiss relation E_{3d} (Cu[2+]) = E_{2p} (O[2−]) has two important effects, one structural and one electronic. The structural effect is the ubiquitous formation of $(CuO_2)_\infty$ square planes (III.3). The electronic effect is the position of defect energy levels, especially those of oxygen vacancies $O^\square$. In general defect energy levels are not pinned near or at E_F, but may be either well above or well below E_F. Because of the high densities of Cu3d and O2p π states and the flexibility of the formal valence of Cu, together with E_{3d} (Cu[2+]) $= E_{2p}$ (O[2−]), narrow $O^\square$ defect bands associated with facile Cu−O charge exchange located near or at E_F seem very likely. These will increase $N(E_F)$ greatly, as shown in Fig. 14(a). However, when we allow for defect-defect interactions, any peak which is pinned at E_F for isolated defects is likely to split into two peaks, reflecting the formation of defect bonds. The scale of the spacing of these two peaks could well be kT_m, where T_m is the metal-semiconductor transition temperature, defined as in III.7 (1). This suggestion is consistent with the observation there that T_m increases as the density of defects increases. This idea is illustrated in Fig. 14(b).

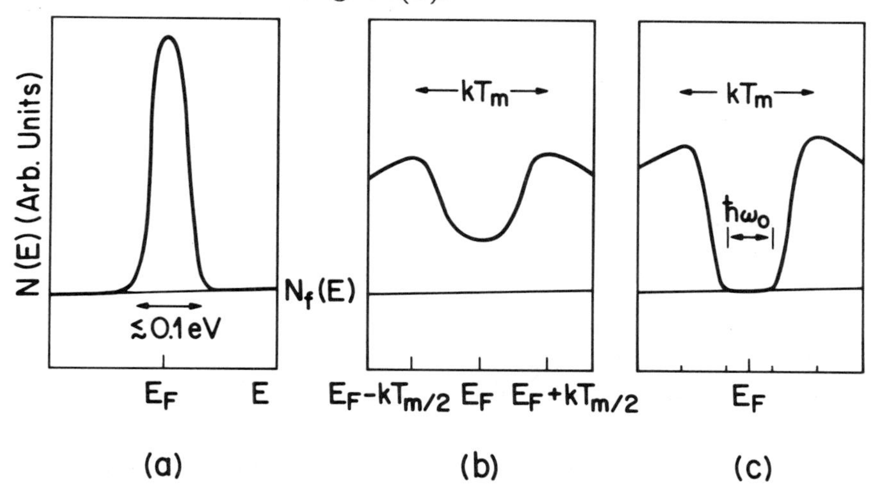

Fig. IV.14. Defect peaks in N(E) may well be pinned at E_F for isolated defects in cuprate superconductors, as in (a), with the peak split into two peaks with a spacing kT_m as the density of defects increases, as in (b). Atomic relaxation using soft phonons with $\hbar\omega_o \sim 40$ cm^{-1} may open a gap in $N_d(E)$ of order $\hbar\omega_o$, as in (c). Broadly speaking this is a kind of Jahn-Teller effect. Note that (b) and (c) are plotted on an enlarged scale.

Next we should consider the effect of atomic relaxation. If very low frequency optic-mode phonons exist of the caterpillar type, with $\omega_o \sim 50\ cm^{-1}$ (III.5, VI.2), then an energy gap can appear in the defect spectrum because of defect relaxation. The magnitude of the gap may be of order $\hbar\omega_o$, as illustrated in Fig. 14(c). More generally, the defect contribution to the density of states $N(E_F)$ can be temperature-dependent, and this temperature dependence can have drastic effects depending on the geometrical rôle played by the defect in completion of electrical paths - for example, intralayer or interlayer. This point will recur repeatedly as an important possibility. If the peak splitting in Fig. 14 is regarded as a generalized Jahn-Teller effect, then this splitting could be reduced to zero near $T_a = 450C$, because this temperature is one commonly used to anneal the oxygen vacancy configurations.

8. Localization and Marginal Dimensionality

One of the most striking aspects of high-T_c oxide structures is that increases in T_c occur as the layered character of the structure becomes more pronounced. The study of electronic conductivity and the nature of electronic states in the presence of substantial disorder has led, since the pioneering early work of Mott[12,30] and Anderson,[31] to the conclusion that the ability of electrons to carry metallic current ballistically (rather than by diffusive hopping) in strongly disordered systems (such as impurity bands) is essentially dependent on dimensionality. Because the present materials are often so close to the metal-semiconductor transition in terms of their normal-state conductivity $\sigma(T)$ and $d\sigma/dT_m = 0$ for T_m close to T_c in many cases (III.7-III.9), these questions of electron localization are of critical importance. Here we are far from good metals ("weak" localization), and are instead concerned with "strong" localization.

The essential state of the theory at present[32,33,34,35] is that there is metallic conductivity for dimensionality $d = 3$, but that all states are localized for $d = 1$. Dimensionality $d = 2$ is marginal, and relatively small structural changes can have drastic effects on the existence of extended electronic states which are capable of carrying ballistic metallic currents. This viewpoint[36] is highly suggestive for layered superconductive oxides, as we shall now see.

In addition to $O^{\square}$ defects in CuO_2 planes and CuO chains, which are intraplanar defects, we can also expect a certain concentration of interplanar defects. In $(La, Sr)_2CuO_4$ the Sr_{La} (i.e., Sr on La sites) are interplanar. They change the formal valence of Cu, and apparently quench the antiferromagnetic Hubbard gap (IV.5). At the same time Sr_{La} or Ba_{La} provides an interplanar electrical path at E_F, as can be seen from $N_{O_z}(E)$ in Fig. IV.9. This interplanar path changes the connectivity of the crystal as seen by electrons near E_F in a fundamental way. Similar remarks apply to Bi_{Ca} in $(Bi, Pb)_2Sr_2$ (Ca, Bi) Cu_2O_9.

In $YBa_2Cu_3O_7$ the interplanar defects are somewhat less obvious, but I believe that interplanar defects are still present, in a concentration of about a few percent, as will be discussed in IV.9. Probably these defects are Ba_Y, with some nearest neighbor O sites occupied adjacent to Ba in the Y plane. Such cation disorder is explicitly present in $(La, Sr)_2CuO_4$ with Sr_{La}, and in $(Bi, Pb)_2Sr_2(Ca, Bi)O_9$ with Bi_{Ca}, but on a level of 10-20%, i.e., three to ten times larger interplanar disorder, which can be recognized by careful diffraction measurements, while cation disorder producing interplanar defects on a scale of a few percent can only be inferred indirectly.

Ordinarily one does not view materials (such as metals) as being strongly disordered when they contain a few percent impurities or defects, but most materials, even those with layered structures (such as the transition metal dichalcogenides), are three-dimensional so far as their electronic properties are concerned.[27] Multilayer oxides which alternate ionic M(2 or 3+) O(2−) layers with metallic CuO_2 or CuO layers produce genuinely electronically two-dimensional cuprate layers in the absence of defects. Then the intra- and inter-planar defects are of critical importance according to localization theory. The intraplanar defects promote localization and drive the marginal conductivity towards the semiconductive limit. They compete with the interplanar defects which tend to increase the dimensionality towards three, where the conductivity becomes metallic. When intraplanar defects dominate, the layers break up into domains or islands separated by insulating lines or domain walls of intraplanar defects. When interplanar defects dominate, current paths bypass these domain walls by jumping from layer to layer.

We can formulate these ideas quantitatively in terms of the densities of electronic states of intraplanar defects $N_p(E)$ and axial interplanar connective defects $N_a(E)$, which we assume are given, together with the density of band states of the defect-free crystalline framework $N_f(E)$. We say that a planar or axial defect contributes actively to the electronic network if the respective conditions

$$N_{p,a}(E_F)/N_f(E_F) > 1 \tag{1}$$

are fulfilled. (This is the condition at $T=0$; the effects of thermal activation can be represented by integrating over an energy width about E_F of order kT.)

The next step is to estimate the average dimension L of a domain bounded by edges of planar defects. Define the concentrations c_a and c_p of active axial connecting or planar blocking defects per conductive CuO_2 planar unit. Then we expect[33,34] that Gaussian disorder makes L^2 proportional to $c_p^{-1} = c_p^{-d/2}$ for dimensionality $d = 2$. If more than about a third of these planar domains are connected by active axial defects, then the material will behave as a metallic conductor, whereas it will behave as a semiconductor if the reverse is true. Thus the condition for metallic conductivity is

$$c_a / c_p \gtrsim b \tag{2}$$

where b is a number of order unity.

One may be inclined, in attempting to visualize planar domains, to associate the domain edges with specifically linear defects such as twin boundaries (X.2). Certainly such boundaries can act as domain edges, but they are not necessary. In a system of degenerate Fermions, $kTN(E_F) \ll 1$, even randomly located impurities can localize electrons in two dimensions.[33] Close to a metal-semiconductor transition (as observed experimentally), with two-dimensional confinement, planar domain formation appears highly likely.

The description of quantum percolation just given is crude but it probably is the best that can be done in the present state of development of localization theory.[32-34] It involves the functions $N_{p,a}(E)$ and $N_f(E)$. While the latter is available from band calculations for the perfect crystal, the former can only be guessed at present, which means that even if (1) and (2) could be refined by computer simulations, little would be gained in terms of analysis of experimental data.

Although the quantum percolation picture is primitive, it still furnishes a basis for discussing anomalies in both normal-state and superconductive properties which is entirely absent in homogeneous models. Our understanding of superconductive properties is based on our description of normal-state properties, and to the extent that we can specify the general features of $N_{p,a}(E)$ from studying normal-state properties we are justified in continuing to use these features to describe the microscopic mechanisms which generate anomalously high values of T_c. At the same time, in analyzing experimental data to identify defects, we must be careful to distinguish defect properties which are intrinsic to metallic superconductive regions from those associated with inadequately oxidized semiconductive regions. Both regions, which can correspond to different superlattice structures, can coexist and intermingle even in single crystals and epitaxial films. In particular, as discussed in IV.5, the simultaneous observation of spin fluctuations and superconductivity in oxygen-deficient samples of $YBa_2Cu_3O_{7-x}$ with $x \sim 0.5$ probably does not indicate that these phenomena are occurring in the same atomic regions. Rather the atomically two-dimensional nature of the cuprate layers suggests phase separation of these effects which minimizes the degree to which the spin fluctuations reduce T_c.

The preceding part of this section was written before the Hall-constant anomalies (which are discussed in the next section) were discovered. Transport data are often explainable with more than model, but certainly the simplest explanation of these anomalies is in terms of the dimensionalities of chains and planes. Specifically the simplest way to relate these anomalies to high-T_c superconductivity is through the marginal dimensionality of CuO_2 planes.

9. Normal-State Transport Properties

Even the earliest data, taken on sintered powder samples prepared in non-optimized ways, showed anomalies in the superconductive resistive transition which could be interpreted as evidence for spatial inhomogeneities associated, for example, with incomplete oxidation. With the improvement of sample preparation methods, including the growth of more nearly homogeneous single crystals, as well as epitaxial films with carefully controlled compositions, many of the broad features observed in electrical transport measurements, or spectroscopically, have narrowed substantially. On these improved samples some anomalies found with sintered powders of variable composition have disappeared, but others have not only remained but have become much more definite and puzzling to conventional models based on idealized, homogeneous, defect-free crystals. These anomalies are of particular interest because they offer us the possibility of separating effects of uniaxial bridging defects from planar blocking defects as described by $N_a(E)$ and $N_p(E)$.

Chronologically the earliest unambiguous anomalies to be studied were those associated with Cu substitution (for instance by Zn or by Ni) in $(La, Sr)_2CuO_4$, $YBa_2Cu_3O_7$, and $EuBa_2Cu_3O_7$, as discussed in III.6. The results of doping $(La_{0.92}Sr_{0.08})_2\ Cu_{1-y}Zn_yO_4$ were surprisingly large, and not explicable merely in terms of compensation of Sr holes by Zn electrons, because $T_c = 0$ for $y = 0.04$, where $[Zn]/[Sr] = 0.25$. Even more surprising were the doping effects in $EuBa_2(Cu_{1-y}Zn_y)_3O_7$. Here even with $0 < y < 0.01$ one already has $T_m > T_c$ and a very rapid crossover to semiconductive normal-state behavior. With the Cu formal valence equal to $[(2.33-y)+]$ and semiconductive behavior expected at $y \sim 0.2 - 0.3$, this result is completely anomalous in the context of homogeneous models. Neither the Zn nor the Ni result can be explained as a magnetic effect, because doping with Zn has a larger effect than doping with Ni, although Zn has a smaller magnetic moment. Quantitatively the orbital scattering potential for 3d states associated with Zn is larger (energy shift ΔE_d about -15 eV) than for Ni (ΔE_d about $+1$ eV), but even if Zn fully blocks 3d orbital currents, for a three-dimensional network this effect would be small.

From the viewpoint of localization theory these effects of Cu substitution are not anomalously large. Suppose the M $= Ni_{Cu}$ or Zn_{Cu} impurities are associated with the interlayer defects, which could be $(Sr_{La})_2$ pairs in $La_{1.85}Sr_{0.15}Cu_{1-y}M_yO_4$ alloys, or Ba_Y in $YBa_2(Cu_{1-y}M_y)_3O_{6.9}$. This association could block the axial interlayer conductivity by moving the peak of $N_a(E)$ away from E_F. Because the dimensionality of the normal-state electrical network depends so critically on $N_a(E_F) - N_p(E_F)$, the very large effects with small y become understandable.

We now turn to the temperature-dependent planar resistivity $\rho_{\|}(T)$ and nominal Hall number $(eR_H)^{-1} = n_{\|}(T)$ in $La_{1.85}Sr_{0.15}CuO_4$ and $YBa_2Cu_3O_{6.9}$. We saw in III.7 that in many measurements at high $T \gtrsim T_c$ both $\rho_{\|}(T)$ and $n_{\|}(T)$ are linear, that is, $\rho_{\|}(T) = a(T_o + T)$ and $n_{\|}(T) = b(T'_o + T)$. It is not difficult to explain the linear dependence of $\rho_{\|}(T)$, so long as $kT_o \sim \hbar\omega_o$, where $\hbar\omega_o$ is a soft optical phonon. Especially in YBCO, where a soft optical phonon band near $\hbar\omega_o = 50\,\mathrm{cm}^{-1}$ has already been observed in a sintered powder by Raman scattering (VI.3), this conventional explanation based on electron-soft phonon scattering which utilizes the linearity of the soft phonon occupation number $n_o(T)$ for $kT \gtrsim \hbar\omega_o$ works well enough. However, with improving sample quality in YBCO single crystals and thin films, both T_o and T'_o tend to zero! This result[35] requires some extension of conventional band theory, which however does explain[24] the positive sign of $n_{\|}(T)$ at high T, and its general magnitude.

The offset temperatures T_o and T'_o correlate with epitaxial film quality as measured by the width ΔT_c and the critical current j_c. The correlation for $\rho_{\parallel}(T)$ is shown[35] in Fig. 15, with the best $YBa_2Cu_3O_{6.9}$ sample being #1. Samples 5 and 6 are Cu enriched and two-phase. We see that $T_o < 0$ and that $|T_o| \to 0$ with decreasing ΔT_c and increasing j_c.

At high temperatures $100K < T < 300K$, $\rho(T)$ is nearly proportional to T because the mean soft phonon occupation number $n_o(T)$ is proportional to T. When a single soft optical mode of frequency $\hbar\omega_o << k\theta_D$ scatters carriers strongly, and we calculate $n_o(T)$ from $n_o(T) = [\exp(\hbar\omega_o/kT) - 1]^{-1}$, we obtain $n_o = (kT/\hbar\omega_o)(1 - \hbar\omega/2kT) = a(T + T_o)$ with $T_o < 0$. With

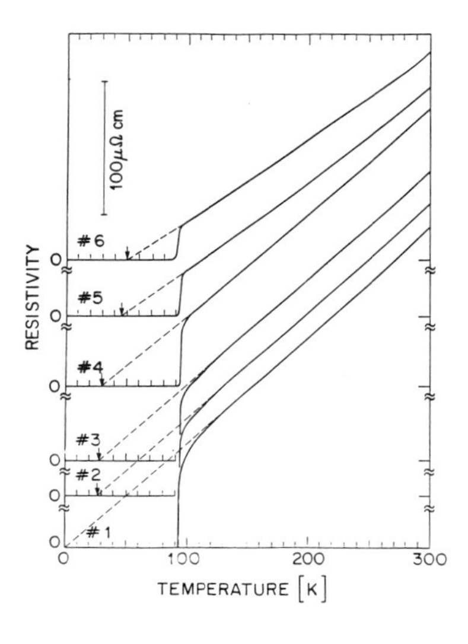

Fig. IV.15. The planar resistivity $\rho_{\parallel}(T)$ for epitaxial YBCO films, highest quality at bottom (ref. 35).

impurities $\rho(T)$ is increased by a temperature-independent residual resistivity $\rho_o = bT_o$. The impurities, however, also increase ω_o through disordering stress fields. In Fig. 15 the most impure (inhomogeneous) samples have the largest $(-T_o)$. This means that the increase in the soft-phonon frequency ω_o due to impurities outweighs their contribution to the residual resistivity. Samples 5 and 6 may be compared to a sintered powder where $\hbar\omega_o/2 \sim 25cm^{-1} \sim kT_o$. With improving sample quality $\hbar\omega_o$ may soften further to $\sim 25cm^{-1}$ or less, and then some compensating residual resistivity can make T_o close to 0.

In a multiply twinned single crystal the planar conductivity is averaged over percolative paths. The single crystal is probably not fully oxidized, so that it consists of three kinds of regions: orthorhombic metallic $YBa_2Cu_3O_{7-x}$ twinned domains rotated by $\pi/2$ with $x \sim 0.1$ and semiconductive domains with $x \sim 0.5$. Each of the three regions occupies less than half the planar area and cannot percolate.

At the low end of the temperature range, $T \sim 100K$, we assume that carriers in the chains and planes are decoupled because the intercarrier interlayer scattering time τ_i is long compared to the intracarrier scattering times τ_1 (CuO chains) and τ_2(CuO_2 planes). In a single orthorhombic domain with chains parallel to the y axis, $v_{x1} = 0$ and $v_{x2} \approx v_{y2}$. This means that only the planes can contribute percolatively to σ_{xx} and σ_{yy} and that $\tau_2(T) \propto (T_o + T)^{-1}$.

The Hall resistivity R_{xyz} for a single domain is given[24] by $E_y/j_xB_z = \sigma_{xyz}/\sigma_{xx}\sigma_{yy}$. Suppose we have only one carrier but the temperature-dependent relaxation time is anisotropic and is described by $\tau_\alpha(T)$ (adequate for orthorhombic symmetry). Then $\sigma_{\alpha\alpha}$ is proportional to τ_α and $\sigma_{\alpha\beta\gamma}$ is proportional to $\tau_\alpha\tau_\beta$, leaving R_{xyz} temperature-independent. This dilemma can be resolved in the context

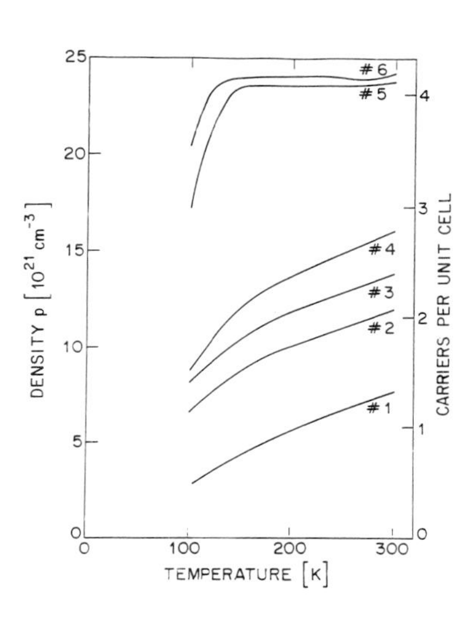

Fig. IV.16. Nominal carrier density $\rho_{\parallel}(T) \propto R_{xyz}^{-1}$ for the films of Fig. 15 (ref. 35).

of a two-carrier (chain and plane) percolative model. For a single domain $\sigma_{xyz} \propto \tau_{x2}\tau_{y2}$ while $\sigma_{xx} \propto \tau_{x2}$ and $\sigma_{yy} \propto (\tau_{y1} + b\tau_{y2})$, where b is a constant. On the chain all the soft modes are x— or z— polarized, and do not backscatter y-polarized chain currents, so that $\tau_{y1} = \tau_o$ (a constant) which can be much larger than $\tau_{y2}(T)$. Hence $R_{xyz} \propto \tau_{y2}(T)/(\tau_o + b\tau_{y2}(T))$ can be proportional to $(T_o' + T)^{-1}$, as observed.[35]

The nominal Hall number $n_{\parallel}(T) = V_o/eR_{xyz}$ (where V_o is the unit cell volume) is shown for the same series of epitaxial films in Fig. 16. We note that $T_o' > 0$ but tends to 0 with increasing sample quality. In the best sample the intercarrier transfer effects gradually set in near $T_b \sim 200K$. Increasing the density of interlayer axial bridging defects reduces T_b and adding a Cu-rich phase (which can short some of the barriers in the percolative process) reduces T_b farther to $\sim$ 120K. The crossover becomes more gradual with improving sample quality, which suggests an intrinsic phonon-assisted interlayer scattering mechanism (VI.3) in the best sample, and an extrinsic, percolatively induced crossover in the two-phase samples. Another model which could explain the freeze-out of the nominal Hall number might involve localization of carriers due to a reduction in the density of axial defect states $N_a(E_F)$ as $T \rightarrow 0$. This would reflect an increasing splitting of the defect bonding-antibonding peaks in Fig. 14, as discussed in IV.7. The freeze-out would begin around $T_a/2$, where $T_a \sim 700K$ is the annealing temperature for oxygen vacancy rearrangement.

The soft phonon percolation model described here is crude, but it is important because it identifies and discusses features of the anisotropic normal state resistivity and Hall effect which correlate well with T_c so far as chemical trends are concerned. Studying these trends one is led to the conclusion that they must be determined by defects, and that a distinction between axial and planar defects, such as is made by Eqn. 8(1)-(2), is necessary. When conduction primarily takes place within a layer, the (possibly thermally assisted) mixing of layer metallic states $\psi_f(E_F)$ with axial defect states $\phi_a(E_F)$ may be of critical significance, because without such mixing there may be no metallic conductivity. For three-dimensional materials (such as the intermetallic compounds discussed in I and II) this mixing is much smaller and alloying effects are correspondingly smaller, as is the temperature dependence of $N(E_F)$.

The theory of marginal dimensionality discussed in IV.8 is concerned with the limiting form of the conductivity in ideal chains and planes as $T \rightarrow 0$. When the chains and planes are weakly coupled by defects and $kT \gtrsim \hbar\omega_s$ for some soft phonons of frequency ω_s, the theory is silent. However, it is reasonable to suppose that one-dimensional freeze-out might occur for $kT \lesssim \hbar\omega_s$, while two-dimensional freeze-out would occur at a lower temperature close to $T = 0$. We saw in III.7 that recent experiments on $Y_2Ba_4Cu_8O_{20-x}$, which contains bilayer chains, show freeze-out, presumably of these chains, near $T = 80K$, which is a typical value for $\hbar\omega_s/k$ (VI.3). Meanwhile single chains in $YBa_2Cu_3O_{7-x}$ may not be conductive, at least across twin boundaries. All of these results are consistent with the general trends expected from dimensional ideas. They imply, moreover, substantial spatial dependence of the superconductive order parameter (gap grading). These morphological problems are discussed further in X.

10. Defect-Enhanced Electron-Phonon Interactions

The usual treatment of electron-phonon interactions in metals assumes that all the electron states near $E = E_F$ are extended states of the Bloch type $\psi_{\mathbf{k}} = \exp(i\mathbf{k}\cdot\mathbf{r})u_{\mathbf{k}}(\mathbf{r})$, where $u_{\mathbf{k}}$ is a function with the periodicity of the lattice. (These states were called framework states ψ_f in IV.8.) The defect states $\phi_d(\mathbf{r})$ which we have been discussing are different. They are like impurity band states in degenerate semiconductors near the metal-insulator transition. Some of them are localized, ϕ_ℓ, and some are extended, ϕ_e. However, because of the marginal quasi-two-dimensionality of the layer geometry, the ψ_f states can carry current only when they mix with the extended defect states ϕ_e to form the expanded set of current-carrying states $\{\psi_f, \phi_e\}$. This situation is basically different from that of defect states not at E_F, which do not alter the current-carrying states or effect Cooper pairing unless they are magnetic. Here the non-magnetic defect states play an essential rôle by increasing the effective dimensionality of the current-carrying states above two so that the set $\{\psi_f, \phi_e\}$ is genuinely metallic.

The electron-phonon interaction operator is $\delta V_\alpha = V_\alpha(\mathbf{r}+\mathbf{r}_o) - V_\alpha(\mathbf{r}_o)$, where V_α is the contribution to the one-electron potential of atom α. In oxides the contribution to λ in (II 1.4) of $|<\psi_f|\delta V_\alpha|\psi_{f^1}>|^2$ will contain the factor $N_f(E_F)$, but in the presence of defects we must also consider terms containing $|<\psi_f|\delta V_\alpha|\phi_e>^2|$. This will be proportional to $N_e(E_F)$, the density of extended defect states at E_F. This may be large because $N_d(E)$ is narrow and pinned near E_F. So long as $|<\psi_f|\delta V_\alpha|\phi_e>|^2$ is comparable to $|<\psi_f|\delta V_\alpha|\psi_{f'}>|^2$, this may lead to large values of λ and large enhancements of T_c.

An interesting feature of this model is that the localized defect states ϕ_ℓ do not become superconducting below T_c, so that although a BCS gap opens in $N_f(E)+N_e(E)$, $N(E_F)$ remains non-zero due to $N_\ell(E_F)$. There is some evidence that this is the case from Raman scattering (VI.2).

The rôles played by axial interplanar defects in contributing to $\{\phi_e\}$ and $N_e(E_F)$ vary significantly with host crystal structure. In $La_{2-x}Sr_xCuO_{4-y}$, for $0 < x < 0.15$ the concentration y of oxygen vacancies $O^\square$ is small and $(Sr_{La})_n$ clusters (probably with $n = 2$) provide the interplanar bridges and make $N_e(E_F) > N_f(E_F)$ (per atom). For $x \gtrsim 0.15$ one finds an increase in y, with association of $O^\square$ with $(Sr)_n$ clusters, for example $Sr_2O^\square$. Probably $N_e(E_F) << N_f(E_F)$ for these complexes, and as the interplanar bridges disappear, the material becomes semiconductive (but not necessarily magnetic), and $T_c \rightarrow 0$. In the magnetic models of high-T_c superconductivity, great emphasis is placed on the narrow semiconductive region with $x \lesssim 0.02$ where antiferromagnetism is observed (IV.5). However, for $x \gtrsim 0.3$, no magnetism has been reported so far. It is possible that the semiconductive behavior is entirely the result in this composition range of two-dimensional localization. The pressure derivative dT_c/dp reaches a maximum near $x = 0.15$ (where T_c is largest) and decreases to zero with increasing x faster than T_c (Fig. III-34). It seems likely that the largest effect of pressure is on the interplanar coupling, and as the latter is reduced the corresponding derivative dT_c/dP may be reduced faster than T_c.

The interplanar coupling in $YBa_2Cu_3O_{7-x}$ is complicated because in addition to the $Cu(2)O_2$ planes one also has the Cu(1)O chains, weakly connected to the planar Cu(1) atoms *via* O4 atoms (Fig. IV-6). As a result current flow can take place either along the chains on the planes, and defects in these units can be bypassed without necessarily utilizing the few percent of Ba_Y defects mentioned in the preceding section. The weak coupling between the Cu(2) planar atoms and the O4 atoms can itself give rise to a large value of $N(E_F)$ and to an enhanced electron-phonon interaction.

The short-range vacancy ordering in the Cu(1) chains in $YBa_2Cu_3O_{6.87}$ illustrated in Fig. III-26 and utilized for the electronic model shown in Fig. IV-10 can be understood if we postulate a repulsive interaction between oxygen vacancies $O^{\square}$, and also assume that presence of $O^{\square}$ on a chain locally give it a positive charge. The total energy is minimized if the chains are alternately charged + (with $O^{\square}$) and − (without $O^{\square}$), as indicated in Fig. III-26. This figure also suggests the possibility[35] that current may flow along the chains without vacancies and still benefit from large electron-phonon interactions with adjacent chains containing $O^{\square}$ phonon modes, in effect combining both favorable features (a) and (b) of Fig. 14.

The model[36] just sketched is but one of several possibilities for utilizing strong-electron phonon interactions and peaks in $N_p(E)$ and $N_a(E)$ for E near E_F to enhance T_c greatly; the complexity of defect electronic structure is so great that more specific conclusions are not easily drawn. It is important, however, to realize that regardless of detailed mechanisms, any model which is consistent with the overall experimental data must include defects, and probably defect peaks in $N_p(E)$ and $N_a(E)$ near E_F. I have stressed particularly the quantitative doping data (III.6) in which Cu is replaced by Ni or by Zn, with effects on T_c which are an order of magnitude larger than those expected from rigid band models ($N_p(E) = N_a(E) = 0$) which attempt to correlate T_c solely with carrier concentration (III, ref. 52). More qualitatively, the extreme sensitivity of T_c in $YBa_2Cu_3O_{6.87}$ to exposure to atmospheric H_2O, suggests that H_2O physisorbtion also shifts the peaks in $N_p(E)$ and $N_a(E)$ away from E_F. These doping effects have fatal implications not only for rigid-band electronic models which contain only the framework density of states $N_f(E)$, but also for the various exotic (non-phonon) models discussed in IV.14.

The idea that peaks in N(E) can not only enhance T_c but also produce variations in T_c which are anomalously rapid, relative to a rigid band picture, is of course not new, and was already familiar in the older models for the A-15 compounds of the Friedel type (II.2). Further extensions of this idea are discussed by Allen and Mitrović in their recent review,[36] especially their Sec. 19. In general, however, such effects are much smaller in metals than those observed in high-T_c oxides.

We can distinguish three stages in the calculation of electron-phonon interactions. In the first stage, appropriate to nearly free electron metals, the matrix elements are described by pseudopotential form factors $<\mathbf{k}|V|\mathbf{k}+\mathbf{q}> = V(q)$, as in BCS. This simple approach fails for transition metals, but it can be replaced by tight-binding models where V(q) is replaced, in effect, by atomic orbital parameters such as $dd\sigma$. This approach has worked well for transition metal-metalloid intermetallic compounds

(II.3), but when it is applied to high-T_c cuprates, with the chief parameter being Cu3d—O2pσ, it fails to explain the anisotropic normal-state resistivity and Hall constant, as well as T_c's much over 40K.

The defect model attempts to resolve these difficulties in two ways. First it supplements $N_f(E)$, the rigid-band density of states for the perfect crystal, with narrow defect peaks pinned near $E = E_F$, because defects are a characteristic feature of oxides in general, and peaks at E_F seem especially appropriate to cuprates. However, as pointed out[37] by Allen and Mitrović (and many others!), such peaks by themselves do not resolve all the anomalies, because there are intrinsic limits to peak widths in a normal quasi-particle model which are set by lifetime effects for poorly conductive material (such as cuprates). Thus a second mechanism, also suggested by the material properties, is invoked, namely, quasi-localization in nearly two-dimensionally confined cuprate layers. Here the electron-phonon interaction should be calculated with a marginal basis set of quasi-particle states whose size may itself be both temperature-dependent and a dynamical variable. This problem is unresolved in the normal state at present, but qualitatively we have seen how it can plausibly explain chemical trends in both normal-state and superconductive properties in the context of quantum percolation. In this model the high T_c's can be understood as the result of strong coupling at the defects. This coupling cannot be calculated by one-electron methods, because the latter at present cannot cope simultaneously with the complexity of defects and those of the crystal structure. At present I regard this latter limitation as primarily technical in nature, and not a problem in principle.

In assessing the contribution of the defect density of states in Fig. 14 to T_c there is one basic point to remember.

If we choose $N_p(E)$ and $N_a(E)$ to be large enough, for example, three times larger than the band density of states N(E) averaged over the region $E_F \pm \hbar\omega_D$, then we can increase λ as much as we wish, in this example by a factor of four assuming constant electron-phonon coupling. We have no direct way to measure $\gamma = N(E_F) + N_p(E_F) + N_a(E_F)$ which would be of much value in practice. In good metals which exhibit a full Meissner effect values of γ can be obtained from the specific heat jump $\Delta C_p(T_c)$, using strong-coupling curves such as Fig. II.8. In high-T_c cuprates, however, where a complete Meissner effect has never been observed, not even in single crystals, the percolative filling factor f_p associated with the superconductive transition is simply not known. As discussed further in X.6, the lower bounds that can be placed on f_p are too small to exclude values of γ which would lead to values of λ fully in accord with the observed high T_c's (even without invoking stronger electron-phonon interactions at the defects). Thus if one is satisfied with a theory based on adjustable parameters, then there are many ways to use a high density of defect states pinned at E_F to explain the observed high T_c's. This point should always be borne in mind when one hears statements that the high T_c's in the cuprates are not consistent with BCS.

11. Coupling Constants and Gap Anisotropy

The very large anisotropy observed in the normal-state resistivity tensor (III.7) suggests that the energy gap itself may be highly anisotropic. Furthermore the "local" energy gap may be different for different Cu planes when these are inequivalent, as for the Cu1 and Cu2 planes in $YBa_2Cu_3O_{7-x}$, as suggested by NQR relaxation data (IX.2). A general formalism exists for the anisotropy problem,[37] and a two-component energy gap has been discussed long ago[38] in the context of the weak-coupling BCS limit (without the additional dynamical refinements provided by the Eliashberg equation[37]). However, actual implementation of these refinements in practice requires the determination of a large number of parameters, which is scarcely feasible in a defect-ridden soft lattice. (For example, the two-component BCS model[38] contains five electronic parameters, $N_A(E_F)$, $N_B(E_F)$, V_{AA}, V_{AB} and V_{BB}, in place of the two electronic parameters $N(E_F)$ and V of a one-component theory). Therefore in this section I consider rough estimates of the coupling parameter λ and the gap ratio E_g/kT_c which neglect these refinements which are probably large and important. So far, however, the overall anisotropy is not believed to be so large as to alter the s-wave pairing[37] (X.3), and so the results estimated here should be qualitatively correct.

With $La_{2-x}Sr_xCuO_4$ values of λ in the range 2-3 and E_g/kT_c ratios between 4 and 5 are reasonably consistent[22,24] with experiment ($x \sim 0.15$) (especially the tunneling experiments discussed in VIII). Tunneling experiments on $Pb_{1-x}Bi_x$ with $x = 0.3$ (highest T_c) have given $E_g/kT_c = 5$ (Fig. II-6), so there is nothing anomalously strong about the electron-phonon coupling in $La_{1.85}Sr_{0.15}CuO_4$. The normalized specific-heat jump (Fig. II-7), $\Delta C/\gamma T_c$, which is 1.4 in the weak-coupling limit, is also maximized near 3 for such materials. Finally one expects the average phonon frequency $\omega_{\ell n}$ to be about $6T_c$ according to this simple model, i.e., near 200K, which is where a soft-phonon bump appears in the phonon spectral density (VI.1). Some of this agreement may be fortuitous, but one may more reasonably conclude that the complications discussed above have a relatively small effect on actual solutions of the gap equation. This is also the experience[37] in a number of intermetallic compounds where the corrections are smaller but the data are also more accurate (more homogeneous samples).

The situation in $YBa_2Cu_3O_{7-x}$, where the Cu1 and Cu2 planes are inequivalent, is more complicated and possibly cannot be entirely resolved at present. There appear to be two energy gaps, with ratios $E_g(\alpha)/kT_c$ bounded above and below by 8 and 2, as observed specifically in NQR relaxation (IX.2) and tunneling (VIII), with the larger gap belonging to the $Cu2O_2$ planes. Because most of the condensation energy is associated with the larger gap, it seems natural[39] to use this gap value to estimate λ. This gives λ in the range 3-10, i.e., super-strong coupling, almost twice as large as has ever been found in an intermetallic compound. The defect and marginal dimensionality theory discussed in the preceding two sections, or anharmonic vibrational models (IV.13), can explain this giant coupling. Note from Fig. II-8 that $E_g/kT_c \sim 8$ implies $T_c \sim \omega_{\ell n}$, i.e., a very soft lattice with $\omega_{\ell n} \sim 100K$. This is consistent with the caterpillar soft phonons observed in $YBa_2Cu_3O_7$ by Raman spectroscopy (VI.2). Finally, such large values, of λ do *not* imply large values of $\Delta C/\gamma T_c$, because according to Fig. II-8, the latter peaks near $\omega_{\ell n} \sim 6T_c$, and declines to $\Delta C/\gamma T_c \lesssim 1$ for $T_c \sim \omega_{\ell n}$.

12. Exotic Theories and Ultimate Causes

The discovery of high-T_c superconductivity soon led to the realization among theorists that standard techniques for estimating the strength of electron-phonon interactions in metals were unlikely to produce coupling strengths large enough to explain T_c's of order 100K. Moreover, the high normal-state resistivities made the materials such bad metals, with resistivities as large or larger than those of strongly disordered A-15 compounds, that alternative interactions (which would be weaker in good metals) now seemed more plausible. At present there are almost no quantitative estimates of the strengths of these alternative interactions, so that their success or failure, especially for specific materials, in not subject to the quantitative criteria which have been applied to electron-phonon interactions. We can say that these mechanisms are vaguely discussed as ultimate causes, with little or no evaluation of proximate causes and material-specific chemical trends. As with cosmology and philosophy, there is no exacting criticism and everything seems possible.

Alternative mechanisms for high-temperature superconductivity were extensively discussed in the 1970's, with the one that attracted the most attention being that of exchange of another kind of boson-specifically, an exciton rather than a phonon. In this model, originally proposed in the context of conductive organic molecules,[40,41] the interaction is purely electronic and the idea is to take advantage of the large exciton energy $\hbar\omega_0 \sim 2$ eV compared to the usual Debye energy of order 0.05 eV. In metals the electronic Coulomb interactions are repulsive and contribute to the Coulomb repulsion parameter μ in formulae for T_c (such as II(4.8)). For excitons to provide an attractive interaction, local field corrections to the dielectric screening function $\epsilon(q, q', \omega)$ must be attractive (overscreening) in a large part of phase space. Little is known about this function, except in a few simple semiconductors (such as Si) and nearly free electron metals, and it is not known for any system which exhibits strong exciton bands and has metallic conductivity.

In the absence of theoretical understanding many experimental studies were made of organic materials, especially bilinear organic compounds, in the hope that this geometry would prove favorable. (One chain was supposed to be metallic, while the other perhaps provided the excitons.) In spite of extensive research efforts, this program failed, and no organic superconductor with $T_c > 10K$ is known.

After many failures theorists suspected that either the basic mechanism could not be activated, or (more probably, since a few bilinear organic conductors are known with $0.1K < T_c < 10K$) that the coupling parameter λ^* for electron-exciton interactions is too small, probably because of the smallness of intermolecular overlap. A geometry which enhances intermolecular overlap is a metal-semiconductor interface, as described by Bardeen and coworkers.[42] They did not attempt to calculate $\epsilon(q, q', \omega)$ in full generality, but instead used a simple physical picture of electrons tunneling from the metal into the semiconductor. Also they isolated only the attractive part of this interaction, estimated it crudely, and largely neglected repulsive corrections. But their worst problem was that of the filling factor - that is, finding a geometry in which the metal-semiconductor interfaces fill a large fraction of the sample volume. Absent such a geometry, their model long seemed to be merely academic.

The discovery of the layer cuprates $(La, Sr)_2CuO_4$ and $YBa_2Cu_3O_{7-x}$ at first seemed to be the perfect realization of the ideal geometry for attractive exciton-mediated electron-electron pairing. Metallic cuprate monolayers alternate with semiconductive M[2+ or 3+] oxide monolayers. As a result, most of the papers at the first theoretical conference[43] on high-T_c cuprate superconductors discussed the exciton mechanism in whole or in part. The idea, however, has not gained general acceptance. Even without detailed calculations it has encountered apparently fatal obstacles in explaining the experimental data, as we shall see shortly. Also the search for positive evidence,[43] in the form of excitons which exhibit interesting chemical trends which correlate well with trends in T_c, has so far not been successful (VII.2).

Another feature of the new materials, which is just as specific as their layer structure, is that they contain Cu ions, and that like so many transition metal oxides, when they are insulating they are often antiferromagnetic. This has led several workers to propose mechanisms based either directly on spin waves[44] (or spin fluctuations) or on crystal field distortions of the Jahn-Teller type.[45] These models suffer from an obvious lack of chemical specificity, that is, they suggest that any transition metal could be used in place of Cu, which seems to be contrary to experiment. As we saw before (IV.4), what probably makes Cu special is the Mattheiss relation $E_{3d}(Cu) = E_{2p}(O)$. This leads to the metallic character, or what Anderson has called "resonating valence bonds." His RVB theory[46] has produced many new ideas, such as the possibility that $e^* = e$ (now discarded), as well as agreeing with magnon or exciton theories which predict zero isotope effect.[47] This seemed to be one of the early successes of the theory, but this success may have little significance, because small isotope effects are expected anyway in strong-coupling superconductors with large defect contributions to $N(E_F)$ (V.1, 2).

One of the main goals of the RVB theory seems to be to find a way in which metallic superconductive and insulating antiferromagnetic properties can coexist in the same spatial region. However, as mentioned again below, all the evidence points to separate phases for these two properties. As occurs very often in oxides, these separate phases can coexist in the same sample, but almost surely in spatially different regions. Thus the RVB idea represents an ingenious and resourceful approach to a problem that (so far at least) seems not to exist.

Several of these theories emphasize electronic interactions only, which in the light of the BCS theory seems "novel" and perhaps appropriate for the novel high-T_c cuprates. Historically however, from the discovery of the phenomenon of superconductivity in 1909 until 1950, *all* microscopic theories of superconductivity utilized only electronic interactions. It was the realization that such interactions are repulsive, and that an attractive interaction is needed to modify the electronic structure near E_F, that led Fröhlich to consider phonon exchange, while Bardeen knew of the first isotope effect data.

The purely electronic theories encounter in principle several problems (apart from the experimental contradictions such as the isotope effect and the absence of structure-specific spin fluctuations (IV.5)). While it is true that $YBa_2Cu_3O_6$ is antiferromagnetic and semiconductive, it is separated from metallic and superconductive $YBa_2Cu_3O_{6.87}$ by a *first-order phase transition.* The question of the proper description of the ground state of the semiconductive antiferromagnetic phase is not only academic but, more importantly, it is *irrelevant* to discussing the ground state of the metallic and superconductive phase. Magnetic fluctuations may be important to quantum percolation, but the most that one can hope is that they will show up as corrections to the defect interactions which alter the marginal dimensionality and hence provide metallic conduction in the normal state. Experimentally these corrections are small and are probably not favorable to superconductivity, according to the Cu substitution (by Ni or by Zn) experiments (III.6). More generally, the phase diagrams for $[La, (Ba, Sr)]_2CuO_4$ and $YBa_2Cu_3O_{7-x}$ show an *inverse* correlation between both spin fluctuations (Fig. IV-9) and magnetic phases and superconductive T_c. An especially good example is Fig. III-16, which shows that the peak in the Meissner volume (as well as T_c) occurs far from the magnetic region, but it coincides exactly with the orthorhombic-tetragonal phase transition, which is the composition not only of maximum lattice instability (with respect to a specific displacive normal mode), but more generally of maximum overall lattice softening and anharmonicity (especially near defects), and hence maximum electron-phonon coupling. Another example which should be mentioned is the $Ba(Pb,Bi)O_3$ and $(K,Ba)BiO_3$ perovskite alloys where T_c has so far been raised up to 30K (III.2). None of these alloys,[4] nor indeed any bismate, has even exhibited unusual magnetic properties other than those appropriate to their itinerant electron conductive character (insulator, semimetal, or metal). In fact $BiFeO_3$ is not magnetic but is ferroelectric (App. C). Here again we have an inverse correlation between oxide superconductivity and magnetism, but a strong positive correlation (through ferroelectricity) to lattice instabilities.

There is a more general objection, which applies not only to magnon exchange models but also to all boson (including exciton) models other than phonons. The central fields of the Cu and O ions couple to electrons through phonons with an ion charge that scales like $Z^* e$, where Z^* is of order 4 (for O) to 8 (for Cu). Exciton and magnon (exchange) interactions couple to single electrons, whose screening contribution to the central atomic fields (by far the largest internal fields) is roughly of order $Z^* \sim 0.3$. Thus if we evaluate electron-exciton or electron-magnon couplings realistically (instead of treating their strengths as adjustable parameters), we find that they are typically 20 times weaker than electron-phonon coupling strengths. It is true that the exchange interactions in cuprates are much stronger than in intermetallic compounds (this is what gives such high Neel temperatures in semiconductive materials such as La_2CuO_4 and YBa_2CuO_6), but the net effect is only an increase of magnon couplings by a factor of 2. This still leaves the electron-phonon interactions stronger by a factor of 10. Unfortunately these relations are often concealed in parameterized model Hamiltonians (such as the Hubbard model) because many of the "large" parameters actually are part of the self-consistent one-electron potential when the latter is properly renormalized. When this is not done, serious misinterpretations of the physics of even simple model problems can easily occur.[48] Even after it is done, the standard approaches to constructing the model Hamiltonians usually involve so many simplifications that little or nothing remains of the crystalline richness and complexity that on purely phenomenological grounds must be present in those rare multinary compounds which yield $T_c > 30K$. Certainly there is little scope in these exotic models for proximate causes of a structural nature.

It might seem at this point as if no mechanism can succeed. However, if we review the material systematics we are unavoidably confronted with mechanical instabilities as the factor which always limits T_c, whether in intermetallic or oxide-based superconductors. This has happened so often that even a mathematician, who has trained himself to be able to believe in five impossible things before breakfast, would be hard put to dismiss the repeated appearance of such instabilities as mere coincidence (Appendix C). It should be noted that given enough effort it has nearly always proved possible to develop preparative methods suitable for growing large single crystals of almost any class of materials, whether they might be semiconductor multi-component alloys, magnetic garnets, ferroelectric insulators, or even solid electrolytes. There is only one exception to this general rule, and that is high-T_c superconductors (X.3). That this is the *only* exception is by itself a startling observation, but it also has far-reaching consequences. It means that at the fundamental operational level of science, that is to say, experiment, we know much less about the atomic structure of these materials than we do of many others. Moreover, we have no justification for assuming that the rules for atomic structure which apply to other materials can be carried over automatically to high-T_c oxide superconductors, because this is a class of materials for which the samples needed to perform the diffraction experiments necessary to fix the structure in quantitative detail are simply so far not available (and this may not be an accident!). In VI we give several examples of phonon frequencies and lattice constants which apparently show anomalous shifts in the cuprates unlike anything seen previously (except for Nb_3Al, end of I). No such anomalies are predicted by purely electronic models.

Our discussion of theory in this chapter has necessarily been of a simplified character, consistent with IV.1. The picture of electron-phonon interactions enhanced by defects in severely anharmonic soft lattices is not very different from what was already anticipated theoretically some time ago.[49] The new element emphasized now in the context of layered cuprates is marginal conductivity. This concept explains many observations, including the absence of superconductivity in more three-dimensional[50] metallic cuprates (Fig. III-35). Further examples appear in later chapters in other experiments.

13. Selective Phonon Condensation

We have seen in IV.11 that giant electron-phonon coupling constants $\lambda \sim 8-10$ seem to be required to explain $T_c \sim 90K$ in $YBa_2Cu_3O_7$. In IV.8 - IV.10 we discussed the possibility that marginal dimensionality and defects might produce such large coupling constants in effect extrinsically. Here we discuss a different model which derives large values of λ intrinsically from an unusual but by no means implausible lattice dynamical model. The advantage of this model is that it resolves several vibrational anomalies identified by Raman scattering as described further in VI.3. The signs of these anomalies are qualitatively the reverse of what is seen in normal intermetallic high-T_c superconductors (such as Nb_3Sn), and they are large in magnitude.

The anomalous lattice dynamical feature is what is usually called a double-well potential. Actual double-well potentials are known for displacive insulating ferroelectrics, which includes many perovskite compounds. Because high-T_c cuprates have structures similar to perovskites, experimental evidence for off-site atomic displacements of Ba atoms in YBCO has been interpreted in terms of such wells.[51] Here I use a different model[51] for such displacements, which I believe is more appropriate for metals. In ferroelectrics the barrier height of the potential between the two wells is related to the Curie temperature. No such order-disorder transformation has been observed so far for the layered cuprates. This means that the detailed form of the effective atomic potential for the lattice softening is not accessible at present. Thus the off-site displacements may be described more simply as fictive phonons, in analogy with the concept of fictive temperature which describes quenched disorder in oxide glasses.

In complex ionic crystal structures the number of constraints imposed by nearest neighbor central forces generally exceeds the number of internal degrees of freedom. If this difference is small (as it is here, with on the average about seven such neighbors, giving 3.5 constraints per atom with 3 degrees of freedom), the internal stress may be relieved partially by static buckling of layers (as observed for the CuO_2 planes in $YBa_2Cu_3O_7$) and partially by anharmonic off-site displacements. These can be described by fictive phonon condensation in a few selected modes localized spatially on a few atoms or bonds in each unit cell. Such fictive condensation of large vibrational amplitudes intrinsically relieves the internal stress by increasing the Debye-Waller factors of the selected atoms. Examples of such distortions have apparently been observed in $BaBiO_3$ and $LaAlO_3$ (III.2). Alternatively one may say that the phonon condensation is extrinsically frozen in at the temperatures (~950C) where the material has been formed. The intrinsic interpretation seems preferable for reasons discussed below. What we assume is that the lattice vibrations are nearly harmonic at the formation temperature $T_f \sim 950C$, and then we use the corresponding high-temperature harmonic oscillator wave functions as a basis set to describe low-T dynamical large-amplitude anharmonic displacements. The main features of these displacements can be described by the occupation numbers n_β of the harmonic high-T phonons. The advantage of using these (high-T) fictive-phonon coordinates at low T over a parameterized double-well potential becomes clear below in the context of numerical estimates.

We can describe this low-T condensation with a fictive phonon Hamiltonian,

$$H_{ph} = \sum_{\alpha \neq \beta} \hbar\tilde{\omega}_\alpha n_\alpha - \hbar\tilde{\omega}_\beta n_\beta + U_\beta n_\beta^2 \tag{3}$$

which is minimized with

$$n_\beta = \hbar\tilde{\omega}_\beta / 2U_\beta \,. \tag{4}$$

It is important to realize that while $\tilde{\omega}_\alpha$ is close to the frequency ω_α observed in single-particle phonon excitations, $\tilde{\omega}_\beta$ is not close to ω_β. We calculate ω_α and ω_β by averaging over atomic distributions as measured by diffraction experiments, and the frozen phonons in the β mode produce dynamical off-site displacements which appear as anomalously large Debye-Waller factors or anomalously large anharmonic amplitudes.

The way in which fictive phonon condensation alters the Bardeen-Fröhlich attractive electron-phonon interaction is simple. Because the electronic band width $W >> |\hbar\omega_\beta|$, the presence of dynamical off-site displacements does not significantly shorten phonon lifetimes, and the lattice vibrations still appear to be harmonic so far as *nonadiabatic* single-particle excitations are concerned.[52] Suppose we calculate (IV.6) the electron-phonon interaction strength λ for $n_\alpha = n_\beta = 0$ in terms of the contributions λ_α of each phonon mode α,

$$\lambda = \sum_{\alpha \neq \beta} \lambda_\alpha + \lambda_\beta \tag{5}$$

and represent each contribution λ_β as the product of electronic and lattice factors:

$$\lambda_\beta = \lambda_\beta^{\circ} \, |<n_\beta|x|n_\beta+1> / <o|x|1>|^2 \tag{6}$$

$$\lambda_\beta = \lambda_\beta^{\circ}(n_\beta+1) \tag{7}$$

where the second factor is the lattice harmonic oscillator dipole matrix element. As $T \rightarrow 0$, for normal bonds $n_\alpha = 0$, but for superfluid bonds one can easily have $n_\beta \sim 5$, depending on the internal stress which has been frozen-in at the quenching temperature. Note that neglecting terms of order $n_\beta \hbar\omega_\beta N(E_F) << 1$, the relation in Eqn. (7) is exact. This is a special case of Migdal's "theorem,"[52] here applied to "fictive" phonons.

A different approach[53] to anharmonicity utilizes the classical double-well potential as if it actually exists as a fixed object which is temperature-independent. The resulting vibrational wave functions can then be calculated numerically. The results are qualitatively similar to (7), namely, a much larger value of $<x^2>$ gives a much larger value of λ. However, it is unlikely that specific atoms actually vibrate in double-well potentials, and a simpler approach is to treat the fictive phonon occupation numbers n_α, n_β as the independent variables, since the phonons are normal modes of the system. When more accurate and complete measurements of temperature-dependent Debye-Waller factors become available, it may be possible to correlate these with the Raman data discussed below to obtain independent estimates of n_α and n_β.

At present there are sufficient uncertainties in both the band-structure calculations (IV.6) and lattice-vibrational models (VI.3) that the relative values of λ_α for various modes cannot be assigned with certainty. However, there is strong evidence that the 336 cm^{-1} Raman mode, assigned[8] to the A_g (347 cm^{-1}) z-polarized O vibrations in buckled CuO_2 planes (VI.3), involves bending of softened Cu-O bonds. All other observed Raman and infrared vibrational frequencies stiffen as T decreases, but this mode reaches a maximum frequency of 340 cm^{-1} at $T_c \sim 90K$ and then drops rapidly to 332 cm^{-1} near $T = 0$ (Fig. VI.11). This 3% reduction can be explained in the present model as the result of an additional *increase* in n_β below T_c which results from the β phonon coupling through λ_β to the superconductive order parameter Δ. It is the right order of magnitude, because $2\Delta/W \sim 60$ meV/2eV $= 0.03$. With $\lambda_\beta^0 \sim 1$, one can easily achieve values of $\lambda \sim 8-10$ which are required to explain the giant values of $E_g/kT_c \sim 8$ observed in $YBa_2Cu_3O_7$.

Other evidence supports the assumption that the 340 cm^{-1} phonon band has condensed. The parity of the condensation is even, and therefore the condensation should not effect the Davydov infrared-active (odd-parity) partner stiffening, as shown by chemical trends as a function of rare-earth (RE) radius in a series of $REBa_2Cu_3O_7$

compounds (VI.3). More generally, the oxygen mobility in $YBa_2Cu_3O_{7-x}$ is anomalously large (comparable to cation mobilities in solid electrolytes), and this is an indicator of the lattice softening that is implied by phonon condensation. The idea that special size ratios are required to produce high-temperature superconductors has been confirmed phenomenologically (Appendix C). These size ratios may facilitate the formation of layer structures, and interlayer misfit may then cause buckling and phonon condensation, that is, both static displacements and anharmonically large vibrational amplitudes.

While the idea of internal stress is intuitively plausible, there has been relatively little done to develop the idea formally. The stress arises from misfit of prototypical lattice constants, for example, of stacked atomic planes. Here by the prototypical lattice constant of an atomic plane we mean the lattice constant which would minimize the energy of the plane in isolation. However this interplanar misfit stress is relieved, whether statically by buckling or dynamically by large vibrational amplitudes, in the presence of a high density of defects, the quenching process will freeze some stress-relieving features that were present at high temperatures. The features are said to be retained at the fictive temperature, which is essentially the temperature from which the sample was quenched.

The frozen-in large vibrational amplitude which is postulated here is a relatively novel idea to most theorists, and it is unlikely that it will gain general acceptance immediately. However, the idea is really a small extension of an old idea, the dynamical Jahn-Teller effect, which is widely known to describe molecular (NH_3) and crystal ($BaBiO_3$) vibrations at high temperatures. Certainly it is unusual to have a flat-bottomed potential at low temperatures (Fig. VI-15), but still this is possible in special structures such as defect perovskites. In fact, as we noted in II.3, epitaxially generated large vibrational amplitudes in a metastable pseudomorphic film of Nb_3Ge enhance its transition temperature to 22° K from the bulk value of 17° K. (Indeed even in that case the substrate is Nb_3GeO_x, so that oxygen has been used as the metastabilizing factor, much like O_x in $YBa_2Cu_3O_6O_x$!) It is true that standard computational procedures are not well-adapted yet to describe fictive phonons, but high-T_c superconductors are not standard materials.

The effect of stress on the mechanical stability of network structures has been analyzed by Thorpe and others[54] in a series of highly original papers which illustrate how quenching can lead to the retention of high-temperature mechanical properties in the presence of internal stress. While their results are not immediately applicable to high-temperature superconductors, their analysis shows many interesting analogies to selective phonon condensation. In particular, they find that the effect of stress is to divide the bonds into three quite distinct classes: nearly normal, abnormally extended, and abnormally compressed. These classes retain their integrity as the network is relaxed, which is a kind of memory or quenching effect. The fictive phonon model is discussed further in VI.3.

REFERENCES

1. L. F. Mattheiss and D. R. Hamann, Phys. Rev. *B28*, 4227 (1983).
2. L. F. Mattheiss, Jap. J. Appl. Phys. *24-2*, 6 (1985).
3. M. S. Hybertsen and S. G. Louie, Phys. Rev. *B34*, 5390 (1986).
4. L. F. Mattheiss, E. M. Gyorgy and D. W. Johnson, Jr., Phys. Rev. *B37*, 3745 (1988); L. F. Mattheiss and D. R. Hamann, Phys. Rev. Lett. *60*, 2681 (1988); R. J. Cava et al., Nature *332*, 814 (1988); Y. J. Uemura et al., Nature *335*, 151 (1988).
5. J. C. Fuggle et al., Phys Rev. *B37*, 123 (1988); A. Bianconi et al., unpublished; B. Lengeler et al., Sol. State Comm. *65*, 1545 (1988).
6. L. C. Smedskjaer et al., Phys Rev. *B36*, 3903 (1987); J. Zaanen, A. T. Paxton, O. Jepsen and O. K. Andersen, Phys. Rev. Lett. *60*, 2685 (1988).
7. L. F. Mattheiss, Phys. Rev. Lett. *58*, 1028 (1987).
8. S. Barisić, I. Batistić and J. Friedel, Europhys. Lett. *3*, 1231 (1987).
9. R. V. Kasowski, W. Y. Hsu and F. Herman, Sol. State Comm. *63*, 1077 (1987).
10. F. Herman, R. V. Kasowski and W. F. Hsu, Phys. Rev. *B36*, 6904 (1987).
11. S. Massidda, J. Yu, A. J. Freeman, and D. D. Koelling, Phys. Lett. *A122*, 198 (1987).
12. N. F. Mott, Proc. Phys. Soc. *A62*, 416 (1949); J. C. Slater, Phys. Rev. *82*, 538 (1951).
13. J. Hubbard, Proc. Roy. Soc. London, *A276*, 283 (1963); *281*, 401 (1964).
14. P. W. Anderson, Phys. Rev. *124*, 41 (1961).
15. J. E. Hirsch, Phys. Rev. *B31*, 4403 (1985).
16. H. Ikeda and K. Hirakawa, Sol. State Comm. *14*, 529 (1974).
17. T. Freltoff et al., Phys. Rev. *B36*, 826 (1987).

18. G. Shirane et al., Phys. Rev. Lett. *59*, 1613 (1987).

19. J. M. Tranquada et al., Phys. Rev. Lett. *60*, 156 (1988).

20. T. Brückel et al., Europhys. Lett. *4*, 1189 (1987); F. Mezei et al., unpublished.

21. W. E. Pickett, H. Krakauer, D. A. Papaconstantopoules and L. L. Boyer, Phys. Rev. *B35*, 7252 (1987).

22. W. Weber, Phys. Rev. Lett. *58*, 1371 (1987).

23. L. F. Mattheiss and D. R. Hamann, Sol. State Comm . *63*, 395 (1987); L. F. Mattheiss (private communication).

24. P. B. Allen, W. E. Pickett, and H. Krakauer, Phys Rev. *B36*, 3926 (1987); Phys. Rev. *B37*, 7482 (1988).

25. W. Weber and L. F. Mattheiss, Phys. Rev. *B37*, 599 (1988).

26. B. M. Klein and W. E. Pickett, in *Superconductivity in d- and f-Band Metals 1982* (Ed. W. Buckel and W. Weber, Kernf. Karls, GmbH, Karlsruhe), p. 477.

27. J. A. Wilson and A. D. Yoffe, Adv. Phys. *18* 193 (1969), see p. 247.

28. R. V. Kasowski, W. Y. Hsu and F. Herman, Phys. Rev. *B36*, 7248 (1987).

29. P. A. Sterne and C. S. Wang, Phys. Rev. *B37*, 7472 (1988).

30. N. F. Mott, Phil. Mag. *6*, 287 (1961).

31. P. W. Anderson, Phys. Rev. *109*, 1492 (1958).

32. E. Abrahams, P. W. Anderson, D. C. Licciardello and T. V. Ramakrishnan, Phys. Rev. Lett. *42*, 673 (1979).

33. J. C. Phillips, Phil Mag. *47*, 407 (1983); _, _ (1988); Sol. State Comm. *47*, 191 (1983); M. Kaveh, Phil. Mag. *B52*, LI (1985).

34. Y. Imry and S.- K. Ma, Phys. Rev. Lett. *35*, 1399 (1975).

35. H. Stormer, A. F. J. Levi, K. W. Baldwin, M. Anzlowar, and G. S. Boebinger, Phys. Rev. *B38*, 2472 (1988).

36. J. C. Phillips, Phys. Rev. Lett. *59*, 1856 (1987).

37. P. B. Allen and B. Mitrović, Sol. State Phys. *37*, 1 (1982).

38. H. Suhl, B. T. Matthias, and L. R. Walker, Phys. Rev. Lett. *3*, 552 (1959).

39. J. C. Phillips, Sol. State Comm. *65*, 227 (1988).

40. W. A. Little, Phys. Rev. *A134*, 1416 (1964).

41. V. L. Ginzburg, Sov. Phys. Usp. *13*, 335 (1970).

42. D. Allender, J. Bray and J. Bardeen, Phys. Rev. *B7*, 1020 (1973).

43. *Novel Superconductivity* (Ed. S. A. Wolf and V. Z. Kresin, Plenum, New York, (1987); I. Bozovic et al., Phys. Rev. Lett. *59*, 2219 (1987).

44. V. J. Emery, Phys. Rev. Lett. *58*, 2794 (1987); J. R. Schrieffer, X.-G. Wen and S.-C. Zhang, Phys. Rev. Lett. *60*, 944 (1988).

45. H. Kamimura, Jpn. J. Appl. Phys. *26*, L627 (1987).

46. P. W. Anderson, Science *235*, 1196 (1987).

47. P. W. Anderson and E. Abrahams, Nature *327*, 363 (1987).

48. L. M. Roth, Phys. Rev. Lett. *60*, 379 (1988).

49. J. C. Phillips, in Douglass II, p. 413.

50. F. Herman, R. V. Kasowski and W. Y. Hsu, Phys. Rev. *B37*, 2309 (1988).

51. P. Marsh, T. Siegrist and R. M. Fleming, L. F. Schneemeyer, and J. V. Waszczak, Phys. Rev. *B38*, 874 (1988); J. C. Phillips, Sol. State Comm. (1988).

52. A. B. Migdal, Sov. Phys. JETP *7*, 996 (1958); G. M. Eliashberg, Sov. Phys. JETP *11*, 696 (1960).

53. J. R. Hardy and J. W. Flocken, Phys. Rev. Lett. *60*, 2191 (1988).

54. H. Yan, A. R. Day and M. F. Thorpe, Phys. Rev. *B* xxx (1988).

V. Isotope Effects

1. Old Materials

One of the simplest predictions (I (2.1)) of the BCS theory (in the weak-coupling limit) is that $T_c \propto \theta_D$, an average phonon energy. Again in the weak-coupling limit where interatomic forces are nearly independent of atomic vibrational amplitudes, θ_D itself scales with average atomic mass $<M>$ like $<M^{-1/2}>$, assuming that electrons interact equally strongly with all atoms. One can define[1] an isotope effect parameter δ by the relation $T_c \propto <M^{-\alpha}>$ with $\alpha = 0.5(1-\delta)$. Experimentally in metals it is found that $\delta \lesssim 0.1 \pm 0.1$ for simple (s-p) metals, as predicted by the BCS theory. Examples[1] of such weakly coupled superconductive simple metals are Zn, Cd, Sn, Hg, Tl and Pb.

The situation for transition metals (Ru, Os, Mo, Zr) and transition metal compounds (Nb_3Sn, Mo_3Ir) is qualitatively different. Here values of δ are much larger, typically $0.3 \lesssim \delta \lesssim 1$. At one time this led Matthias and coworkers to suggest that some mechanism other than the electron-phonon interaction must be responsible for superconductivity in metals containing d bands. Careful analysis shows, however, that all of these "anomalies" are merely the result of the fact that d bands are narrower than s-p bands, with lower cutoffs for the Coulomb repulsion interactions when the gap equation is solved. Garland's early analysis[1] of these band effects was carried out in the absence of detailed band calculations, but when these later became available they not only confirmed his early qualitative reasoning but also provided striking quantitative support for it. For example, in Nb_3Sn the isotope effect is very small, that is, $\delta = 0.84 \pm 0.04$, but as we notice from Fig. II-3, there is a very narrow peak in N(E), the electronic density of states, for E near E_F, and this narrow peak effectively reduces the isotope shift. The same conclusion is reached by a similar route in the Friedel model (II.2) because of the narrow peak in N(E) for E near E_F. Actually what these anomalies really demonstrate is that often strong electron-phonon interactions accompany peaks in N(E), and therefore since the latter often suppress the isotope effect ($\delta \sim 1$), it is just in the strong-coupling case that we expect small or nearly zero isotope effects. Paradoxically zero-isotope effect, coupled with high T_c, is often the signature of strong electron-phonon coupling.

The anomaly in metals which attracted the most attention was the reverse isotope effect in PdA, where $T_c(PdH) \sim 9K$ and $T_c(PdD) \sim 11K$. At first this was thought to show that the electron-boson interaction involved magnons instead of phonons, on the grounds that Pd was almost ferromagnetic. This nearly ferromagnetic character is chemically reasonable (Pd lies just below Ni in the periodic table) and Fe impurities (but of course not H!) in Pd were known to have very large local moments. However, once again a careful analysis[2] of the BCS equation showed that this reverse anomaly could be understood if the Pd-A interatomic force constant k was renormalized by the zero-point vibrational amplitude of the interstitial H. Ganguly estimated that a 12% increase in k(PdH) compared to k(PdD) would account for the observed "anomaly". Subsequent neutron scattering studies[3] inferred that the actual increase was about 20%. Considering the difficulties that surround even the most careful experiments involving interstitial (rather than substitutional) impurities, this is very good agreement indeed, and it shows that there is no need to discuss magnon exchange as anything more than a minor factor.

2. Cuprates

Experimental determinations of δ are normally difficult because (except in the case of H) isotopic shifts in M are small and sharp transitions are needed to measure δ accurately. In the case of $Pd(H, D)_x$ the isotopic shift in M is large but great care must be exercised to obtain homogeneous samples with the same value of x for H and D. For high-T_c cuprates the ease of oxygen exchange in $YBa_2Cu_3O_{7-x}$ makes a *partial* isotope effect measurement relatively easy, and this was the method used in the earliest experiments.[4] These showed little or no isotope shift ($\delta = 1$, that is, an even smaller value than in Nb_3Sn), and not surprisingly they were hailed by some[5] as evidence that "eliminates conventional strong coupling between electrons and phonons" as "unthinkable." Others, however, realized that these early experiments, while interesting, were inconclusive for several reasons.[6] These are: (1) At the exchange temperature $T_{ex} \sim 500C$ only about half of the ^{16}O was actually replaced by ^{18}O, (2) If defects are present, and there are large peaks in the defect density of states $N_{a,p}(E)$ for E near E_F, the observed[4] isotopic changes in the framework vibrational spectrum (θ_D in the simple BCS formula) may be compensated by changes in $N_{a,p}(E_F)$ or even by changes in the local vibrational $\theta_{a,p}$ frequencies of the defects, and (3) More generally, in a soft lattice, atomic relaxation and complex configurational rearrangement may be induced with isotopic substitutions.

Many questions about the isotope shift in $YBa_2Cu_3O_7$ have been raised because of the incompleteness of the isotope exchange in some of the earlier experiments,[7] but these have largely been laid to rest in recent experiments[7] in which samples were synthesized from starting materials with different ^{16}O and ^{18}O isotopic weights. The more recent results for YBCO and some earlier results for other oxides are listed in Table 1. They show complex chemical trends which are most easily explained in terms of differences in $N_{a,p}(E)$ and not in terms of the isotope shift of the framework θ_D.

Compound	T_c(onset) (K)	ΔT_c (K)	Meissner effect (%)	Isotope shift (K)	BCS shift (K)	^{18}O (%)
$BaPb_{0.75}Bi_{0.25}O_3$	11.0	6	21	0.6	0.63	60
$La_{1.85}Ca_{0.15}CuO_4$	20.6	10	23	1.6	1.14	75
$La_{1.85}Sr_{0.15}CuO_4$	37.0	14	12	1.0	2.10	75
$YBa_2Cu_3O_7$	90.5	10	82	0.2	5.21	90

3. Fictive Phonons and the Isotope Effect

While it is true that a small isotope effect can be explained by a variety of electronic effects, it is nevertheless interesting to work out the consequences[8] of the fictive phonon model (IV.13) and (VI.3) of quenched-in large vibrational amplitudes for the 330 $cm^{-1} = \bar{\omega}$ vibrations of CuO_2 planar O atoms transverse to these planes. This model is especially interesting because it contains no adjustable parameters and the predicted isotope effect can be calculated easily, including the strong-coupling corrections described by II (4.7), rewritten here for convenience:

$$\tilde{T}_c = 0.25\bar{\omega}[e^{2/\lambda} - 1]^{-1/2} \tag{1}$$

where $\tilde{\lambda}$ has been approximated by

$$\tilde{\lambda} = \lambda_o(1+\tilde{n}_f) \tag{2}$$

$$\tilde{n}_f = (\exp(\hbar\bar{\omega}/kT_f) - 1)^{-1} \tag{3}$$

and T_f is the formation temperature (950C). With $\tilde{n}_f = 2.5$ we now find that in addition to the weak-coupling isotope shift associated with the prefactor $\bar{\omega}$ in (1) there is a strong-coupling correction associated with the term in brackets which cancels approximately half of the isotope shift predicted by the prefactor.

Because we now know that the CuO_2 planar $\bar{\omega} = 330\ cm^{-1}$ phonon plays a unique rôle in the superconductivity of YBCO, it seems natural to measure $\delta\bar{\omega}$ directly by Raman scattering in samples with ^{16}O replaced by ^{18}O. This has recently been done,[9] and it is found that the $330\ cm^{-1}$ band shows the same fractional isotope shift as the $500\ cm^{-1}$ band (corresponding respectively to the A_g modes labelled 347 and 515 cm^{-1} in Fig. IV-10). What this means is that the equations (1)-(3) do not describe all the factors associated with determining T_c. The most important of these factors may be the difference between $\tilde{T}_c$ (the transition temperature of an unclamped isolated grain, which is correctly described by (1)-(3)), and the resistivly measured T_c in a single crystal or compacted powder sample clamped at the grain contacts which determine T_c. As is shown in VI.4 and Fig. VI-9, this difference may be of order $\Delta T_c = \tilde{T}_c - T_c \sim 30K$. The shape memory effects discussed there are likely to be especially significant at clamped grain contacts. It may be that microscopically this shift ΔT_c depends mainly on interplanar coupling by defects, and that the most important of these defects are the Cu(1) O chains themselves, whose vibrational frequencies seem to exhibit negligible isotope shift.[9] Thus the observed small shift may reflect compensation of the Cu(2) O_2 isotope shifts discussed with (1)-(3) by Cu(1) O non-isotope shifts, an effect which is difficult to calculate because little is known about the electronic defect mechanism responsible for the latter. Single-crystal specific heat data indicate[10] that such defects are present in metallic YBCO.

There is a hidden point[6] concerning the isotope effect in all high T_c cuprates which requires careful thought. The materials are always subjected to complex preparation procedures which include various annealing stages to increase the oxygen content and optimize the internal site occupancy configuration. Some of the consequences of these procedures are discussed explicitly in X.7, but these preparation procedures (which are essentially determined by trial-and-error) must effect many internal degrees of freedom in ways that will lie beyond direct measurement for the foreseeable future. Thus it seems that we should not be surprised if optimized samples exhibit little isotope effect.[6] For example, if the Fermi energy is pinned by a defect peak, and this peak position can be tuned to E_F, and the peak width is $\lesssim k\theta_D$, then little isotope effect is expected. (It would be second order in $\delta M_O/M_O$, rather than first order, as expected from a näive model without defects.) In fact, this kind of possibility may well explain much of the scatter observed even in metals (V.5).

4. Lattice Instabilities and Anharmonicity

We have discussed several effects of anharmonicity, on T_c and on isotope shifts. In general anharmonicity of O motion is very difficult to measure because the O contribution to X-ray intensities is small compared to the cation contributions, while neutron scattering structural determinations require large, homogeneous single crystals which are not so far available. However, from cation motion there is evidence for anharmonicity. This evidence results from a careful and complex determination of cation anharmonic thermal motion in $YBa_2Cu_3{}^{17}O_{7-x}$ by dual wave length x-ray diffraction on a single crystal with a small amount of twinning.[11] The crystal structure including thermal ellipsoids is shown in Fig. 1. The largest

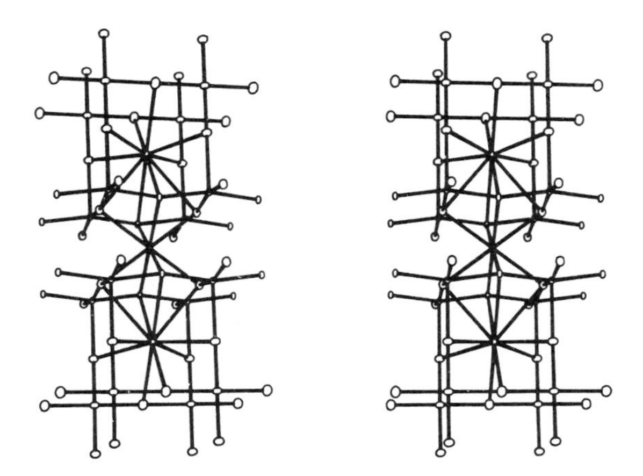

Fig. V.1. Crystal structure of $YBa_2Cu_3O_7$ with thermal ellipsoids. Note the ten-fold coordinated Ba atoms and the O buckling in the $Cu2O_2$ plane.

anharmonicity is found not for the Cu1 chain atoms but for the ten-fold coordinated Ba atoms, which actually move in a double-well potential corresponding to a z-axis Ba-Ba breathing mode, corresponding to the electron density map shown in Fig. 2. It seems very likely that the major source of the Ba anharmonicity is associated with the short Ba-O bond length, where the O atoms are nearly coplanar (relative to z) with Ba. This squeezes the Ba away from the chains towards the planes, which in turn causes the latter to buckle, as shown in Fig. 1. It could easily be the case that the degree of buckling of the Cu2 planes is strongly effected by the O isotopic mass through the Ba coupling. This in turn implies strong electron-phonon coupling for the Cu2 planar O vertical bond-bending modes. However, more generally, it shows that isotopic substitution can have a drastic effect on interplanar coupling either intrinsically (through the Ba double-well configurations) or extrinsically (through defects, e.g., in the Y plane that couple to the Cu2 planar O buckling configurations, or through isotope-dependent growth kinetics).

5. Physical Implications of Isotope Shifts

Because so much significance has been attached to the smallness of the isotope effect in $YBa_2Cu_3O_7$, we show in Fig. 3 the known isotope shifts[1] for simple metals and for transition metals and their compounds as two separate groups, together with the experimental data[4,7,12] for $YBa_2Cu_3O_7$, $La_{1.85}Sr_{0.15}CuO_4$, and $Ba(Pb,Bi)O_3$. Crudely speaking, the general trend in the oxides somewhat resembles that in the metals, but with T_c increased by about a factor of 10. However, this statement has little real

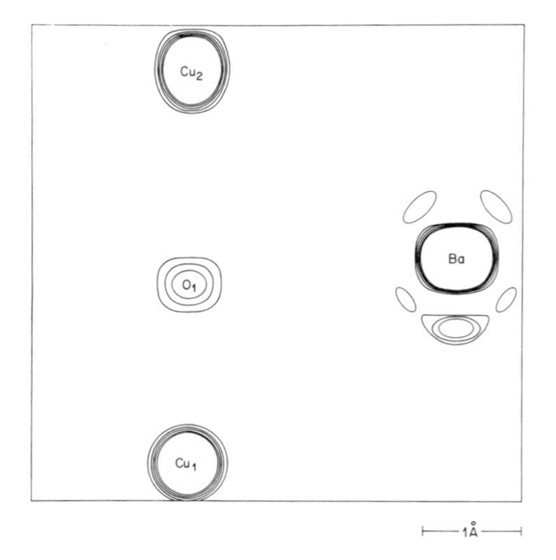

Fig. V.2. An electron density map of an $[(x+y), z]$ plane of $YBa_2Cu_3O_{7-x}$ including Cu1, Cu2 and Ba. The density contours span a range of 5 and clearly show both anharmonic skewed potential (z^3) and double well (z^4) Ba motion.

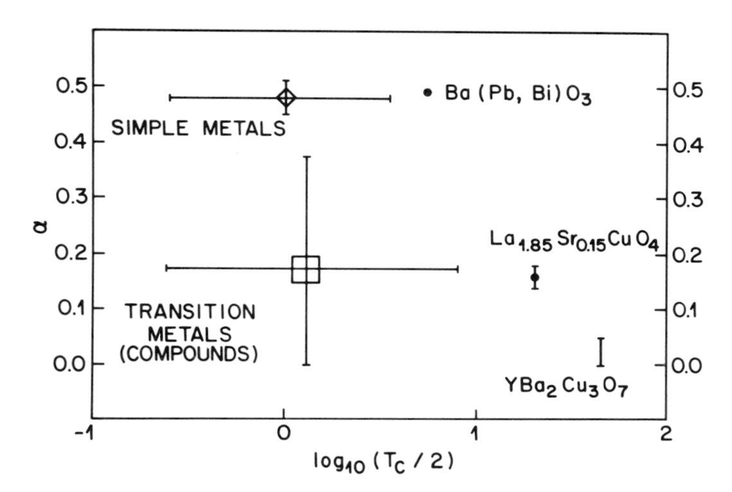

Fig. V.3. Isotope shifts α for metals and for oxides. The metals are separated into two groups, s-p simple metals (Al, Pb, ... and transition metals and their compounds (Ru, Nb_3Sn, $\cdots$) with the ranges shown for α and T_c: note that for the latter group the range of α is large. The large range in α spanned by the oxides is thus not surprising, especially considering the high defect density in $YBa_2Cu_3O_9O^{\square}_{2+x}$.

significance because there is so much scatter and the data are so sparse. For example, $\alpha \approx 0.0$ not only in $YBa_2Cu_3O_7$ ($T_c = 92K$), but also in Ru ($T_c = 0.5K$). If one concludes from $\alpha = 0$ in the former that superconductivity there is not mediated by electron-phonon interactions, then the same conclusion should apply to the latter.

It is now generally believed[1] that $\alpha = 0.0$ in Ru because of its complex electronic band structure. The electronic band structure of "ideal" $YBa_2Cu_3O_7$ is supposed to be simple, but in real samples with oxygen vacancies, stacking faults, and cation disordering, this simplicity is probably lost. If a peak in N(E) near E_F is responsible for the smallness of α in Ru and Nb_3Sn, then a similar peak, associated with defects and located near E_F because of the Mattheiss relation $E_{3d}(Cu) = E_{2p}(O)$, can easily explain the smallness of α in $YBa_2Cu_3O_9O^{\square}_{2+x}$, where the formula emphasizes the high concentration of oxygen vacancies $O^{\square}$ in the latter defect perovskite compound compared to perovskite $Ba(Pb, Bi)O_3$. These defects are responsible for the precursor and memory effects associated with structural relaxation in $YBa_2Cu_3O_{7-x}$ which are discussed in VI.4.

REFERENCES

1. J. W. Garland, Jr., Phys. Rev. Lett. *11*, 114 (1963); R. Sharma, K. S. Sharma, and L. Dass, Phys. Stat. Sol. (b) *133*, 701 (1986).
2. B. N. Ganguly, Z. Physik. *265*, 433 (1973).
3. A. Rahman, K. Sköld, C. Pelizarri, S. K. Sinha, and H. Flotow, Phys. Rev. *B14*, 3630 (1976).
4. B. Batlogg et al., Phys. Rev. Lett. *58*, 2333 (1987).
5. P. W. Anderson and E. Abrahams, Nature *327*, 363 (1987).
6. J. C. Phillips, Phys. Rev. *B36*, 861 (1987); Phys. Rev. Lett. *59*, 1856 (1987).
7. H. -C. zur Loye et al., Science *238*, 1558 (1987); E. Garcia et al., Phys. Rev. *B38*, 2900 (1988).
8. J. C. Phillips (unpublished).
9. M. Cardona et al., Sol. State Comm. *67*, 789 (1988); H. Katayama-Yosida et al., Jap. J. Appl. Phys. *26*, L 2085 (1987).
10. S. von Molnár, A. Torressen, D. Kaiser, F. Holtzberg and T. Penney, Phys. Rev. *B37*, 3762 (1988).
11. P. Marsh, T. Siegrist, R. M. Fleming, L. F. Schneemeyer and J. V. Waszczak, Phys. Rev. *B38*, 874 (1988).
12. B. Batlogg et al., Phys. Rev. Lett. *59*, 912 (1987); T. A. Faltens et al., ibid., *59*, 915 (1987).

VI. Lattice Vibrations

1. Metallic, Covalent and Ionic Forces

The traditional methods of treating lattice vibrations in superconductors are designed for good metals where covalent and ionic forces can be neglected. However, high T_c cuprates are bad metals and we expect ionic forces to be especially large because the screening of the internal electric field by cuprate holes is good only at distances large compared to unit cell dimensions. At present no method exists for treating metallic and ionic lattice forces simultaneously and joining them to an SCF energy-band calculation to obtain electron-phonon interactions. Such a method could be obtained in principle by generalizing the Varma-Weber method (II.3) for metals to include an internal macroscopic field **E** which is screened by a selected group of "metallic" carriers. We now discuss a model which carries out the ionic part of this program in the case where there are no metallic carriers. It is possible that eventually it will prove feasible to combine ionic and metallic forces in a systematic way involving few or no parameters.

The most advanced method at present for treating lattice vibrations in ionic solids is the potential-induced breathing (PIB) model.[1] This model is basically quantum-mechanical, but it contains one classical feature that greatly simplifies the calculation and makes possible the analysis of complex ionic crystals such as multinary cuprates. This is the introduction of a neutralizing fictitious uniform positive spherical charge density at each anion site with the anion assigned its formal valence, here O[2−]. The atomic charge densities of the anions are calculated quantum-mechanically in the presence of this fictitious stabilizing potential, which is allowed to vary or "breathe" during atomic vibrations. There is interatomic charge density overlap, and the energy associated with this overlap is treated by the Thomas-Fermi method, which (as is often the case) works (not so surprisingly) well. The method is designed to correct the most serious weakness of classical ionic models, namely ionic rigidity, and it does this reasonably well, although the spherical approximation of necessity excludes covalent effects. Probably the most serious limitation of the method is that the stabilizing anionic potential is independent of cation-anion interactions and so works best for small, rigid cations (like Mg) and worst for large, polarizable cations (like Ba).

Some recent results obtained by the PIB and related models include discussions of ferroelastic and ferroelectric effects in perovskite halides.[2] The lattice vibration dispersion curves of MgO and BaO were calculated and compared with neutron scattering data, as shown in Figs. 1

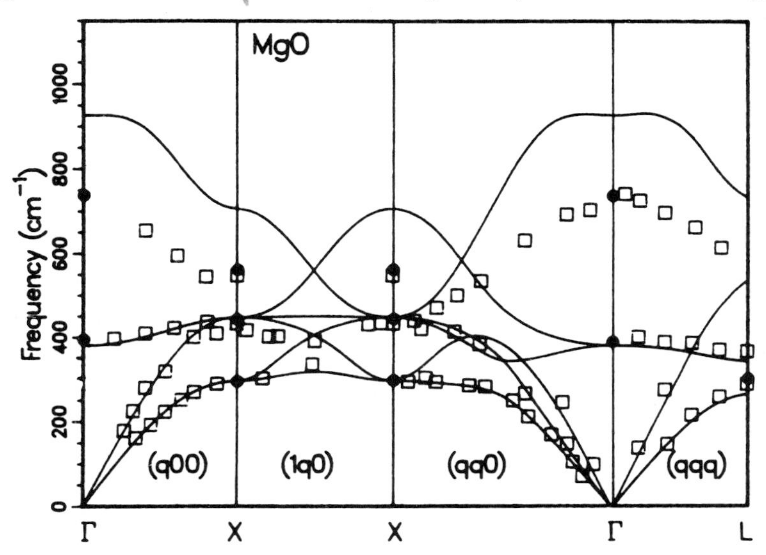

Fig. VI.1. Theoretical (PIB) lattice vibration curves for MgO compared to neutron scattering data (ref. 2).

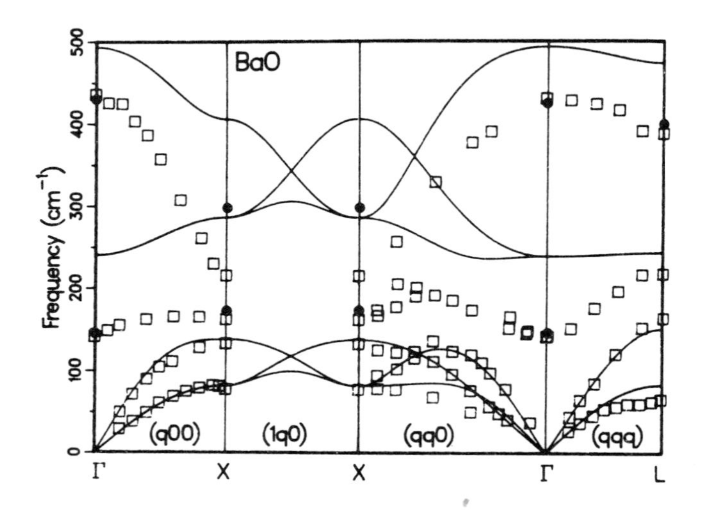

Fig. VI.2. As in Fig. 1, for BaO (ref. 2).

and 2. The longitudinal optic (LO) modes given by the theory are not accurate, but the LO-TO splitting at $\mathbf{k} = 0$ is reasonable. The TO mode in MgO is quite accurate, but there are substantial errors for the TO mode which increase monotonically in the series MgO-CaO-SrO-BaO and the **k**-dependence of the LO-TO splitting is described increasingly poorly (Fig. 2). Clearly the corrections associated with cation polarizability and covalency are large and important in BaO.

Because of its simplicity (similar to that of tight-binding method) and accuracy (the atomic charge densities and wave functions breathe radially), the results obtained[3] when the PIB model is applied to a cuprate are of great interest. First it is instructive to compare the sphericalized ionic charge density with that obtained by a full SCF calculation. The major differences in La_2CuO_4 are found only in the CuO_2 plane, where it is seen that SCF charge moves out of the Cu-O overlap regions, minimizing electron-electron repulsive energies. Because the antibonding $d_{x^2-y^2}$ states are partially empty in the SCF calculation, the SCF calculation transfers charge from the planar $Cu-O_{x,y}$ overlap region to the overlap region of the axial (long bond) $Cu-O_z$. Finally there is a small SCF bond charge at the $Cu-O_{x,y}$ contacts.

The PIB model also predicts a tetragonal distortion ratio $c/a = 3.01$ compared to the experimental 3.49. (Because of the way ionic effects are introduced in the PIB model, it would predict a similarly smaller ratio for La_2NiO_4 (III.3), and in this case the agreement with experiment would be almost perfect. This leads to the inference that the c-axis elongation, or planar a-axis compression, is a covalent effect in La_2CuO_4.) An orthorhombic distortion of 1.6% is also predicted compared to the experimental distortion of 0.8%. (Again this distortion is ionic in nature and is overestimated because of the suppression of covalent effects by atomic sphericalization.)

The predictions of the PIB model for La_2CuO_4 in the tetragonal and orthorhombic structures are shown in Fig. 3. As we shall see, the errors are similar to those found for SrO, that is, the theoretical frequencies are too high by about a factor of two. It is clear that the model succeeds in including the effects of oxygen polarizabilities, but not the Sr or Cu polarizabilities. The eigenvalues ω^2 may be negative, corresponding to lattice instabilities (shown as negative ω in Fig. 3). Most of these tetragonal instabilities, Fig. 3(a), are removed by the orthorhombic distortion, Fig. 3(b), but one remains, corresponding to the monoclinic distortion which is discussed in VI.2. The orthorhombic instability is identified with the octahedral tilting mode

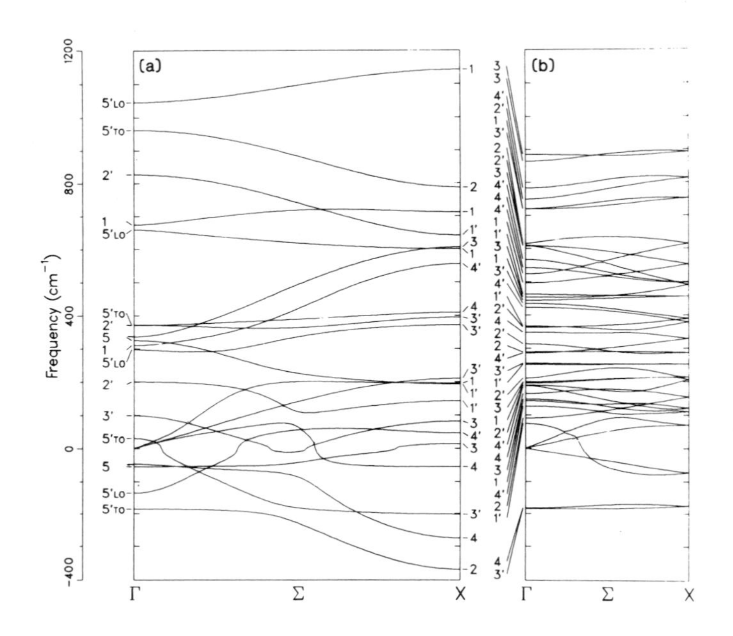

Fig. VI.3. Theoretical (PIB) lattice vibration curves for La_2CuO_4 in (a) tetragonal and (b) orthorhombic structures (ref. 3).

(similar to $BaBiO_3$), in good agreement with experiment.

One comment should be made here. In Weber's non-ionic model of (IV.6) of electron-phonon interactions, the octahedral tilting mode does not couple to metallic electrons, and this has sometimes been used to argue that the coincidence of the lattice instability with maximum T_c in $La_{2-x}Sr_xCuO_4$ alloys (Figs. III 13, 14) is accidental. However, it is important to remember that the $Sr_{[La]}$ impurities not only dope the material by upsetting the stoichiometry, thereby moving away from the antiferromagnetically insulating half-filled valence band phase, but they also provide electrical bridges between the layers (axial interplanar defects, IV.8). They are therefore essential to the three-dimensional metallic character of the material, and at these impurities the local symmetry is not that of the crystal. At the same time the local or resonant modes associated with the Sr defects are not easily identified in the experimental spectra, except insofar as they alter the orthorhombic-tetragonal phase transition mode directly. At the impurities the metallic electrons specifically will have strong electron-phonon interactions with all these localized modes, some of them of a multiphonon nature involving combinations of the soft phonons with other phonons, most probably z-polarized O transverse vibrational modes.

2. Lattice Instabilities and Phonon Spectra

In (I.2) we saw that lattice instabilities have been observed to correlate well with T_c enhancement in intermetallic compounds. In the new materials (so far all cuprates) such correlations are also observed, but they may have a special added significance, because of the special rôle played by defects in both normal state properties and T_c enhancement (IV.7-IV.10). Soft lattices offer many configurational choices for defects which can contribute to screening of the ionic field $\mathbf{E}$ with defect state electronic energies E pinned at or near $E = E_F$. However, because of the large number N of atoms per unit cell in these materials, the lattice vibration bands must be quite complex. (For example, La_2CuO_4 with $N = 7$ is the *simplest* case in this family, yet it has $3N = 21$ vibrational bands.)

When we look for lattice instabilities the situation we have in mind is similar to that shown in Fig. 4. In the unrelaxed lattice a specific vibrational mode is unstable, $\omega_\alpha^2(\mathbf{k}_\alpha) < 0$. Usually $\mathbf{k}_\alpha \neq 0$ and often it lies at a Brillouin zone boundary of the unrelaxed lattice, as in Fig. 4(a).

After relaxation the same mode often appears as in Fig. 4(b). Now $\omega^2(\mathbf{k}) > 0$ for all $\mathbf{k}$, as expected, but it is nearly flat in the region where it was negative before. Thus the relaxation involves many small correlated rearrangements of the atomic site configurations to the extent that these can be made in a fashion consistent with crystal symmetry.

The flat dispersion over a large portion of the Brillouin zone suggests also isolated defects (Fig. I.4), and often if is experimentally difficult to separate such defect bands from those of a relaxed and reconstructed lattice. In some of the older superconductive materials, such as NbN_x, anion

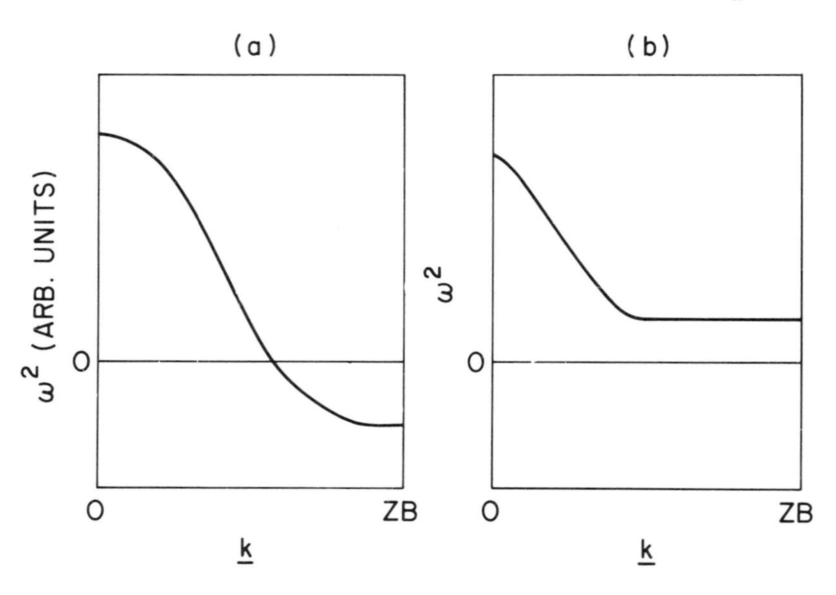

Fig. VI.4. An example of an unstable lattice mode (a) before relaxation, and (b) after relaxation. The maximum instability is assumed to occur near the Brillouin zone boundary (ZB).

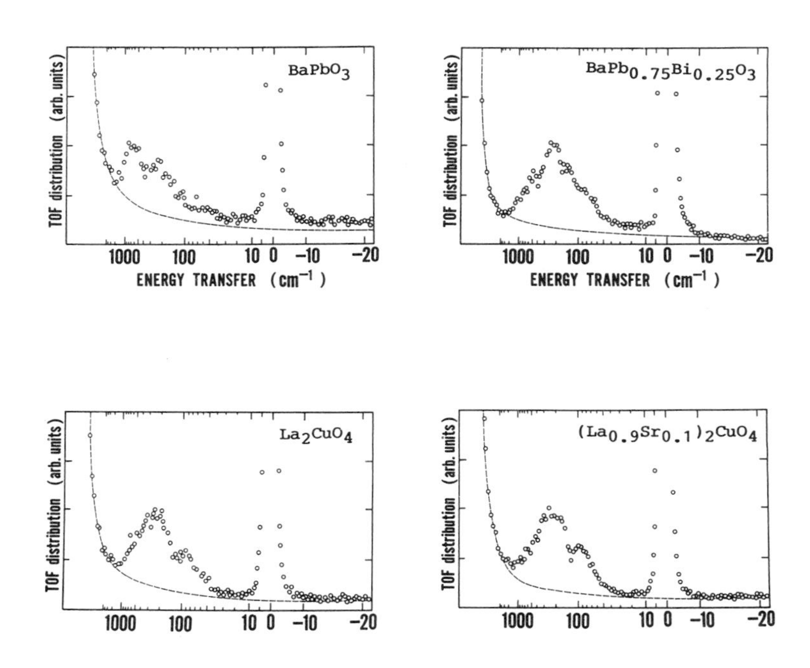

Fig. VI.5. Phonon spectra for pseudo-perovskites (ref. 4).

vacancies almost appear to be native to the material: decades of materials research on NbN_x have improved x from 0.8 to 0.9 to (presently) 0.98, but truly stoichiometric NbN has never been made.

While it is easy to recognize unstable bands of the kind shown in Fig. 4(a) in theoretical plots (an example is the Γ_5 (380 cm^{-1}) to X_4 (280 i cm^{-1}) band in Fig. 3), even in the presence of many nearly flat crossing bands, spotting something like the stabilized bands of Fig. 4(b) in experimental data requires some faith that such relaxed and reconstructed bands must be present. As examples we may consider the phonon densities of states $F(\omega)$ measured by inelastic neutron scattering. These have been measured for a series of pseudoperovskite high-T_c compounds, and the chemical trends within these series show the softening effects implied by Fig. 4(b).

Measurements of $G(\omega)$, which is $F(\omega)$ weighted by neutron scattering cross sections, which should vary slowly with ω, are shown[4] for the series $BaPbO_3$-$BaPb_{0.75}Bi_{0.25}O_3$, La_2CuO_4 and $La_{1.8}Sr_{0.2}CuO_4$, in Fig. 5. These were obtained by TOF (time of flight) techniques, so that the low

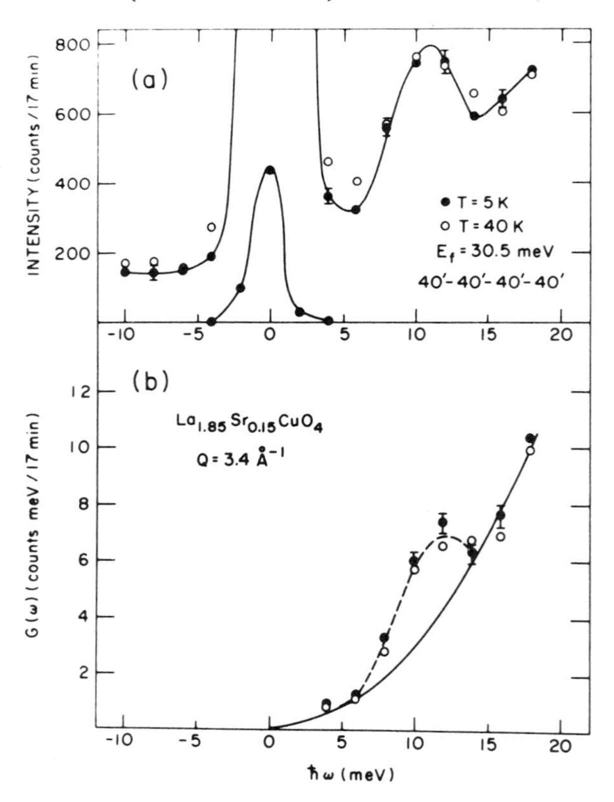

Fig. VI.6. Neutron scattering spectrum for $La_{1.85}Sr_{0.15}CuO_4$ (a) as measured and (b) after subtraction of an energy-independent background. The acoustic component, varying as ω^2, is shown by the solid line (ref. 5).

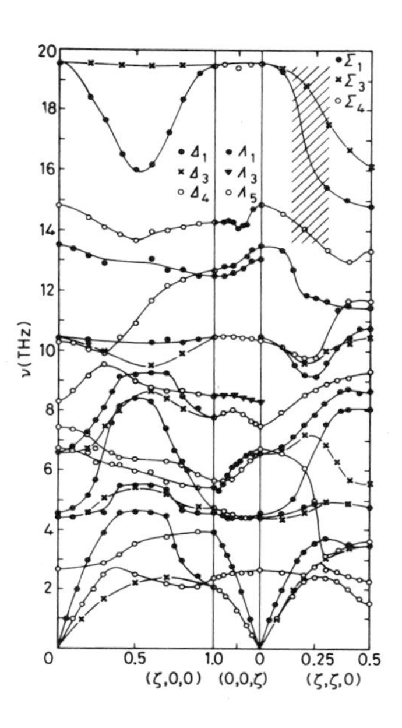

Fig. VI.7. Vibrational spectra of La_2NiO_4 (ref. 6).

frequency region (of greatest interest) is well resolved. Note the development of a soft-mode peak in $G(\omega)$ near 80 ± 20 cm^{-1} as we go from $BaPbO_3$ ($T_c \sim 0K$) to $BaPb_{0.75}Bi_{0.25}O_3$ ($T_c \sim 12$ K). Similarly, the strength of the soft-mode peak near 100 cm^{-1} in La_2CuO_4 ($T_c \sim 0K$) is noticeably enhanced in $La_{1.8}Sr_{0.2}CuO_4$ ($T_c \sim 30$ K). The soft-mode peak at $\sim 10-12$ meV ($80-100$ cm^{-1}) is recognized easily[5] in fixed-Q scattering from $La_{1.85}Sr_{0.15}CuO_4$ after background subtraction, as shown in Fig. 6.

While it has not so far proved possible to grow single crystals of La_2CuO_4 large enough to obtain complete vibrational spectra by neutron scattering, such crystals are available for semiconductive and antiferromagnetic La_2NiO_4, which is a reasonably satisfactory analogue compound.[6] These spectra are shown in Fig. 7. Note that near the (110) zone boundary the lowest optic mode phonon frequency is about half that at the zone center.

Experimental data on $\omega(q)$ in $La_{2-x}Sr_xCuO_4$ are so far incomplete,[7] but they show considerable similarity overall to the La_2NiO_4 spectrum shown in Fig. 7, except that near the Brillouin zone boundary the Σ_4 (octahedral tilting) mode frequency goes to zero, in accordance with the tetragonal-orthorhombic transition. It is suggested[7] that the frequencies are similar[8] to those of Sr_2TiO_4, after scaling by a mass factor m_{Sr}/m_{La}. However, there is no theoretical basis for choosing only this mass ratio, and too few modes are known from Raman scattering to justify the few correlations proposed between these two A_2BO_4 crystals with A and B valences reversed.

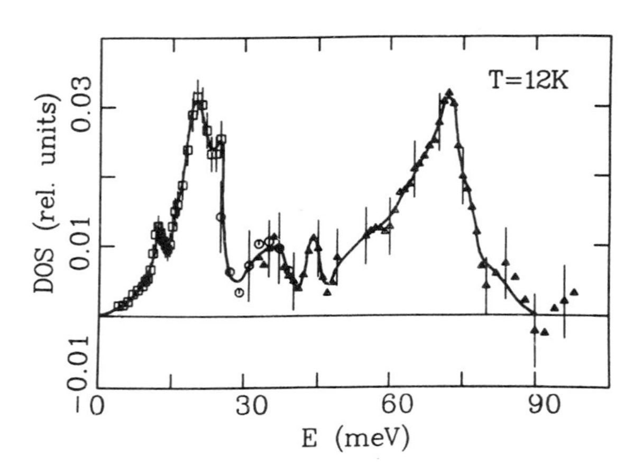

Fig. VI.8. Phonon spectrum of $YBa_2Cu_3O_7$ (ref. 9).

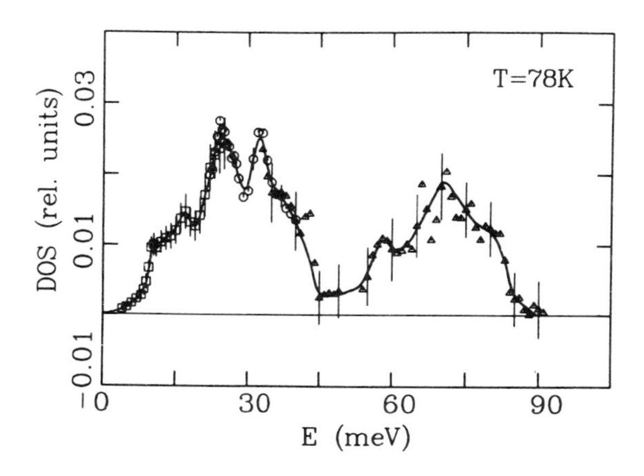

Fig. VI.9. Phonon spectrum of $YBa_2Cu_3O_6$ (ref. 9).

The situation for $YBa_2Cu_3O_{7-x}$ is more complicated.[9] By comparing (Figs. 8 and 9) the phonon spectra for $x = 0 (T_c \sim 90$ K) and $x = 1$ $(T_c \sim 0)$, one can see that the transverse (in-plane) chain vibrational modes are concentrated below 45 meV, and that removal of the chain O atoms on going from $x = 0$ to $x = 1$ increases the fraction of modes below 30 meV. It is possible that some of these modes are relaxed by the formation of O vacancies.

3. Infrared and Raman Lattice Vibration Spectra

Experience with large molecules has shown that infrared and Raman spectra normally yield the most information on vibrational modes and frequencies. Symmetry considerations simplify the spectra considerably, yielding narrow bands (because of the $\mathbf{k} = 0$ selection rule) for crystalline modes and broader bands for defect modes. In addition crystalline selection rules predict only relatively few allowed bands (by factor group analysis) for polarized infrared and Raman spectra on single crystals. By now

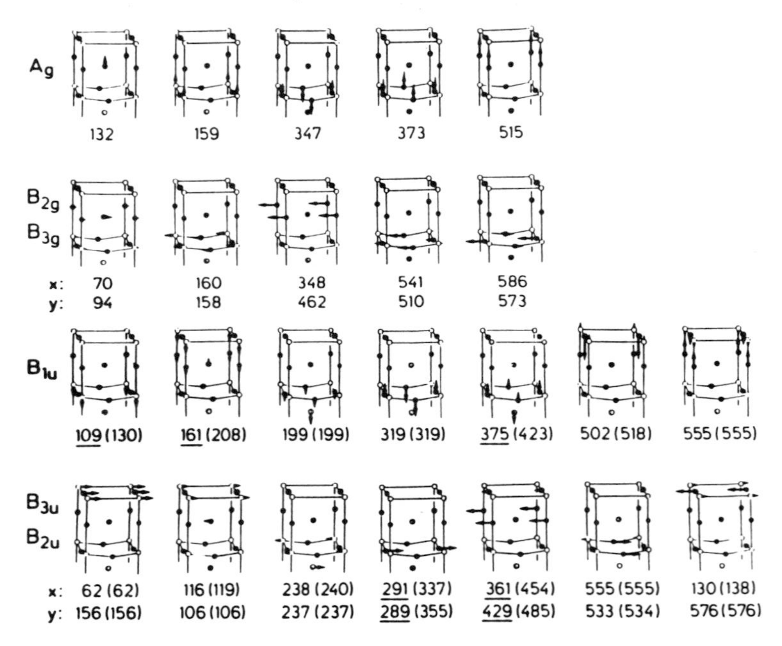

Fig. VI.10. Normal modes and frequencies for $YBa_2Cu_3O_7$ (ref. 11).

there is general agreement on the labelling of these bands in $(La, Sr)_2CuO_4$ and $YBa_2Cu_3O_{7-x}$. The salient features of the analysis are reviewed here briefly.

Two excellent theoretical studies[10,11] have used 16 central force constants to calculate frequencies and normal modes of the 36 optical vibrations of the 13-atom unit cell of $YBa_2Cu_3O_7$. The force constants were determined by comparison with a variety of materials[10] or by comparison only with perovskite compounds.[10] The most important difference between the two models, however, is that the later model utilized parameters adjusted to fit polarization selection rules inferred from single-crystal data,[11] while the earlier model utilized only powder data. As expected, the later model fits the experimental data somewhat better. Force constants and bond lengths for the earlier model are published,[10] while normal modes and frequencies for the later model are shown in Fig. 10.

Certain general characteristics of the phonon modes shown in Fig. 10 are quite interesting. The Raman-active (g) modes involve atoms in the BaO and $Cu2O_2$ planes, while the infrared-active (u) modes involve atoms in all three planes, Cu1O, $Cu2O_2$, and BaO. Experiments on single crystals so far have primarily utilized xx and xy polarizations to identify orthorhombic A_g modes by Raman scattering. The preferred correspondence between experiment[11,12] and (theory) for the A_g modes in $YBa_2Cu_3O_{6.9}$ is (in cm^{-1}) 116 (159), 150 (132), 340 (347), 440 (zz, not well resolved) (373), and 504 (515). The agreement between (theory) and experiment (as measured by $\omega^2_{exp} - \omega^2_{theory}$) is very good, except for the (373) mode, which involves the buckled O atom modes in the $Cu2O_2$ plane, which may not be well described by central forces only.

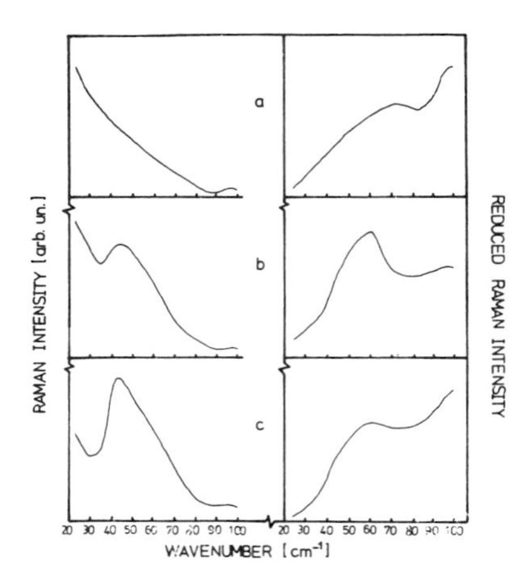

Fig. VI.11. Softest optic mode phonon in $YBa_2Cu_3O_{6.9}$ powder observed by micro-Raman scattering (ref. 13).

As we saw in IV.9, soft phonons with $\hbar\omega \lesssim 100\,cm^{-1}$ are of special interest in normal-state transport properties. In Fig. 10 we see a B_{3u} soft phonon corresponding to in-plane transverse vibrations of Cu1O chains, with a calculated frequency of 62 cm^{-1}. It appears that this mode has been observed by micro-Raman spectroscopy[13] near 40-50 cm^{-1}, as shown in Fig. 11. The mode is nominally only infrared active, but the selection rules for such low-frequency modes are easily broken by defect fields. Note also that the frequency of the (132) Ba mode, 150 cm^{-1}, is very low.

Several features of the Raman spectra have attracted considerable attention.[11,12] Except for the 340 cm^{-1} band, all other bands stiffen with decreasing temperature, but this one stiffens as T decreases to T_c and then softens dramatically (by $\sim 3\%$) below T_c, as shown[12] for single-crystal $YBa_2Cu_3O_{7-x}$ with x near 0.1, in Fig. 12. For this sample the (x,x) and (x,y) polarized spectra are shown at $T = T_c$ and T near O in Fig. 13. The phonon spectra interfere with an electronic background which peaks near $\omega = 470\,\text{cm}^{-1}$ (58 meV) at low T. This is surely a superconductive energy gap, and it agrees well with the maximum gap $E_g^{max} \sim 60\,\text{meV}$ seen in tunneling experiments (VIII.2) as well as the $Cu2O_2$ plane local gap seen by NQR relaxation (IX.2). However, there is a smooth electronic continuum extending to below $100\,\text{cm}^{-1}$ which decreases almost proportionally to ω. It is possible that this continuum represents a distribution of energy gaps arising from crystalline defects. It is important to realize that even this single-crystal sample, with its relatively sharp resistive transition at $T_c = 90K$, $\Delta T = 15K$ in magnetic

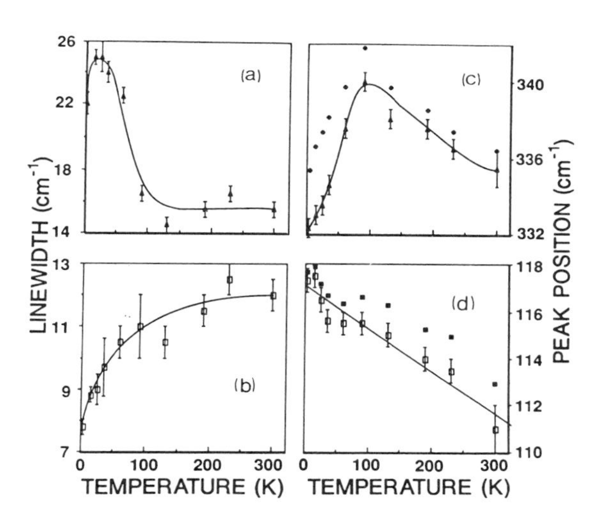

Fig. VI.12. Linewidths and peak positions for the 116 (159) and 340 (347) modes of the z-polarized normal modes of the Cu2 and O atoms of the $Cu2O_2$ plane (ref. 12).

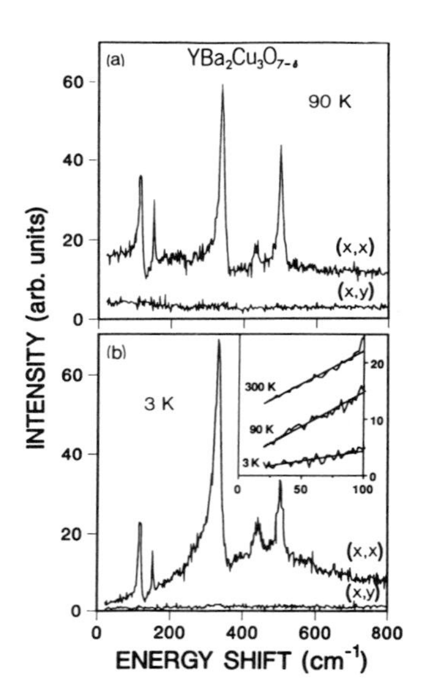

Fig. VI.13. Polarized ($(x, x) = A_g$) and depolarized ($(x, y) = B_{1g}$) spectra for single crystal $YBa_2Cu_3O_{7-x}$ at 90K and 3K. The inset shows the small frequency linear regime, after dividing by the thermal Bose factor, offset at 90 (300)K by 2 (10) units (ref. 12).

susceptibility, may still not be homogeneously and fully oxidized, and that this could produce a distribution of energy gaps, as could cation disorder (X.7).

The asymmetric Fano interference with the electronic background is large for the 116 and 340 cm^{-1} bands in Fig. 13 and assigns them to the Cu2 and O modes of the metallic $Cu2O_2$ plane. The electronic background is unusually large and suggests interlayer scattering from the $Cu2O_2$ to the Cu1O planes.[12] This scattering process is the key to layer localization, as seen in NQR relaxation (IX.2), and it is central to the apparent freeze-out of the nominal Hall carrier concentration (IV.9).

The anomalous softening of the 340 (347) band shown in Fig. 12 below T_c may be caused by an order-disorder transition of dynamical off-site displacements of the buckled O atoms in the $Cu2O_2$ plane. These atoms are buckled in part by their interactions with the double-well Ba atoms (V.4) and they must also couple strongly to the carriers which develop a large superconductive gap $E_g \sim 8kT_c$, according to NQR data (IX.2). In rare-earth substituted samples $MBa_2Cu_3O_7$ the 340 (347) Raman band softens with increasing M ionic radius, but its Davydov partner infrared bands stiffen.[14] This supports the double-well interpretation, since the even parity of the double well would interact oppositely with the even (g) Raman - and odd (u) infrared-active modes. As the buckled atoms order, the electron-phonon coupling and superconductive condensation energy may increase. The strength of this band is increased by a factor of 10 above the orthorhombic-tetragonal phase transition (670C) where oxygen vacancies are created, which suggests that the strength of this band may be defect-enhanced and that the same defects may enhance T_c. This observation provides direct evidence in favor of defect enhancement of T_c (IV.10).

A striking feature of the data shown in Fig. 12 is that not only does the buckling mode frequency decrease below T_c, but also its band width increases. This is very surprising because 330 cm^{-1} $<$ 470 cm^{-1}, which means that at low temperatures the phonon lies inside the energy gap. Such phonons are energetically incapable of decaying by excitation of electron-hole quasi-particle pairs, and so their band widths should decrease once they lie below the gap. This effect was studied[15] in Nb_3Sn and there it reduced the phonon band width by a factor of two, as shown in Fig. 14. Thus the present observation of a band-width doubling is anomalous, but it is plausible in the context of the fictive phonon model discussed in IV.13. It may also arise as the result of incomplete condensation of the superconductive phase, so that the "broad" band actually consists of two unresolved bands.

The spectra shown in Fig. 12 were obtained using a very sensitive detector, which is important because the Raman scattering intensity from a metal is much weaker than from an insulator because it is confined to a surface layer determined by the skin depth. With a less sensitive

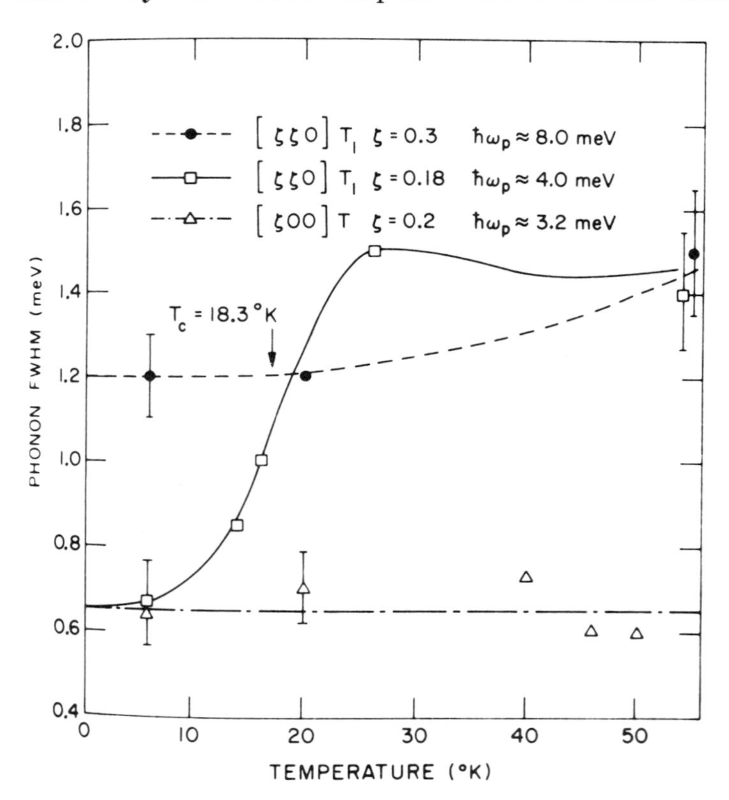

Fig. VI.14. When the phonon energy lies outside the gap in Nb_3Sn, its band width decreases gradually with decreasing T. When it lies inside the gap but is strongly coupled to the electrons [ςς0], a rapid decrease is found below T_c (ref. 15).

detector but with a sample which was 90% orthorhombic (not tetragonal) in one orientation, [12] the anomalous softening of the 335 cm^{-1} phonon occurs within a temperature region of less than 10K below T_c, compared to a region of nearly 100K in Fig. 12. This suggests that the width observed in Fig. 12 reflects the presence of defects or inhomogeneities coupled in some way to the twin boundaries. More specifically, it may be that most of the interlayer coupling occurs at the twin boundaries, and when these are far away the very rapid temperature dependence of order parameters characteristic of two dimensions (Fig. IV.7) is observable.

Studies of infrared and Raman spectra as a function of rare-earth (RE) radius in a series of RE $Ba_2Cu_3O_7$ compounds[14] have shown that with increasing RE radius the frequency of the infrared-active Davydov partner of the 335 cm^{-1} Raman band increases, as expected, but that the Raman band itself softens, and it is the only band which shows this anomalous behavior. This is consistent with both the double-well and fictive phonon models, because both describe the even-parity part of the effective atomic potential seen by the O atoms in the CuO_2 plane.

In discussing the coupling of the 340 cm^{-1} band to the superconductive order parameter Δ and its softening below T_c, I have made no effort to show that in an ideal crystal a transverse mode of this type actually would couple strongly to holes near E_F. The reason for this neglect is that there is a very high density of defects in the sample which will probably erase all selection rules (or quasi-selection rules) associated with dipole transitions and crystal symmetry. The situation is analogous to $\mathbf{k} = 0$ selection rules for optical vibrational modes which are valid in compound semiconductors (where a line spectrum is observed), but not for alloyed compound semiconductors (where band spectra, roughly mimicking the density of vibrational states for all modes and all **k** are easily obtained). Misuse of such selection rules in highly defective samples can be quite misleading.

At this point it is instructive to compare the strong coupling of the 340 cm^{-1} mode in YBCO, which is obvious from its temperature dependence, (Fig. 12(c)), with the disappointing lack of observation of similar effects in $La_{2-x}Sr_xCuO_4$, which has puzzled several experimental groups.[7] As we mentioned above, the vibrational modes associated with replacement of less than 10% of the La with Sr atoms are very difficult to identify in observed spectra, and quite possibly they are not even localized, so that their strength is distributed merely as a broadening and small shift of La_2CuO_4 modes (except for the soft Σ_4 zone-boundary mode). This is what one would expect theoretically, so there is no reason to infer (as many have done) that strong Fermi surface electron-phonon coupling is absent in $La_{2-x}Sr_xCuO_4$. Such arguments (which have been used to justify various exotic models) seriously underestimate the complexity of electron-phonon coupling in a crystal with a complex, open structure, and many vacancies, substitutional impurities, and other defects.

The great advantage of the fictive phonon model is its simplicity. In IV.11 a strong-coupling model with $\tilde{\omega} \sim 100$ cm^{-1} (a transverse chain phonon) and $\lambda \sim 8-10$ explained $E_g = 60$ meV and $T_c = 90K$ in YBCO. The Raman data just discussed show, however, that the correct fictive phonon frequency is $\tilde{\omega} = 330$ cm^{-1} (transverse planar phonon), and with $E_g = 470$ cm^{-1}, then according to Fig. II-8 and the interpolation formula II(4.7), the best fit to the data is obtained with $\tilde{T}_c = 125K$, $E_g/k\tilde{T}_c = 5.5$, and $\lambda = 3.2$. (In this case there is no choice of λ that will fit both T_c and E_g). This value of $\tilde{T}_c$ is perfectly acceptable; as we have discussed in IV.9 and IV.10, the measured value of T_c reflects not only the true intralayer $\tilde{T}_c$ but also the effects of interlayer coupling by defects, which could easily reduce T_c from $\tilde{T}_c = 125K$ to the measured value of 92K. With $\mu \sim 0.1-0.2$ this gives (II (4.8)) $\lambda \sim 3.5-3.8$. From (IV.13(7)) with $\lambda_\beta^0 = 1$, this gives $n_\beta+1 \sim 3.5$. The thermal equilibrium value of n_β at the formation temperature $T_F \sim 950C$ is 2.1. Considering the complexity of the problem, this is very good agreement indeed. It means that the large vibrational amplitudes are quenched in by defects at the formation temperature, and that the effect of these amplitudes on λ is determined once the formation temperature and the fictive phonon frequency $\overline{\omega}$ are known. This means that the fictive phonon model, unlike the various exotic models discussed in IV.12, gives a good estimate for $T_c \sim 100K$ *without adjustable parameters.*

A word should be added here about the rôle of defects. As discussed at length in IV.10, because of localization and marginal dimensionality (IV.8), the number of extended orbits with E near E_F can depend very sensitively on defects, especially if these produce states near E_F. Even if most of the transverse O vibrations of the CuO_2 plane do not have anomalously large vibrational amplitudes, the effective electron-phonon coupling for extended orbits can still depend on $n_\beta(T_F)$, because the detect configuration is a metastable one which is quenched in near T_F.

We mentioned above in connection with the shifting and broadening of the 340 cm^{-1} band that an additional unresolved satellite band, centered at low $T \lesssim 20K$ near 330 cm^{-1}, may grow in scattering strength below T_c and cause the observed shift and broadening. Because of the z polarization of the 330 cm^{-1} band it is tempting to associate this band with interplanar defects, but then its proximity to the 340 cm^{-1} band, which is thought to be intrinsic, would be accidental. A problem here is that Raman scattering strengths, unlike optical oscillator strengths, do not satisfy the f-sum rule. As a result Raman scattering strengths may vary by orders of magnitude from one vibrational mode to the next, and defect modes, often have exceptionally large scattering strengths. Thus if the band is split, it most probably indicates that only parts of the sample actually become superconducting. The different behavior could arise from partial ordering of oxygen vacancies $O^{\square}$ on the Cu(1) O chains which would have little broadening effect on the vibrational frequencies of the Cu(2) O_2 planar modes, and cause part of the shift from 340 cm^{-1} to 330 cm^{-1}. These vacancies can also provide impurity centers which pin E_F, as illustrated in Fig. IV-10.

While it is strictly correct only to discuss the fictive phonon number n_β, it may be helpful in clarifying the physical picture if we sketch an "effective" atomic potential for the transverse planar oxygen motion. This is done in Fig. 15. Because so many perovskites (like $BaTiO_3$) are ferroelectrics, we compare these potentials (which are not observables) for ferroelectric insulators and for cuprate

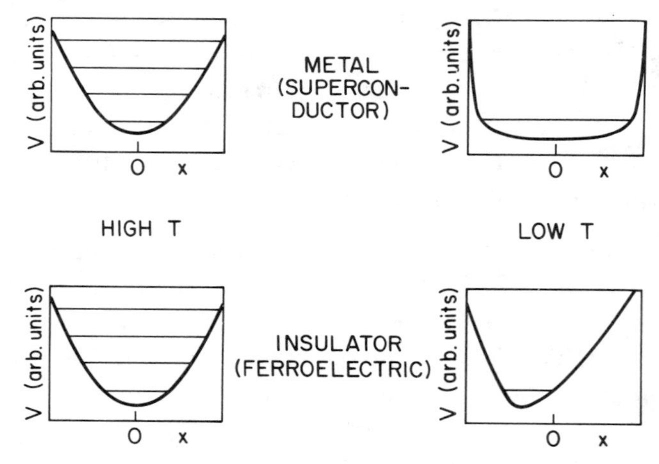

Fig. VI.15. A comparison of "effective" atomic vibrational potentials for a ferroelectric insulator and a superconductive, metallic layered cuprate.

superconductors, at high T (~1000C) and low T (~100K). One can explain the ferroelectric instabilities of ABO_3 perovskites in part by noting that the unit cell contains one adjustable parameter (the lattice constant), which must effect a compromise between two bond lengths, A-O and B-O. In the cuprate case a single planar lattice constant must compromise between the Cu-O bond length and the other cation bond lengths (Y-O and Ba-O in YBCO). The compromise fails for Y and the O sites in the Y plane are vacant, while the Ba-O bond length is also unsatisfactory, leading to anharmonic displacements for Ba, as discussed in V.4. At present, the anharmonic transverse planar O vibration can only be inferred from Raman spectra, because direct diffraction measurements of the O motion would require large (~1 $cm^{3)}$ single crystals for neutron diffraction. Further crystal chemical arguments consistent with fictive phonons are given in Appendix C.

In semiconductive samples of $YBa_2Cu_3O_{7-x}$ ($x \gtrsim 0.7$, $T_c = 0$) at frequencies much higher than those of one-phonon Raman bands, additional Raman scattering bands have been observed.[16] These are attributed to spin fluctuations because of their similarity both in frequency and polarization selection rules to the spin-flip spectra of $La_{2-x}CuO_{4+\delta}$, which is known to be antiferromagnetic from polarized neutron scattering studies (IV.5). The dependence of these spectra on x and T_c is shown in Fig. 16. In (a), $x \gtrsim 0.7$, $T_c = 0$ and a strong, poorly resolved doublet is observed near (2400, 2800) cm^{-1}. In (b), $x \sim 0.4$, $T_c \sim 60K$, and the doublet is weaker and more widely spaced (2000, 3000) cm^{-1}. In (c), $x \sim 0.2$, $T_c \sim 88K$, and the doublet has become very weak but its spacing has not changed.

The trends shown in Fig. 16 are just the ones that would be expected from a decreasing level of oxygen vacancies which are probably concentrated inhomogeneously into

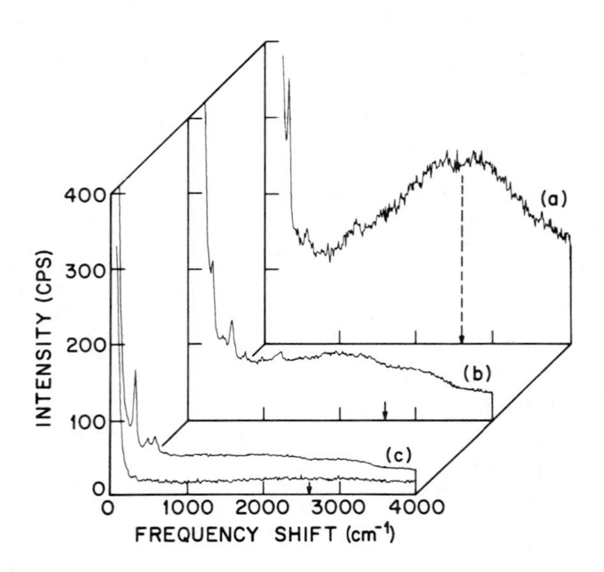

Fig. VI.16. Magnon polarized Raman spectra of (a) $T_c = 0$, (b) $T_c \sim 60K$ and (c) $T_c \sim 88K$ single crystal $YBa_2Cu_3O_{7-x}$ samples. The background curve in (c) is for non-magnon active polarization (ref. 16).

semiconductive or semimetallic pockets whose filling factor scales roughly with x. These pockets could be responsible for the distribution of energy gaps shown in Fig. 16. If the pockets are small, so that their interfacial volumes with the metallic regions are large, these boundary regions could also account for the Fano interference effects shown in Fig. 16. The gap could be graded linearly normal to such interfaces, and this could account for the linear distribution of gap states shown in Fig. 16.

Spectra such as those shown in Fig. 16 are sometimes optimistically mentioned as evidence that spin fluctuations contribute significantly to high-T_c superconductivity. Regardless of the details of the mechanism, the interaction strength will depend on the product $N_e(E_F)n_B\, g_{Be}$, where $N_e(E_F)$ is the extended metallic electron density of states, n_B is the Boson (exciton, magnon, or phonon) number, and g_{Be} is the electron-Boson coupling strength (nearly constant). In the $YBa_2Cu_3O_{7-x}$ alloys, $N_e(E_F)$ is roughly proportional to $(0.7 - x)$, while from Fig. 16 the number of magnons is roughly proportional to x^2. If electron-magnon interactions enhanced T_c, then the maximum of T_c would correlate well with the maximum of $(0.7 - x)x^2$, which occurs far from the observed maximum near $x = 0$. Thus the data shown in Fig. 16 actually constitute strong evidence *against* a magnon mechanism for high T_c.

4. Sound Velocities, Internal Friction and Lattice Instabilities

One of several ways of studying mechanical instabilities macroscopically is the vibrating reed method, which can be used with superconductive powders. This method is very sensitive, and it has been used by many workers to search for elastic phase transitions in $(La, Sr)_2CuO_4$ and $YBa_2Cu_3O_{7-x}$. Some of the most interesting data[17] have been taken on samples of $YBa_2Cu_3O_{7-x}$ and $EuBa_2Cu_3O_{7-x}$. The former data resemble those of other workers while the latter data, taken on much denser samples with narrower resistive transitions ($\Delta T_c = 2K$ for the Y sample, $\Delta T_c = 1K$ for the Eu sample), show much narrower anomalies.

The relative change in the sound velocity v_E measured by the vibrating reed method at audio frequencies on $YBa_2Cu_3O_{7-x}$ agree well with measurements at ultrasonic frequencies four orders of magnitude higher. Both experiments show a much larger (100 times) increase in v_E below T_c than can be explained by electronic interactions only.[18] This large increase can be explained by pressure dependence of the frozen-in vibrational amplitude (fictive phonons), i.e., by a term dn_β/dP which may also couple to dv_E/dP. This increase is shown for $EuBa_2Cu_3O_{7-x}$ in Fig. 17.

One of the advantages of the vibrating reed method is that it provides a very accurate measurement of internal friction (sound damping). This quantity shows three peaks in both materials, with narrower peaks in the denser and more homogeneous Eu samples,[17] as shown in Fig. 18. The maximum of the peak near T_c occurs near 87.5K, and this T agrees well with that of the break in slope of v_E shown in the inset of Fig. 17. If we assume that these samples are inhomogeneous, then the (linear) percolative resistive

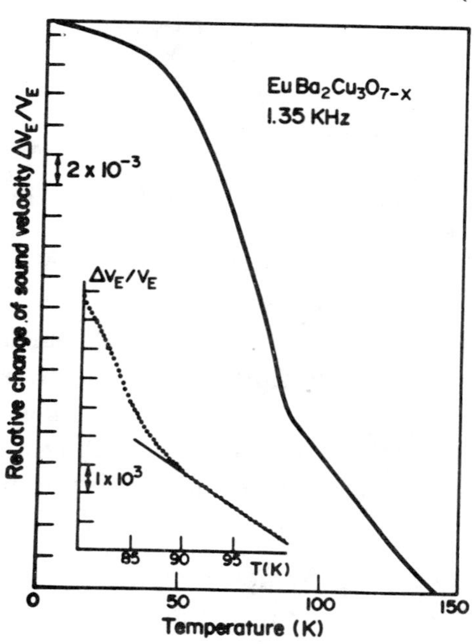

Fig. VI.17. Relative change of the Young's modulus sound velocity $v_E(T)$ (ref. 17).

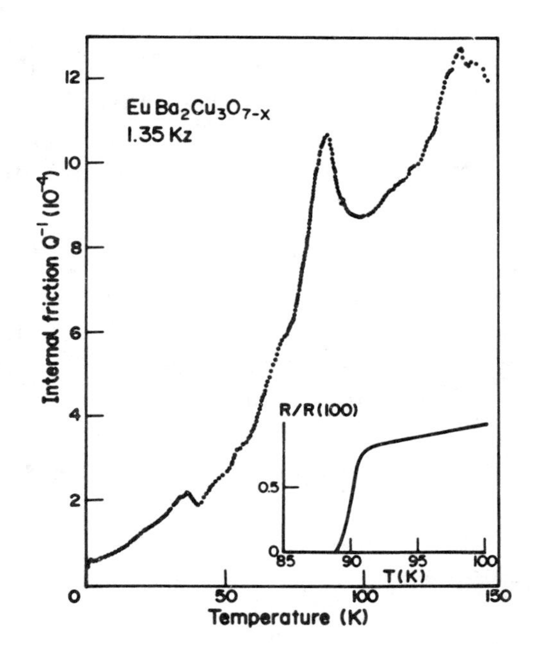

Fig. VI.18. Internal friction $Q^{-1}(T)$ (ref. 17).

threshold may occur at slightly higher T than the planar v_E threshold. The latter may measure the onset of the c-axis coupling of planes by defects or fictive phonons, which causes percolatively broadened first-order discontinuities in n_β and v_E. This is consistent with the idea discussed in IV.9, 10 and VI.3, that the intraplanar T_c may be $\sim$ 125K.

The simplest manifestation of a mechanical instability is a lattice constant anomaly. Because $YBa_2Cu_3O_{7-x}$ is obtained by oxidation of $YBa_2Cu_3O_6$, to observe such anomalies special care must be taken to insure a homogeneous distribution of oxygen throughout the sample. This is probably best done with a powder of uniformly small grains. In any case, inhomogeneities are expected to clamp the grains and to reduce the magnitude of the observed anomaly. Thus it appears that the best experiment is the one that observes the largest effect. Lattice constant anomalies of about -0.2% have been observed[19] in $YBa_2Cu_3O_{7-x}$ powders centered at $T = 110K$ and extending from 90K to 125K. The anomalies exhibit thermal hystersis, as expected from their strong coupling to defects, as shown in Fig. 19.

When the anomalies shown in Fig. 19 are viewed on a larger scale.[19], including the linear temperature dependence above and below the oscillation, it is clear that as T decreases the lattice constants initially drop, reaching a local minimum near $T = 110K$, and then almost completely recover to their extrapolated linear values near $90K < T < 100K$. This recovery suggests a microscopic model which can reconcile these large powder anomalies to single-crystal data, which do not exhibit oscillatory anomalies. We suppose that the fine-grained sample is

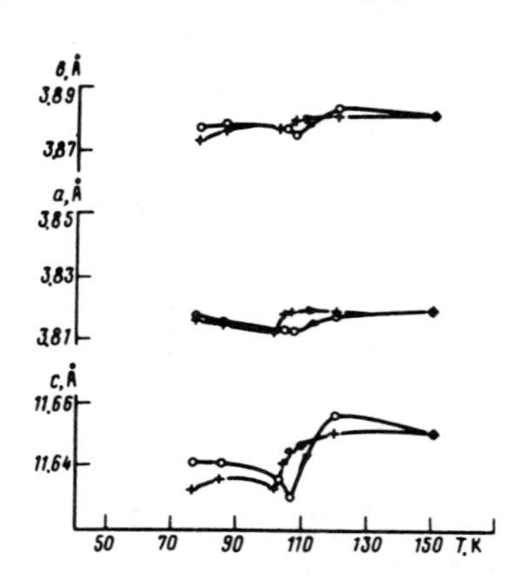

Fig. VI.19. Temperature dependence of orthorhombic lattice constants of $YBa_2Cu_3O_{7-x}$: open circles, cooling, crosses, heating (ref. 19).

textured and densely compacted, so that the c axes of the ellipsoidal grains are locally nearly parallel. As individual grains become superconducting, there is a large change ($\gtrsim$0.2%) in the ratios c/(a+b) of individual grains. The T_c's of the nearly unclamped grains are distributed between 90K and 125K, and as T decreases below 125K the change in some c/(a+b) ratios generates a large intergranular compressive internal misfit stress, which reduces all the lattice constants. Half of the granular volume is transformed near T = 115K. Below 115K the transformed volume fraction increases to near unity, and as nearly all grains have nearly the same c/(a+b) ratio again, the internal stress is relieved and the anomalous decrease in lattice constants is erased. This is an example of the shape memory effect, which is well known in martensitic transformations of metallic alloys.[20] The observed lattice constant changes (0.2%) are a lower bound for the actual changes which must occur for isolated fine grains. This in turn implies a very large electron-phonon coupling which is enhanced below T_c, as in the fictive phonon model. It also means that the clamping characteristic of single crystals or epitaxial films may mask many of the relaxation effects which can occur in powders. Finally, intergranular contacts are clamped much more than the free surfaces of grains, and these same mechanical contacts also form the electrical bridges between grains which determine onset values of T_c in resistive measurements. In IX.1 this anomaly is discussed further and it is suggested that it arises from a surface or grain boundary phase transition which is especially sensitive to shear.

REFERENCES

1. R. E. Cohen, L. L. Boyer and M. J Mehl, Phys. Rev. *B35*, 5749 (1987).
2. J. W. Flocken, R. A. Guenther, J. R. Hardy, and L. L. Boyer, Phys. Rev. *B31*, 7252 (1985).
3. R. E. Cohen, W. E. Pickett, L. L. Boyer and H. Krakauer, Phys. Rev. Lett. *60*, 817 (1988).
4. A. Masaki et al., Jap. J. Appl. Phys. *26*, L405 (1987).
5. A. P. Ramirez et al., Phys. Rev. *B35*, 8833 (1987).
6. L. Pintschovius et al, Europhys. Lett. *5*, 247 (1988).
7. P. Böni et al., Phys. Rev. *B 38*, 185 (1988); B. Renker et al., Z. Phys. *B67*, 15 (1987).
8. G. Burns, F. H. Dacol, and M. W. Shafer, Sol. State Comm. *62*, 687 (1987).
9. J. J. Rhyne et al., Phys. Rev. *B36*, 2294 (1987).
10. F. E. Bates and J. E. Eldrige, Sol. State Comm. *64*, 1435 (1987).
11. R. Liu et al., Phys. Rev. *B37*, 7971 (1988).
12. S. L. Cooper, M. V. Klein, B. G. Pazol, J. P. Rice and D. M. Ginsberg, Phys. Rev. B *37*, 5920 (1988); R. M. Macfarlane, H. Rosen and H. Seki, Sol. State Comm. *63*, 831 (1987); C. Thomsen, M. Cardona, B. Gegenheimer, R. Liu, and A. Simon, Phys. Rev. *B37*, 9860 (1988).
13. M. P. Fontana, B. Rosi, D. H. Shen, T. S. Ning and C. X. Liu, Phys. Rev. *B38*, 780 (1988).
14. M. Cardona et al., Sol. State Comm. *65*, 71 (1988); R. Nishitani, N. Yoshida, Y. Sasaki and Y. Nishina, Jap. J. Appl. Phys. *27*, L1284 (1988).
15. J. D. Axe and G. Shirane, Phys. Rev. Lett. *30*, 214 ((1973).
16. K. B. Lyons, P. A. Fleury, L. F. Schneemeyer and J. V. Waszczak, Phys. Rev. Lett. *60*, 732 (1988).

17. C. Durań, P. Esquinazi, C. Fainstein and M. N. Regueiro, Sol. State Comm. *65*, 957 (1988).

18. D. J. Bishop et al., Phys. Rev. *B36*, 2408 (1987).

19. A. I. Golovashkin et al., JETP Lett. *46*, 410 (1987); L. Sun, Y. Wang, H. Shea, and X. Cheng, Phys. Rev. *B38*, 5114 (1988).

20. M. B. Salamon, M. E. Meichle and C. M. Wayman, Phys. Rev. *B31*, 7306 (1985).

VII. Optical Spectra

1. Samples and Surface Preparation

Because the energy gap E_g plays such a central role in the BCS theory, substantial efforts have been made to observe this gap directly by optical methods. Compared to tunneling, optical methods have the advantage that direct electrical contact (possibly through a surface dead layer) need not be made. It appears that the energy gap is observed in the single-crystal Raman spectra in Fig. VI-13(b). However, ideally one would also like to observe gap structure in the infrared reflectance spectrum with different spectral weighting factors.

In the exciton variants of exotic theories (IV.12) it has been hoped that an exciton might be resolved in the infrared with an excitation energy $E_{ex} \lesssim 1\,eV$. If such an excitation could be resolved, then one could hope that chemical trends in $E_{ex}(x)$ in $YBa_2Ca_3O_{7-x}$ might correlate with $T_c(x)$ and $E_g(x)$.

A number of early efforts to resolve E_g reported small values in the range 2-3 kT_c for powder samples. Gradually it was realized that the strong anisotropies found in electrical properties of these materials are continued into the infrared where the optical properties are also strongly anisotropic. Disentangling these anisotropies by reflectivity measurements on isotropic powders presents formidable difficulties. Thus the measurements which have obtained the most clear-cut results are those on single crystals. In what follows, we review data on composite samples as well as single crystals. The problems of interpretation are reduced for the single-crystal samples but many difficulties still remain. We discuss primarily measurements on $YBa_2Cu_3O_{7-x}$, because thick large single-crystal samples, suitable for optical reflectivity experiments, so far have been synthesized only for this material.

Another important point is surface preparation. Most samples, composite or not, do not have highly reflecting specular surfaces. When such surfaces are obtained by polishing, the resulting damage in the surface layer probably reduces T_c substantially in that region. It may also reduce the normal-state conductivity, because close correlations between the two are often found as a function of composition in d.c. resistivity measurements. Thus a standard technique measures reflectivity differences between the normal and superconducting state, and describes these with Drude currents for the former and BCS currents for the latter. A danger in this approach is that

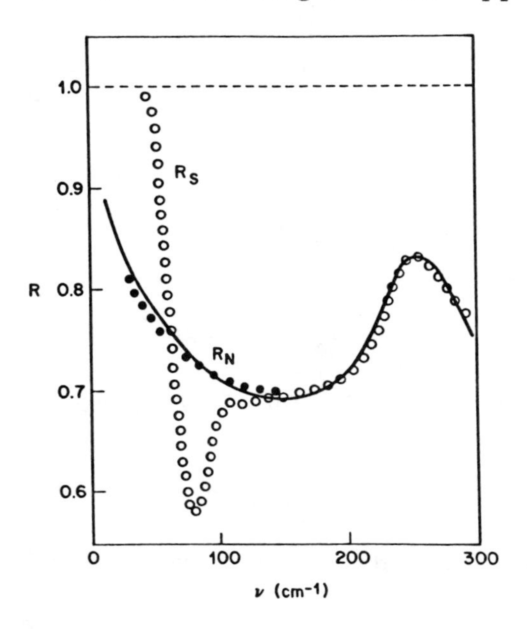

Fig. VII.1. Reflectivity $R(\nu)$ in the superconductive state $(T/T_c = 0.3)$ and normal state $(T > T_c)$ in $La_{2-x}Sr_xCuO_4$ with $x = 0.175$ (ref. 1).

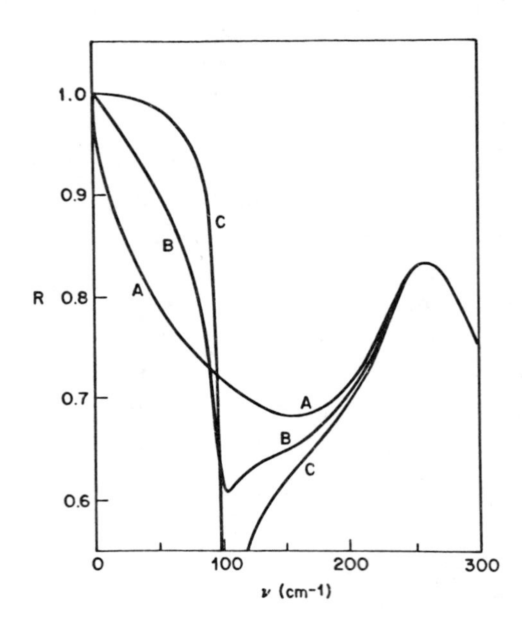

Fig. VII.2. Theoretical reflectivity for normal state Drude model (A) and superconductive BCS model with fraction f = 0.5 (B) or 0.1 (C) of material superconducting (ref. 1).

the degree of anisotropy may change substantially between the normal and superconductive state, which again complicates the interpretation of data taken on composite samples.

2. Composite Reflectance Data

Instructive data taken on a polished powder composite of $La_{z-x}Sr_xCuO_4$ with x = 0.175, shown in Fig. 1, exhibit two simple features.[1] These are a crossover of normal and superconductive reflectivities R_n and R_s, with a maximum dip in the latter near 80 cm^{-1}, and a strong peak common to both R_s and R_n near 240 cm^{-1}. The crossover and dip are explained[1] by a Drude model for R_n and a BCS model at $T = 0$ for R_s, as shown in Fig. 2. Strictly speaking, R in Fig. 1 has not been measured absolutely, so that the Kramers-Kronig transform was performed to obtain R theoretically, as in Fig. 2, using a BCS gap value, $E_g = 2\Delta = 3.5\,kT_c \approx 85\,cm^{-1}$. The shift from $T=0$ to $T = 10K = 0.3\,T_c$ is small in the BCS theory. Thus the actual value of E_g/kT_c obtained by comparing Fig. 1 and Fig. 2 is about 3.0. Considering the effects of sample polishing, which should reduce T_c and E_g near the surface, and the effects of gap anisotropy, which should reduce E_g for c axis polarization, this value represents a large lower bound for E_g/kT_c. Smaller values have often been reported, and this large value probably is an indication of more careful surface preparation.

The surprising aspect of the phonon structure shown in Fig. 1 is that only one phonon peak (near 240 cm^{-1}) is observed. Between 100 and 300 cm^{-1} many phonons are observed in the reflectivity of La_2CuO_4, as shown in Fig. 3.

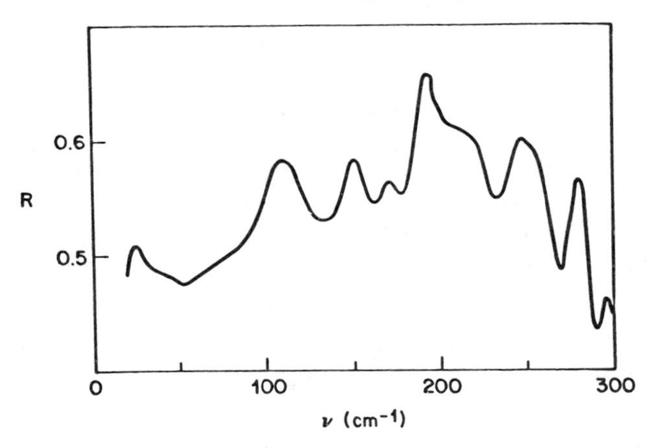

Fig. VII.3. Reflectivity of insulating La_2CuO_4 (ref. 1).

Moreover, the fitted oscillator strength of the 240 cm^{-1} peak in Fig. 1 is exceptionally large, corresponding to an effective charge $Z^* \sim 10$. It is suggested[1] that this might indicate a charge density wave, but there is no evidence for such a wave in La_2CuO_4 (where Fermi surface nesting would be favorable in the absence of antiferromagnetic localization). In the light of subsequent data (VI.3) which show giant coupling between a CuO_2 planar transverse O mode near 330 cm^{-1} and Δ in the superconductive state of YBCO, it is tempting to assign this phonon to such a mode here, but bound to Sr impurities. The reduced valence of Sr^{2+} compared to La^{3+} would lower the O transverse vibrational frequency and might produce, through the valence difference, large anharmonic quenched-in vibrational amplitudes (fictive phonons, VI.4) bound to the Sr impurities. These impurities also act as interplanar defects, providing electrical bridges between CuO_2 planes, and if dynamically marginally stable they could easily generate very large values of Z^* (~ 30) per impurity because of the large anharmonicity. This is just another way of describing an anomalously large vibrational amplitude.

3. Single Crystal YBCO Reflectivities

For YBCO the first reliable data[2] used a small partially oriented polycrystalline mosaic sample and measured the ratio R_s/R_n to obtain the results shown in Fig. 4. Some of the oscillations (those near 500 cm^{-1}) are instrumental, but they do not mask the broad trend, which is a broad maximum at a value of $E_g = 2\Delta$ which is near 500 cm^{-1} at low T, as shown by the BCS theory (dashed line) with $E_g(T)$ shown in the inset. This result is in excellent agreement with later Raman data (VI.3) on a larger single-crystal sample with possibly a slightly reduced T_c and $E_g = 470\,cm^{-1}$.

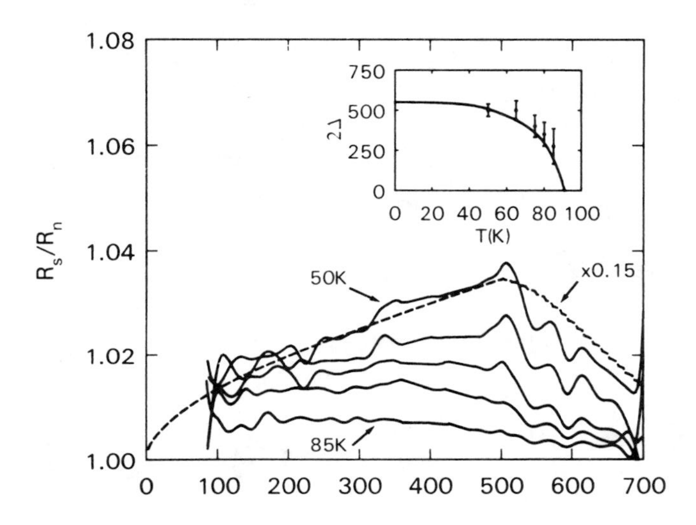

Fig. VII.4. Measured and calculated (dashed curve, T = 50K) reflectivity ratios R_s / R_n for YBCO (ref. 2).

The crystallites used to prepare the mosaic used to obtain the data shown in Fig. 4 were selected because of their good specular surfaces normal to the crystallite c axes. It is not possible, however, to measure R absolutely with such a mosaic, and it is difficult to eliminate interferometric grating oscillations associated with the mosaic itself. So far, no one has been able to grow large single crystals of $YBa_2Cu_3O_{6.9}$ (T_c = 90K) with specular native surfaces, but it has proved possible to obtain such samples of $YBa_2Cu_3O_{7-x}$ with larger x and T_c = 50 and 70K. (This is another illustration of the general rule that high T_c means greater lattice instabilities and poorer crystalline morphologies). Even these samples, with their improved surfaces, need not be homogeneous with regard to their oxygen deficiency x, because of lower diffusivities and the internal stresses which build up in larger crystals.

With these reservations in mind the measured absolute reflectivities[3] for those two samples shown in Fig. 5 are dramatic. Up to $h\nu \gtrsim 100\ \text{cm}^{-1}$ the value of R is 1.000(5), demonstrating beyond doubt the presence of a BCS-like energy gap. (This apparently rules out the various theoretical models of high-T_c superconductors based on boson condensation without an energy gap.) At the same

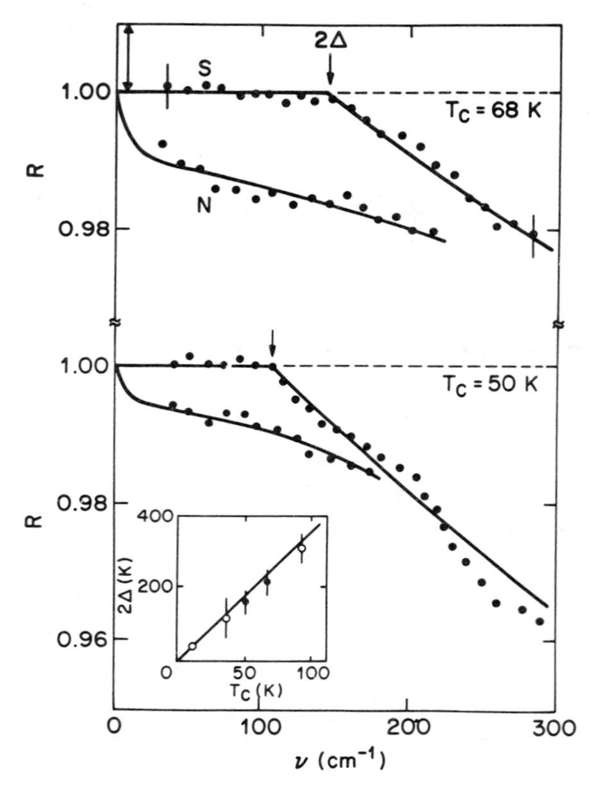

Fig. VII.5. "Absolute" reflectivities for two large single crystals of $YBa_2Cu_3O_{7-x}$ ($x \sim 0.3$ to 0.5) with nearly specular a-b planar surfaces, in the normal and superconductive states (ref. 3).

time the value of $2\Delta/kT_c$ (marked by the arrows in Fig. 5, and corresponding to the limiting range of R = 1.00) are much smaller, ~3.5, than the largest values obtained on other samples by other methods (reflectivity, Fig. 4, tunneling, VIII, Raman VI.3, and NMR, IX). The key point which distinguishes the estimates of 2Δ shown in Fig. 5 from these other estimates is that the surface current give *perfect* reflectivity, R = 1.00(1), whereas the other estimates show a distribution of energy gaps which could easily be spatially graded.

In an inhomogeneous sample with a spatial distribution of energy gaps and transition temperatures, it is in general misleading to compare an energy gap defined by propagation of a plane wave (optical or acoustical) with a transition temperature defined by percolation of a dendrite, as in an electrical resistance measurement. The two cases actually correspond to different kinds of percolation (called

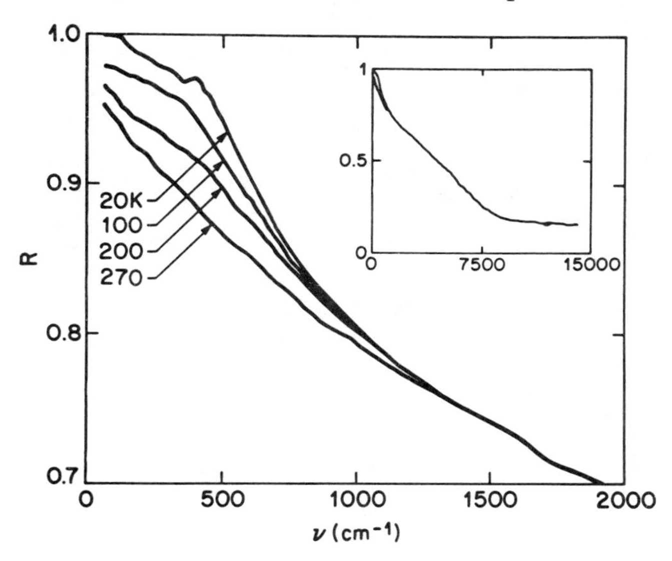

Fig. VII.6. Reflectivity of the T_c = 50K sample in Fig. 5 as a function of T (ref. 3).

scalar and vector for the one- and two-dimensional propagating units, respectively[4]). A more consistent definition of E_g/kT_c relative to the data of Fig. 5 is obtained by using T_c as measured by the vector percolation maximum in attenuation of acoustical waves (IX.1). While T_c (electrical) $= 92K$, T_c (acoustical) $= 65K$. Taking E_g from Fig. 5 and $T_c = 65K$, the ratio E_g/kT_c is close to 5.0, which agrees reasonably well with the results obtained by other methods, especially after allowance is made for clamping effects in the single crystal which may inhibit complete oxidation.

The conductivities of these samples, as contained in $R(\nu)$, vary rapidly in the range up to $\nu = 500\ cm^{-1}$, as T increases from 100K to 270K on the sample with $T_c = 50K$, as shown in Fig. 6. This rapid variation has been attributed,[3] through a Kramers-Kronig analysis, to electron-phonon coupling with $\lambda \sim 10$. To carry out such an analysis one must assume that the sample is homogeneous. For inhomogeneous samples where the width of the range of the inhomogeneity in σ is temperature-dependent, and decreases with increasing T, this may lead to overestimates of λ. In any case these data present evidence for very strong electron-phonon coupling in the phonon spectral region $\lesssim 500\ cm^{-1}$.

REFERENCES

1. G. A. Thomas, A. J. Millis, R. N. Bhatt, R. J. Cava, and E. A. Rietman, Phys. Rev. *B36*, 736 (1987).
2. Z. Schlesinger, R. T. Collins, D. L. Kaiser and F. Holtzberg, Phys. Rev. Lett. *59*, 1958 (1987).
3. G. A. Thomas et al., Phys. Rev. Lett. *61*, 1313 (1988).
4. J. C. Phillips and M. F. Thorpe, Sol. State Comm. *53*, 699 (1985).

VIII. Tunneling

1. Point Contacts and Dead Layers

In I and II we saw many examples of successful superconductive tunneling studies of intermetallic compounds which permitted direct measurements of $\alpha^2(\omega)$ which defines the electron-phonon coupling strength λ through eqn. I (2.2). Quantitative measurements of this kind are obtained with planar evaporated metallic electrodes deposited on thin oxide interfacial layers on planar metallic sample substrates. Qualitative measurements of energy gaps can be made with point contacts, but these have seldom yielded quantitative results for $\alpha^2(\omega)$. In practice the anisotropy of the energy gap in layered cuprates turns out to be so large as to prevent resolution of α-fine-structure in the tunneling characteristic of electron-phonon interactions. Thus we must be content with more modest tunneling information which can be compared with results from other experiments.

The preparation of the tunneling interface in a metal-interface-superconductor sandwich (MIS) is known to be of crucial significance for intermetallic superconductors. The interfacial barrier is usually an oxide (such as Al_2O_3) obtained by oxidizing a thin metal layer (such as Al) deposited on the superconductor. Here the cuprate itself may be insulating at its surface, and this appears to be the case for "as-grown" samples. These have native dead layers which may be too thick to form suitable barriers for superconductive tunneling. One can try to avoid these dead layers in various ways. For example, one can cleave the sample at low temperatures, for instance, in liquid He. However, the cleavage process itself produces substantial local heating and may easily create a new (but thinner) dead layer at the surface.

With point contact tunneling one can drive the point through part of the as-prepared dead layer and obtain tunneling characteristics (see Fig. 1) which show an energy gap.[1] Many experiments of this type have been reported, and it was generally assumed that the largest gap observed would represent the "true" gap — that is, the one which involved the smallest damage to the cuprate structure near the contact. However, one can also explain the wide range of observed gaps in a different way, as the result of anisotropy, so that the measured gap reflects the local interfacial orientation near the contact.

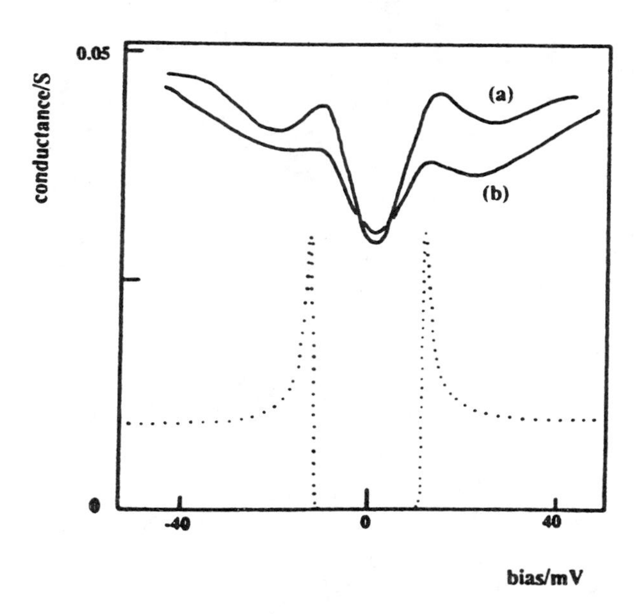

Fig. VIII.1. Conductance (dI/dV) spectra for two YBCO junctions. The dotted curve shows theory for an isotropic energy gap with a small thermal broadening, $T/T_c = 0.05$ (ref. 1).

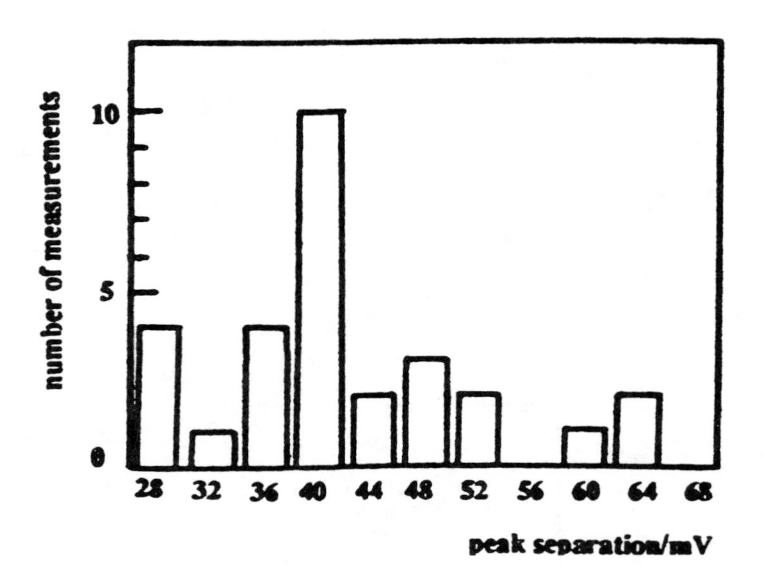

Fig. VIII.2. Histogram (4mV boxes) for 29 contacts where the conductance peak could be clearly resolved (ref. 1).

The distribution of gaps measured by point-contact tunneling on $YBa_2Cu_3O_7$ samples with tunneling characteristics of similar quality[1] is shown in Fig. 2. The largest gap has a value corresponding to $E_g = 2\Delta \sim 60-64$ meV. This is in good agreement with the largest gap seen by Raman scattering (polarization $\perp$ c) from single crystals (IV.3). It corresponds to $E_g/kT_c = 8$, while the most commonly observed gap ~ 40 meV or $E_g/kT_c = 5.5$. This is the largest gap observed in YBCO using a high-field scanning tunneling microscope (STM), where the feedback circuits drive the tip deep into the sample.[2] Some selected data obtained in this way are shown in Fig. 3. It is possible[1] that the gaps larger than 40 meV in Fig. 2 contain additional contributions from Schottky barriers in series with the superconductive barrier, but then one is surprised to see that the cutoff in the distribution agrees so well with the maximum gap in the Raman spectrum (VI.3) and infrared reflectance (VII.2).

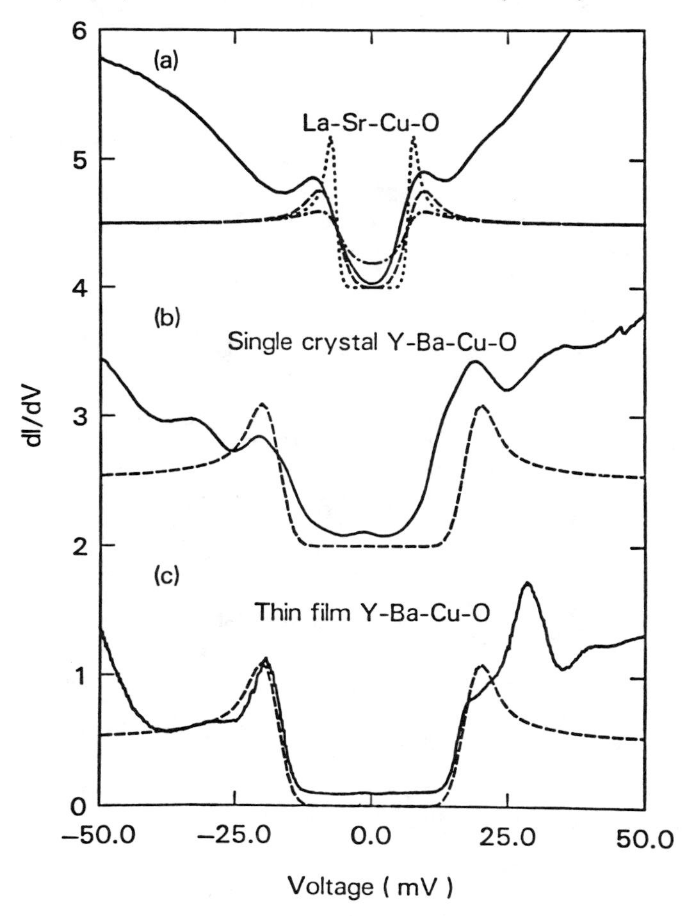

Fig. VIII.3. STM conductance spectra for samples as indicated. In (a) the dotted curve is theory with thermal broadening, the dot-dash curve includes lifetime broadening. Dashed curve corresponds to Gaussian gap distribution (ref. 2).

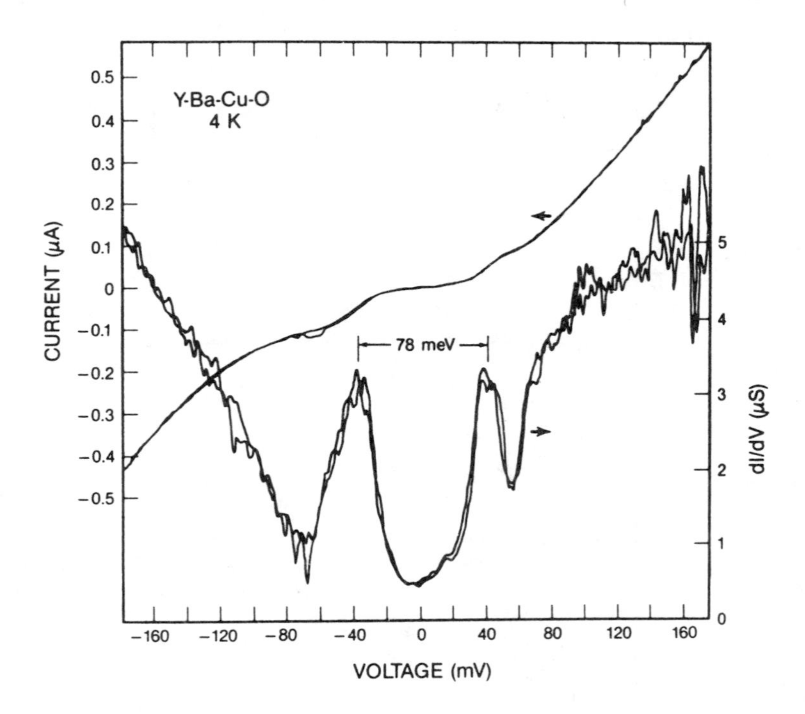

Fig. VIII.4. A tunneling conductance for YBCO-YBCO interface measured by the break junction method. The gap marked here was the maximum gap observed and presumably corresponds to current flow in the a-b planes (ref. 3).

2. Break Junctions

The dead layer which is usually an obstacle to forming good tunneling contacts may arise because of lattice collapse to form a metastable surface phase. Because of their open structure oxygen-deficient perovskites such as $YBa_2Cu_3O_{7-x}$ (or in the perovskite notation, $YBa_2Cu_3O_{9-y}$) are especially likely to suffer this kind of surface degradation. Surface degradation can be minimized by preparing a freshly cleaved surface by breaking a sample under liquid He and then joining the cleaved interfaces. Occasionally this method, called the break-junction technique,[3] yields a good tunneling characteristic from which the gap can be estimated. The maximum gap measured in this way[3] is shown for YBCO in Fig. 4. It gives $2\Delta \sim 40$ meV, which corresponds to the most frequently measured gap (Fig. 2) using the point-contact method, and an E_g/kT_c ratio of about 5.5.

3. Gap Anisotropy of Epitaxial Thin Films

One way to stabilize an unstable oxygen-deficient perovskite is by growing it as a thin film on a lattice-constant matched stoichiometric stable perovskite substrate such as $SrTiO_3$ (X.2). One then breaks the thin film at a performed substrate groove, as shown in Fig. 5, again in liquid He to minimize the local damage and lattice reconstruction associated with the heat of breaking. This technique has yielded very interesting results.[4] Note a special feature of Fig. 5. After the oriented edge has been broken, it is brought into contact with a Pb electrode, which is not a point contact. (It is more like a line contact.) The break-edge method has certain features in common with the bulk break-junction method,[3] but there are important mechanical differences as well as the obvious ability to control the anisotropy of the interface by using an oriented substrate. Specifically it is possible that the thickness of the dead layer at the edge is a function of the distance from the substrate-film interface and is at a minimum value (~ 20 Å) near this interface. Then the tunneling could take place primarily into the film near this interface, where the YBCO film is stabilized mechanically by the substrate. It is also worth noting that Tsai et al. mention that the broken edge is not a grain boundary, whereas this could easily be the case for the break-junction method.[3] All these factors make it possible to contact the film with the electrode over a larger region then a point contact would use, and without the kind of isotropic averaging over several point contacts implied by the break-junction method. At the same time the substrate provides mechanical stabilization which reduces the likelihood of deformation of the YBCO by the pressure exerted at the contact by the Pb electrode. Indeed pressure variations alter the junction resistance by a factor of 10^5 without changing the measured gap values,[4] which is not the case with other junction geometries (especially point contacts).

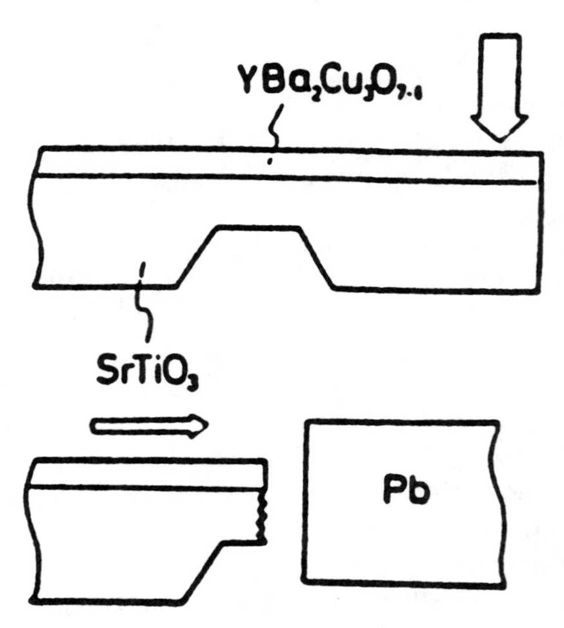

Fig. VIII.5 The thin-film geometry used to form a broken edge junction between YBCO and Pb. The YBCO film is oriented by deposition on a oriented crystalline $SrTiO_3$ substrate (ref. 4).

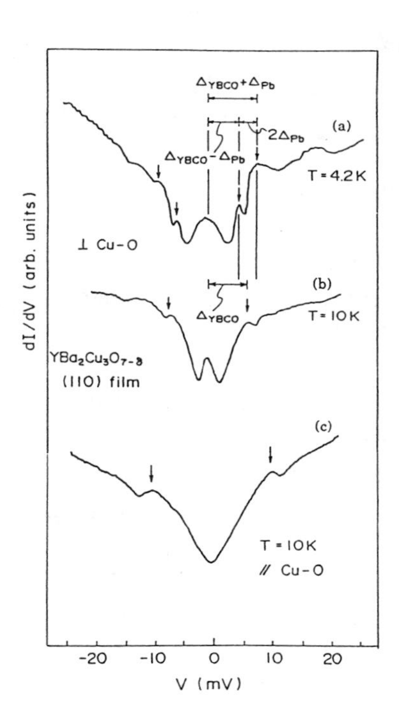

Fig. VIII.6. Tunneling characteristics for the junction geometry shown in Fig. 5. In (a) and (b) the current is flowing perpendicular to the YBCO a-b planes, and in (c) parallel to these planes (ref. 4).

The tunneling conductance shown in Fig. 6 exhibited the usual peaks at $\Delta_{YBCO} \pm \Delta_{Pb}$, but the value of Δ_{YBCO} was found to be strongly anisotropic. The current above the tunneling gap consists of resistive quasi-particles with conductivities weighted by the anisotropy of the normal-state conductivity which may be as much as 10^4 times greater in the a-b plane than normal to it. Thus the apparent value of $\Delta^{\perp}$ may be too large, as it may contain a contribution from $\Delta^{\parallel}$. At the same time, some damage and some misorientation must occur even for tunneling current flow nominally in the a-b plane, so the measured value of $\Delta^{\|}$ may be too small, relative to the maximum bulk value. Another important factor is the effect of break damage on the local value $T_{c\ell}$ of T_c at the junction. Care was taken to determine T_c *not* from a separate resistivity measurement but directly from $\Delta(T) = \Delta_{YBCO}(T)$, as shown in Fig. 7. Here the *measured local gap* refers to current tunneling through an edge containing the c-axis into the a-b plane. It gives $\Delta^{\|}(o) = 20$ meV, in good agreement with the maximum value obtained by the break-junction method, as shown in Fig. 4. However, according to Fig. 7, at the junction one has $T_{c\ell} = 76(4)$K, not 92K, as was used previously[3] to estimate $E_g^{\|}/kT_c$. With this value one has $E_g^{\|}/kT_c = 5.9(2)$. When the c axis was normal to the edge, the ratio $E_g^{\perp}/kT_c = 3.6(2)$. Both of these ratios appear to be independent of $T_{c\ell}$ as shown in Fig. 8.

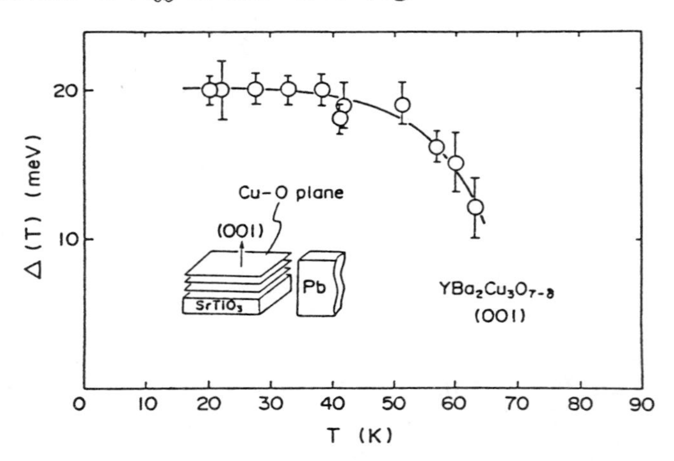

Fig. VIII.7. Measurement of $\Delta^{\|}(T)$ to determine T_c for point #6 in Fig. 6 (ref. 4).

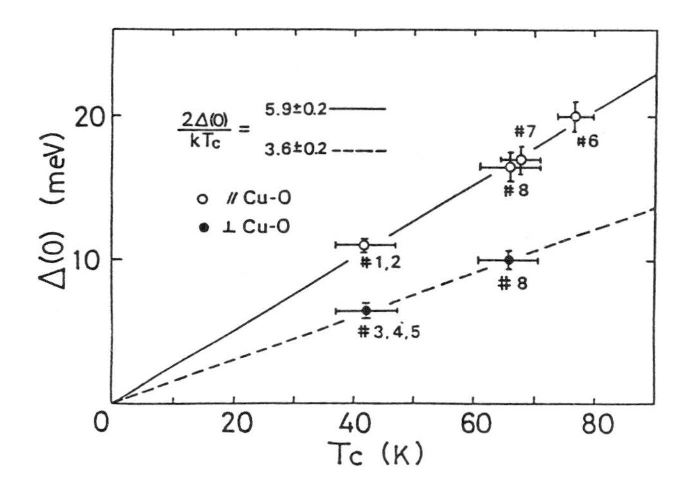

Fig. VIII.8. Plots of $\Delta^{\parallel,\perp}$ for current flow $\parallel, \perp$ to the CuO_2 planes as a function of the local $T_c = T_{c\ell}$ as determined by the method shown in Fig. 7 (ref. 4).

In discussing Fig. 8 it is important to emphasize that T_c as shown is the local value $T_{c\ell}$ at the edge where the junction is formed, which in general is less than the resistive value of T_{cr} for the entire film. If $T_{c\ell}$ were constant over a microscopically homogeneous film, then variations in $T_{c\ell}$ would change the ratio $2\Delta(o)/kT_{c\ell}$, in a way qualitatively similar to Carbotte's theory (Fig. II-8). One reason why this might not be observed is that the film is microscopically inhomogeneous, with a certain fraction f superconductive with a fixed intraplanar $\tilde{T}_c \sim 125$K and an intraplanar $\tilde{\Delta}^{\parallel}$ such that $2\tilde{\Delta}^{\parallel}/k\tilde{T}_c = 5.5$ (or 5.9(2), according to Fig. 8). Even when f = 1, $\Delta^{\perp} = 3.6\tilde{\Delta}^{\parallel}/5.9$, because the planes are coupled by defects of some kind where $\Delta^{\parallel,\perp}$ are reduced, and $T_{c\ell} < \tilde{T}_c$. As this fraction f decreases, a smaller volume of the sample is superconductive , and superconductive islands are formed with decreasing $T_{c\ell}$ and decreasing $\Delta^{\perp}$ and $\Delta^{\parallel}$. It is an interesting problem to see whether the constancy of $\Delta^{\perp}/T_{c\ell}$ and $\Delta^{\parallel}/T_{c\ell}$ can be explained theoretically, taking inhomogeneities into account.

4. Physical Model of Cuprate Junctions

Tunneling experiments with high-T_c cuprate samples have provided data which is qualitatively quite different from that obtained for the best junctions between the older superconductive metals (II). Enormously variable results have been obtained, especially with point contacts (Fig. VIII-2), while the break junction technique is in practice even more difficult. To experts accustomed to the sophisticated level achieved with good junctions of the older metals, this situation has proved quite disappointing, and one often hears it said that tunneling cannot provide reliable data for cuprates. However, there is considerable consistency between the data discussed earlier in this chapter. Here we discuss some junction models which utilize this consistency to explain the variability which has often been observed from one junction to the next.

First we notice that gap anisotropy in a powder sample means that a point contact, driven through a surface dead layer, will probably contact several grains, and that these grains will in general have different orientations and quite possibly different local $T_{c\ell}$'s. As a result, in addition to the jump in I(V) at the largest gap V_o (which corresponds to the orientation with current flow most nearly in the highly conductive a-b planes), there will be smaller "leakage" currents occurring for $V < V_0$. The situation is similar to the resistivity $\rho(T)$ in a multiphasic sample, which may show a drop, but not to zero, near $T = T_c$ of a minority phase with a high T_c. Moreover, as the point contact pressure is varied, the relative weights and even orientations of the grains relative to the tip may change drastically. It is noteworthy that in contrast to point contacts the break-edge junctions (VIII.3) give gap energies which are independent of junction impedance over five orders of magnitude.[4]

So far the largest gap observed in YBCO (bulk or film) break junctions is 40 meV, whereas the largest gap observed by the point-contact method is 60 meV. The ratio 40/60 = 2/3 is already a familiar one in the context of $T_c(YBa_2Cu_3O_{6.6})/T_c(YBa_2Cu_3O_{6.9}) = 60/90 = 2/3$, so it is tempting to suppose that cleavage produces the oxygen-deficient phase at the junction, but whether oxygen deficiency or cation disordering is involved here is hard to say. In any event, the primary gap distribution shown in Fig. VIII-2 also seems to have a cutoff at 40 meV, with only a secondary tail extending to 60 meV.

We have already argued that this secondary tail is genuine, and not an artifact resulting from Schottky barriers (which would not be symmetrical between $\pm V$ anyway) because the 60 meV cutoff has been observed in so many different experiments, some involving tunneling, others involving surfaces without contacts (infrared and Raman optical studies), and some involving only bulk probes (NQR relaxation). Still the question remains why the secondary tail exists and so far has been reported only for point contacts. It may be that most of the point contacts damage the junction in much the same way as cleavage does (40 meV primary cutoff in Fig. VIII-2), but that a small fraction of the junctions suffer less or even no damage. This would explain the secondary tail.

REFERENCES

1. A. Edgar, C. J. Adkins, and S. J. Chandler, J. Phys. *C20*, L 1009 (1987); similar results, with $1.7 \leq E_g/kT_c \leq 7.5$ in YBCO, have been reported by N. V. Zavaritsky, V. N. Zavaritsky, S. V. Petrov and A. A. Yurgens, Pisma JETP, Suppl. *46*, 23 (1987).
2. J. R. Kirtley et al., J. Vac. Sci. Tech. *A6*, 259 (1988).
3. J. Moreland et al., Phys. Rev. *B35*, 8856 (1987).
4. J. S. Tsai et al., Physica *C 153-155*, 1385 (1988) and unpublished.

IX. Relaxation Studies

1. Sound Velocity and Attenuation

The wave length of sound waves is of order 10^{-1} cm at frequencies of order MHz, which means that in a polycrystalline sample most of the scattering will take place at grain boundaries. These are of interest because in a powder sample with grain sizes of order 10 μ the grain boundaries are the weak links which determine critical currents and may contribute a large part of the normal-state resistivity. In examining ultrasonic velocity and attenuation data on powders the possibility that some of the observed anomalies are associated with the grain boundaries and not with their bulk-like interiors should always be considered. In single crystals of metals (such as Sn) it is possible to identify electronic effects associated with the superconducting energy gap in the temperature dependence of the longitudinal sound wave attenuation below T_c, although even with the best crystals there are still substantial corrections due to scattering from dislocations.[1]

In a powder sample, which is an isotropic elastic medium, there are two independent elastic moduli, the bulk modulus B and the shear modulus G, which are determined by the longitudinal and shear sound velocities.[2] For a second-order phase transition thermodynamics shows that the jump in the specific heat ΔC_p at the transition requires a jump ΔB in the bulk modulus which is proportional to ΔC_p and $(\partial T_c/\partial P)^2$. The latter factor is so small in $YBa_2Cu_3O_7$ that ΔB is negligibly small, but the thermodynamic relation has been confirmed by several workers for $La_{2-x}Sr_xCuO_{4-y}$ alloys.[2]

Comparison of the longitudinal v_L and shear v_s velocities in powders shows that in addition to kinks near T_c another

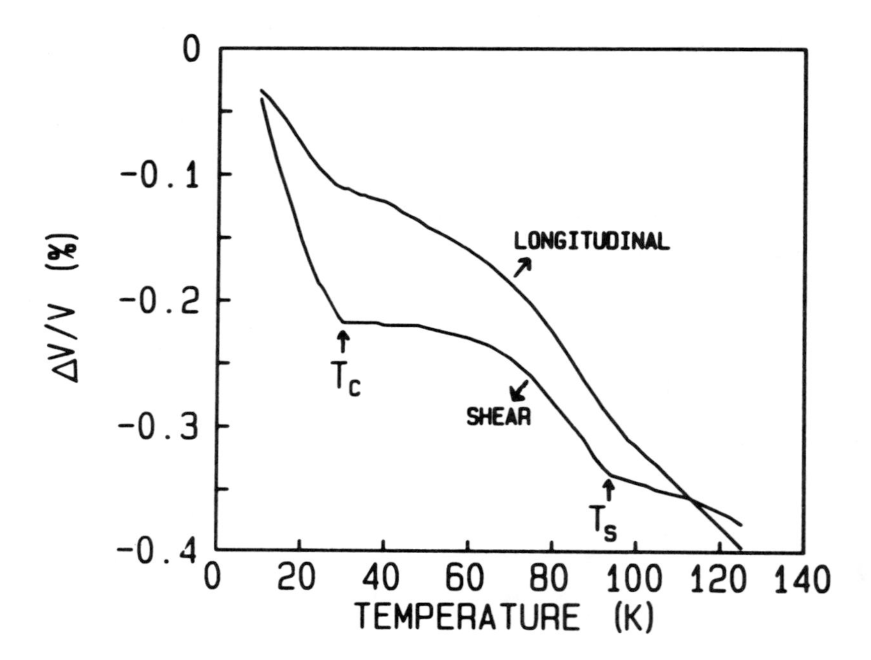

Fig. IX.1. Temperature-dependent sound velocities in $La_{1.8}Sr_{0.2}CuO_{3.9}$, with kinks at T_c and T_s (ref. 2).

kink can be observed near a second temperature $T_s > T_c$, especially in the shear velocity. This second kink is not associated with the bulk phase transition, but it can be resolved on carefully prepared powders, as is shown in Fig. 1 for $La_{1.8}Sr_{0.2}CuO_{4-y}$ with $y = 0.5$, and in Fig. 2 for $YBa_2Cu_3O_{6.9}$. In both materials the softening anomaly at T_s is significantly larger in v_s than in v_L. This anomaly occurs near $T_s = 120K$ in $YBa_2Cu_3O_{6.9}$ which is where the lattice-constant powder anomaly (discussed in VI.4) begins with decreasing T. It seems likely that this is a phase transition of the grain boundary which changes the clamping condition of the material which is ferroelastic. Such a change primarily alters the ratios of grain dimensions while keeping the grain volume fixed. It is much more sensitive to a shear wave than to a longitudinal wave for this reason.

We notice in Figs. 1 and 2 that although v_s is continuous at T_c, below T_c there is a steep increase with decreasing T. This arises from a change in screening of the ion-ion interaction as the energy gap increases, as well as coherent changes in the electron-phonon scattering matrix elements as independent electron wave functions are replaced by Cooper pairs. The gap alone would account for the sign of dv_s/dT in the cuprate examples shown here, but in some metals (such as V)[3] for $T < T_c$ one has $dv_s/dT > 0$ for $T < T_c$. It is important to remember here

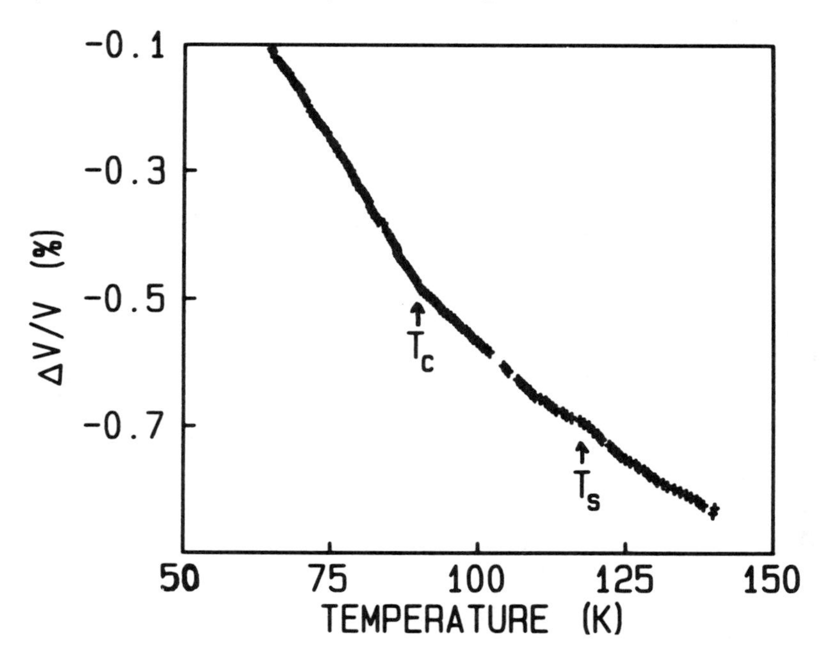

Fig. IX.2. Transverse sound velocity in $YBa_2Cu_3O_{6.9}$ (ref. 2).

that shear waves are screened only by local field or Umklapp effects, and that these are complex, whether in a transition metal like V or in cuprates. The overall magnitude of these gap effects should scale, however, with E_g, which is about 20K in V compared to about 700K in $YBa_2Cu_3O_7$. The fractional shift in v_s in V between O and T_c is about 10^{-4}, while in $YBa_2Cu_3O_{6.9}$ it is about 10^{-2}, reflecting primarily the larger value of E_g in the latter material.

Sound attenuation in $YBa_2Cu_3O_{6.9}$ shows two maxima,[2] one at $T_s \sim 120K$ and one at $T' \sim 65K \sim 0.7\ T_c$, with a similar pattern in $La_{1.8}Sr_{0.2}CuO_{3.9}$. We expect that T_s is associated with the grain boundaries, but that T' is somehow associated with T_c. The latter in a percolative system occurs when a large enough volume fraction has become superconducting so that superconductive filaments traverse the sample. Plane waves, however, generally require a larger volume fraction to percolate than do dendrites. Technically the electrical and acoustical cases are known as scalar and vector percolation, respectively, and quantitative results for the active volume fractions are known in simple models. These models cannot be quantitatively compared to these experiments because the volume filling fractions are not known accurately enough. However, the distinction between scalar and vector percolation is fundamental, and so I have used T' as an estimate for the vector T_c in (VII.3) to calculate E_g (vector)/kT_c (vector), obtaining a ratio in reasonably good agreement with other experiments.

2. Cu NMR Relaxation Rates in YBCO

In elemental metals the temperature-dependent nuclear spin-lattice relaxation time $T_1(T)$ closely follows[4] the BCS theory, which predicts not only a freeze-out of the relaxation rate T_1^{-1} as the energy gap opens up but an actual enhancement for T just below T_c because of coherence factors associated with Cooper spin pairing. Moreover, when there is more than one inequivalent Cu site per unit cell (such as the Cu(1) chain and Cu(2) plane sites in YBCO), these are generally associated with separate resonances. The relaxation rate at each site probes the local BCS gap at that site, and reflects anisotropies which necessarily arise when these sites are in series (current flow along c axis) or in parallel (current flow in a-b plane). However, just as is the case for acoustic attenuation, at low temperatures these relaxation rates can depend sensitively on defect configurations which are little known and which have different effects on different sites. At this early stage there are still several obvious uncertainties in interpretation of these data, which nevertheless appear to be in good overall agreement with the anisotropic normal-state transport data (III.7) and anisotropic tunneling data (VIII.3). While nuclear spin-lattice coupling to defects is complex, compared to most other methods of measuring gaps NMR has the advantage of being a bulk average which is not surface-sensitive and does not require external contacts, as in electrical or acoustical transport experiments. Unlike most metallic superconductors, the oxides have low-lying optic modes with $\hbar\omega_0 \sim kT_c$, and these may affect the magnetic character of defects. While no one experiment may be said to be decisive, NMR has made informative contributions primarily because of its unique ability to explore specific sites.

In $YBa_2Cu_3O_{6.9}$ NMR spin-lattice relaxation times can be studied[5] in zero magnetic field through nuclear quadrupole resonance (NQR) which shows two lines near 22 and 301 MHz. Below T_c both lines exhibit relaxation rates which freeze out exponentially (Fig. 3), corresponding to local energy gaps E_g such that E_{g_1} (lower line)$/kT_c = 2.4$ and E_{g_2}(higher line)$/kT_c = 8.3$, with $T_c = 92K$. It seems

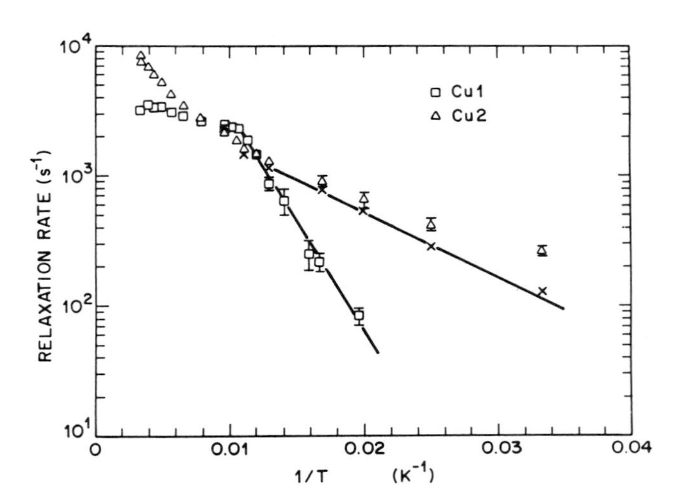

Fig. IX.3. Arrhenius plot of relaxation rates T_1^{-1} vs. T^{-1} for Cu(1) (triangles) and Cu(2) (squares) sites in $YBa_2Cu_3O_{6.9}$ (T_c = 92K) powder. Crosses denote Cu(1) data after correction for constant background associated with oxygen vacancies on Cu(1)O chains (ref. 5).

likely that E_{g_2} describes the intrinsic BCS gap, while $E_{g_1} \ll E_{g_2}$ may arise because of a proximity effect.

Two methods have assigned these two lines, lower and higher, $E_g/kT_c = 2.4$ and 8.3, to chains and planes respectively. In a magnetic field of about 80 kG at T = 100 K measurements of the NMR spectrum of an aligned single-crystal platelet mosaic sample[6] showed that the higher line has nearly axial symmetry about the c axis, as expected for the Cu(2) planar site, while the lower line does not (its axial symmetry is about the a axis). The width of the lower line is some four times greater than that of the higher line, as expected from oxygen vacancies on the Cu(1)

O chains. Although crystal symmetry predicts that $|\partial^2 V/\partial z^2|$ should be zero at the Cu(1) sites, a non-zero value is measured (about 2% of the Cu(2) values), possibly reflecting stacking faults.

A second method is to study the temperature dependence of the NQR lines in $GdBa_2Cu_3O_{6.9}$ compared to $YBa_2Cu_3O_{6.9}$. The magnetic moment of Gd greatly alters the relaxation rate at the nearly Cu(2) sites while having much less effect on that of the Cu(1) sites. The results[7] shown in Fig. 4 clearly show a much larger change for the higher band, both in magnitude and temperature dependence, confirming the single crystal assignment.

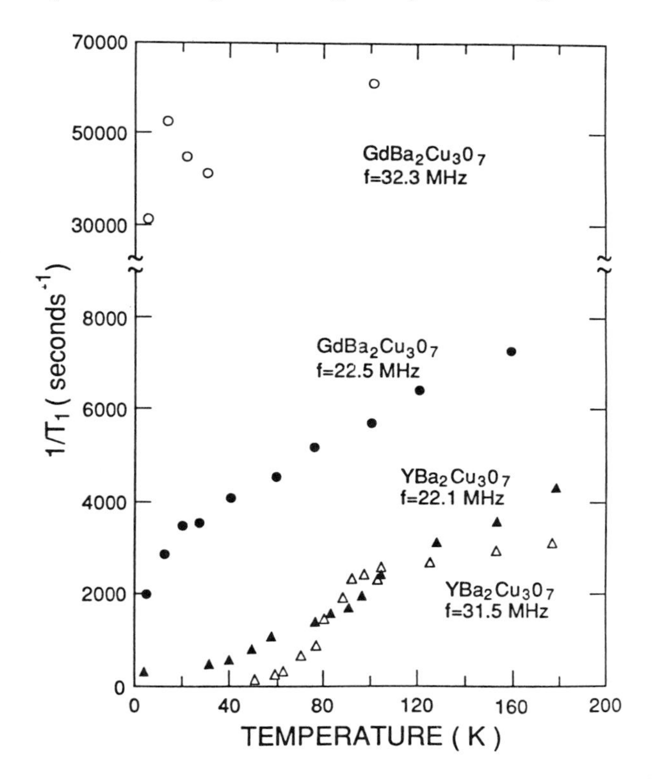

Fig. IX.4 Comparison of relaxation rates for Cu(1) and Cu(2) NQR lines in $GdBa_2Cu_3O_{6.9}$ and $YBa_2Cu_3O_{6.9}$ powders (ref. 7).

In the normal state the relaxation rate of the lower or Cu(1) line shows the linear temperature dependence of a normal metal, which is consistent with the two-carrier model for the normal-state transport properties described in IV.9. The relaxation rate of the Cu(2) line, however, shows much less temperature dependence, and this, together with differences in the relaxation rate of nearby Y nuclei, has led to the suggestion that the Cu(2) electron spin fluctuations in $YBa_2Cu_3O_{6.9}$ are antiferromagnetically correlated.[8] In view of the antiferromagnetic ordering of the Cu(2) O_2 planes in $YBa_2Cu_3O_6$ (IV.5), this suggestion would be attractive if it were not for the polarized neutron diffuse scattering data (Fig. 8a and VI) which show that in the frequency range appropriate to nuclear spin relaxation, less than 1% of the Cu(2) spins carry local magnetic moments. Instead it seems more likely that these anomalies are associated with some degree of cation substitutional disorder either at the Cu(2) or Y sites, which alter the local spin-orbit coupling.

At this point the site-specific Cu spin relaxation data all seem to be in generally good agreement with other data on $YBa_2Cu_3O_{6.9}$. However, a surprising change occurs for the tetragonal $YBa_2Cu_3O_{6.7}$ phase (T_c = 60K). Compared to $YBa_2Cu_3O_{6.9}$, the Cu(2) normal-state spin-lattice relaxation rates are reduced by more than three orders of magnitude, while the Cu(1) rate is little changed.[9] Because the additional oxygen vacancies are concentrated mainly on the Cu(1) O chains according to diffraction experiments (III.4.5), this is just the opposite of what we would have expected. The $YBa_2Cu_3O_{6.7}$ phase is superconductive with T_c = 60K and one would naturally assume that this superconductivity is associated, as in the T_c = 92K $YBa_2Cu_3O_{6.7}$ phase, primarily with the CuO_2 planes, with T_c reduced by weaker interlayer coupling. However, if the Cu(2) spin-lattice relaxation rate is so greatly reduced as to be comparable to its value in the insulating $YBa_2Cu_3O_6$ phase, then the CuO_2 planes should be insulating and possibly even antiferromagnetic. This would mean that the $T_c = 60K$ superconductivity is associated with the Cu(1)O chains only, which are widely separated by insulating planes including the Y planes. This result is so unexpected that it must be confirmed by independent experiments before it can be discussed further.

One additional point concerns the "coherence factor" enhancement of T_{ls}^{-1}/T_{ln}^{-1} which was observed[4] to be about a factor of two in Al for $T/T_c \sim 0.9$. The magnitude of the enhancement in Al depends on several factors, including the electron spin relaxation broadening. Here if we examine the original $YBa_2Cu_3O_7$ data in Fig. 2 carefully near 90K, we see that the break in slope of the 22 MHz Cu(1) line seems to occur between 100K and 120K, while the 31 MHz Cu(2) line shows a break close to $T_c \sim 90K$. The Cu(1) break may reflect the effects of strain associated with the precursor effects on the powder lattice constants, as discussed in VI.4 and IX.1 and the postponement (as T is reduced) of the break in the Cu(2) rate may actually reflect some coherence factors for the Cu(2) plane, as more magnetic defects (and greater electron spin relaxation broadening) are expected in the Cu(1) O chains. These differences in prefactors of an Arrehenius behavior may be of only marginal significance, however.

REFERENCES

1. W. P. Mason, in *Physical Acoustics* (ed. W. P. Mason, Academic Press, N.Y., 1966), Vol. IVA, p. 299; R. W. Morse, H. V. Bohm, and J. D. Gavenda, Phys. Rev. *109*, 1394 (1958).

2. S. Bhattacharya et al., Phys. Rev. Lett. *60*, 1181 (1988); Phys. Rev. *B37*, 5901 (1988).

3. G. A. Alers, in *Physical Acoustics* (ref. 1 above), p. 292.

4. L. C. Hebel and C. P. Slichter, Phys. Rev. *113*, 1504 (1959); A. G. Redfield and A. G. Anderson, Phys. Rev. *116*, 583 (1959).

5. W. W. Warren, Jr., R. E. Walstedt, G. F. Bronnert, G. P. Espinosa and J. P. Remeika, Phys. Rev. Lett. *59*, 1860 (1987); *High Temperature Superconductivity*, MRS Symp. Proc. (1987). The labelling of Cu(1) and Cu(2) sites in these papers should be reversed.

6. C. H. Pennington et al., Phys. Rev. *B37*, 7944 (1988).

7. P. C. Hammel, M. Takigawa, R. H. Heffner and Z. Fisk, Phys. Rev. *B38*, 2832 (1988).

8. R. E. Walstedt et al., Phys. Rev. *B* (Nov. 1, 1988).

9. W. W. Warren, Jr., et al., Phys. Rev. Lett. xxx.

X. Materials Morphology

1. Intergranular Weak Links

Because of mechanical instabilities the most convenient and most natural morphology for high-T_c cuprate materials is that of a fine-grained polycrystalline powder. The small dimensions (typically $\lesssim 2-20\mu$) of fine grains limit the accumulation of internal stress, or put differently, the natural grain dimensions are small because the accumulation of internal stress eventually limits homogeneous grain growth. Chemical factors are also important. With a multiplicity of cations homogenization of the source oxides Y_2O_3, BaO and CuO, for example, can be achieved much more easily on a small scale which should not be much larger than a cation diffusion length during laboratory times at the formation temperature. The final stage of additional oxidation (for instance, from $YBa_2Cu_3O_6$ to $YBa_2Cu_3O_7$) also can take place more completely with fine grains.

Even when the grains are small, however, they are unlikely (X.7) to be completely homogeneous, especially near the surface where cation disproportionation is thermodynamically favorable (surface segregation). This can explain the dead layers observed in point-contact tunneling experiments (VIII.1). Intergranular weak links are also expected because of anisotropic conductivities (III.7) and poor mechanical connectivity.

The effects of weak links on the critical current density J_c are drastic. Intragranular values of $J_c \gtrsim 10^5$ amp/cm^2 are expected on the basis of thin-film data, but in sintered powders $J_c \lesssim 10^3$ amp/cm^2 even at $\mathbf{H} = 0$. With increasing H, J_c decreases by a factor of 10 or more[1] at H = 100 gauss and T = 77K, as shown in Fig. 1.

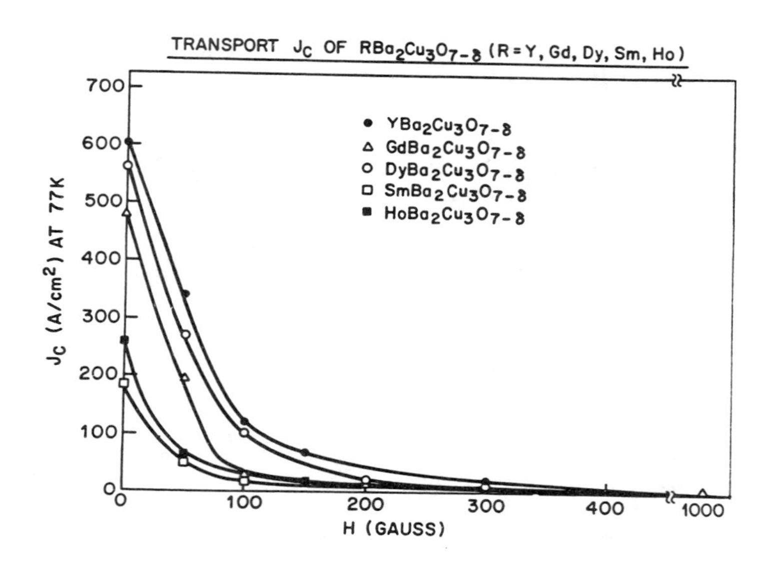

Fig. X.1. Critical current for sintered $RBa_2Cu_3O_7$ powders at 77K vs. magnetic field (ref. 1).

Many of these problems are greatly reduced by orienting (texturing) the grains, either with a magnetic field, or by melting the powder and regrowing the solid in the presence of a temperature gradient (melt-texturing). Some results for $J_c(H)$ obtained so far in this way[2] are shown in Fig. 2.

2. Epitaxial Films

The problem of grain boundaries can be made much less serious by growing thin ($\sim 1\mu$) films on substrates with lattice constants close to those of the basal plane of a tetragonal cuprate.[2] This method has worked well for (La, Sr)CuO_4, $YBa_2Cu_3O_7$, and the bismates and thallates. It produces oriented samples with high critical current densities ($\gtrsim 10^5 - 10^6 A/cm^2$, about 10^3 larger than found for powder samples). The obvious drawback is the small sample volume and small critical current. Less obvious is that the much thicker substrate will often unavoidably contribute to the physical properties of the sample.

The substrates which have been most effective in practice have good lattice constant matching to $YBa_2Cu_3O_7$ ($(a+b)/2 = 3.85A$); they are $SrTiO_3$, $a = 3.90A$ and Yttria-stabilized zirconia, $a = 5.16A = \sqrt{2}(3.65A)$, a $\pi/4$ rotation of the basal plane, while films grown on MgO substrates ($a = 4.20$) appear to be inadequately epitaxial. If lattice-constant matching were the only criterion, $SrTiO_3$ would be the substrate of choice, but in practice during the high-temperature anneal in O_2 which converts the film from insulating to metallic, cation exchange (Ba, Sr) with the substrate can degrade the superconductive film properties.

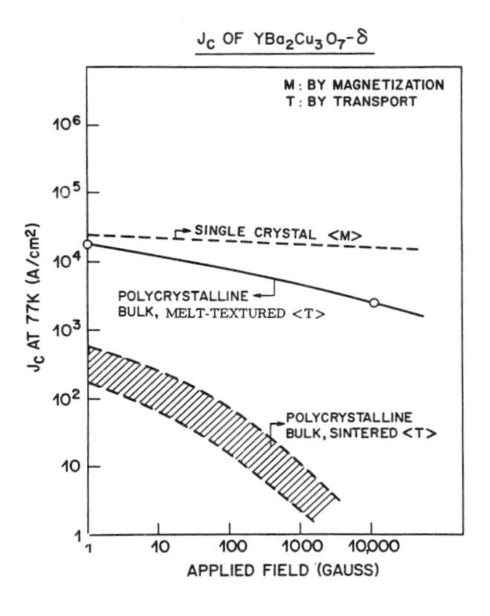

Fig. X.2. Comparison of sintered and melt-textured critical currents vs. H compared to single-crystal values (ref. 2).

Several methods have been used to produce the thin film material; most successful to date have been evaporation from multiple sources, sputtering from one or more sources, and laser ablation from one source. Depending on the source of material and the substrate, different temperatures and pressures are used for the final oxidation, which should be done in situ, because contamination (by H_2O vapor especially) is a serious problem for thin film oxides with an open structure.[2] These epitaxial films have been especially valuable for studies of transport properties.

3. Single Crystals

So long as an oxide is insulating, even it contains an odd electron/unit cell so that the nonmetallic behavior is caused by spin density waves, as in La_2CuO_4, single crystals with dimensions of order 1 cm can be grown.[3] When the crystals become metallic and superconductive, growth of large single crystals with narrow transition widths ΔT_c becomes much more difficult. The highest quality single crystals of $BaPb_{0.7}Bi_{0.3}O_3$, grown hydrothermally and in a gold capsule,[4] had $\Delta T_c = 1.8K$ and gave a rod with a diameter of about 3 mm. Large crystals of $La_{1.92}Sr_{0.08}CuO_4$, volumes $\sim 1cm^3$, have been grown,[5] but only with $T_c < 20K$, and with superconductivity confined to the sample surfaces probably because of slow or self-limited oxygen diffusion.

For $YBa_2Cu_3O_{7-x}$, oxygen diffusion is facile at $T \sim 1000C$, so long as $x > 0$. Thus single crystal growth from the melt depends primarily on the ternary phase diagram of this temperature, which is now believed to be approximately as shown[6] in Fig. 3. The region of partial melting is shifted towards CuO but almost touches $YBa_2Cu_3O_7$. Slow cooling of certain of these partially melted compositions has produced free-standing crystals[7,8] on the edge of the solidified charge with surface area approaching $1cm^2$ and thicknesses (along the tetragonal

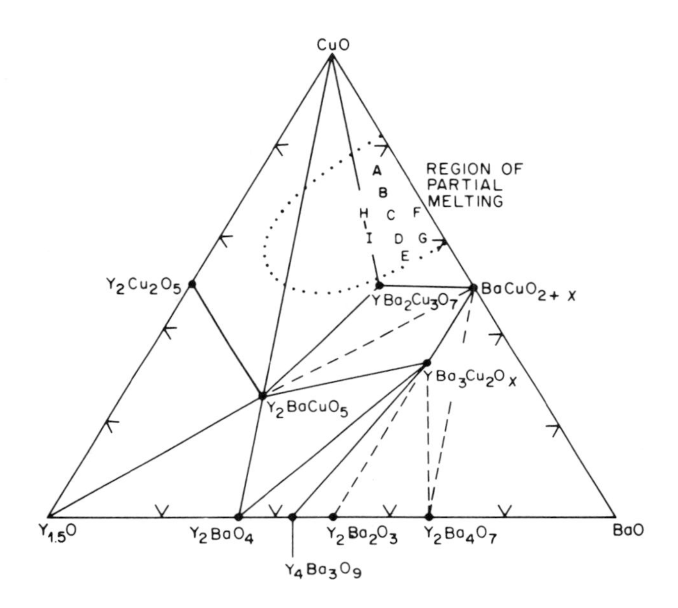

Fig. X.3. Ternary phase diagram for Y-Ba-Cu oxide (ref. 6).

axis) of order 100μ.

Crystals have also been grown of the bismate $Bi_{2.2}Sr_2Ca_{0.8}Cu_2O_{8+\delta}$ using both CuO fluxes[9] (as in $YBa_2Cu_2O_7$) and alkali chloride fluxes.[10] Here maximum dimensions so far are much smaller ($8mm \times 300\mu$), as one might expect from the rigid mechanical structure of the Bi_2O_2 layers which are lattice-constant matched to the CuO_2 layers (XI) and which probably account for the micaceous cleavage properties. Still thinner black platelet ($t \sim 0.1\mu$) crystals have been reported for the Tℓ-Ba-Ca-Cu-O system.[11] Quite consistently maximum crystal dimensions decrease as T_c increases in the pseudoperovskite family from BaPbBiO to tℓBaCaCuO. (While the dimensions of the more recently discovered materials are expected to increase, this technical improvement should not alter the overall trend.) This point is illustrated in Fig. 4, which indicates a correlation between T_c and mechanical instabilities and thus points to electron-phonon interaction as the underlying mechanism responsible for high-T_c superconductivity.

It is instructive to compare Fig. X.4 and Fig. V.4. The absence of the isotope effect in $YBa_2Cu_3O_7$ can be interpreted as evidence against the electron-phonon interaction only if both the electronic structure is simple and the lattice structure has almost no defects which could pin the Fermi energy. It is very implausible that this is the case (Figs. IV-10 and IV-14). By contrast, unlike this

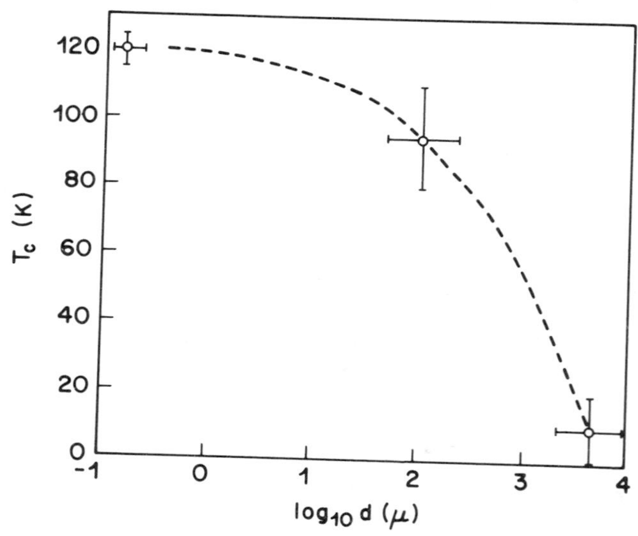

Fig. X.4. Trend of T_c vs. smallest crystalline dimension for largest samples grown so far. The three points correspond to thallates, $YBa_2Cu_3O_7$ and bismates, and $(La,Ba,Sr)_2CuO_4$ and $Ba(Pb,Bi)O_3$. The crystalline dimensions might be increased by more elaborate preparation, but the general trend is unlikely to change significantly.

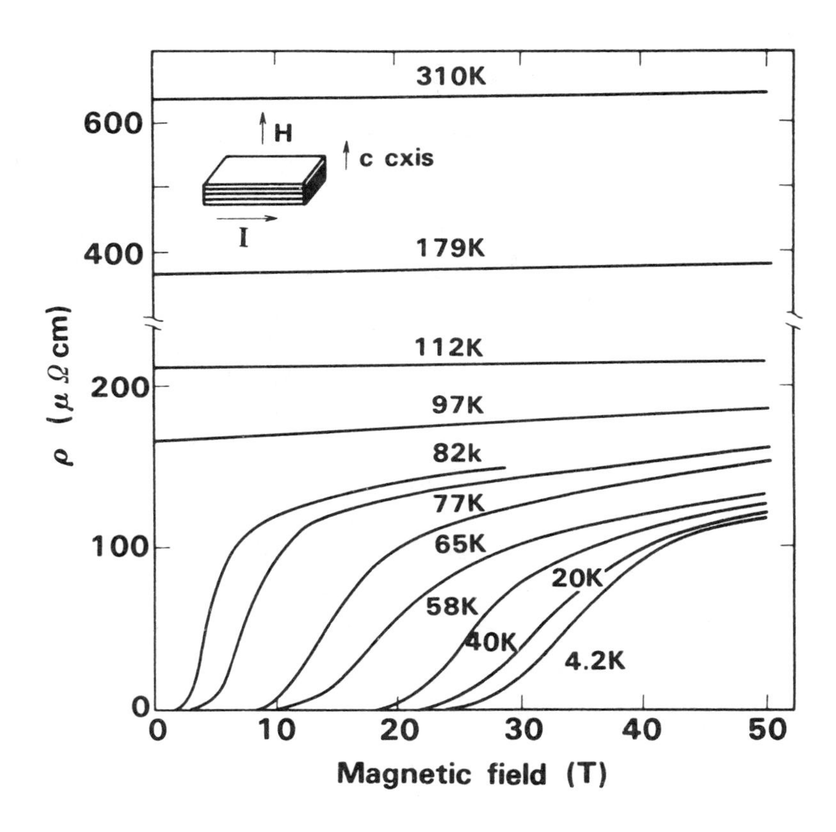

Fig. X.5. Perpendicular magnetoresistance in $EuBa_2Cu_3O_7$ (ref. 13).

reasoning which is so fragile that it collapses in the face of a realistic model for the crystal structure, the correlation shown in Fig. X.4 is robust in the sense that it not only survives in the presence of defects and Fermi level pinning, but actually thrives in this more realistic context. This is because as the defect density increases the hysteretic structural properties (such as the shape memory effect, VI.4) become more serious and more effective in limiting the size of single crystals.

4. Critical Field Anisotropy and Macroscopic Parameters

The critical field $H_{c2}(T)$ determines several important macroscopic parameters, as described for the macroscopic anisotropic Landau-Ginsberg theory in Appendix A. The general form of $H_{c2}(T)$ is well-known from extensive numerical studies[12] on limits of various models. To test the theory one must have resistivity data on single crystals at very high magnetic fields ($\gg 10T = 100$ kgauss). These data have recently been obtained[13] on a high-quality single crystal of $EuBa_2Cu_3O_{7-x}$ with $T_c = 94K$ and $\Delta T_c < 1K$, for H both $\parallel$ and $\perp$ to the basal plane. The in-plane magnetoresistivity is shown in Fig. 5. With increasing field the resistive transition is broadened, as shown for T near T_c in Fig. 6. (An explanation for this broadening in terms of giant flux creep is given in X.6.) This leads to two definitions of H_{c2} shown in Fig. 6, for $\rho = 0.05(5)\rho_0$, or for

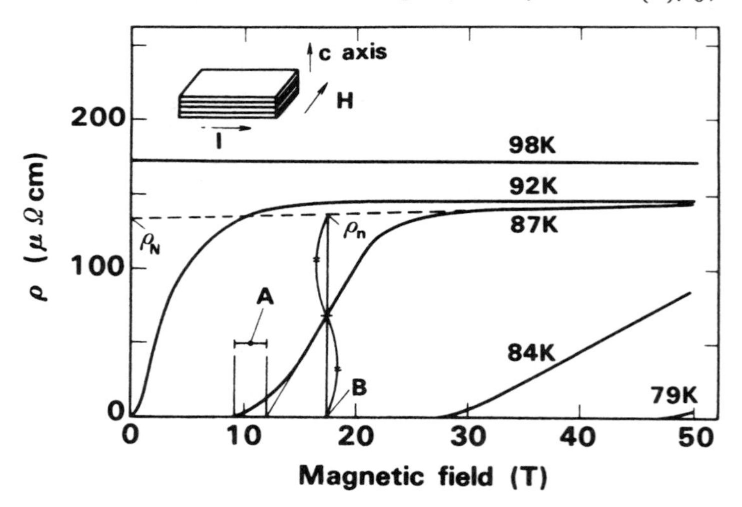

Fig. X.6. Parallel magnetoresistance in $EuBa_2Cu_3O_7$, with two definitions of $\rho_{\parallel}(T)$ indicated by the geometrical constructions A and B (ref. 13).

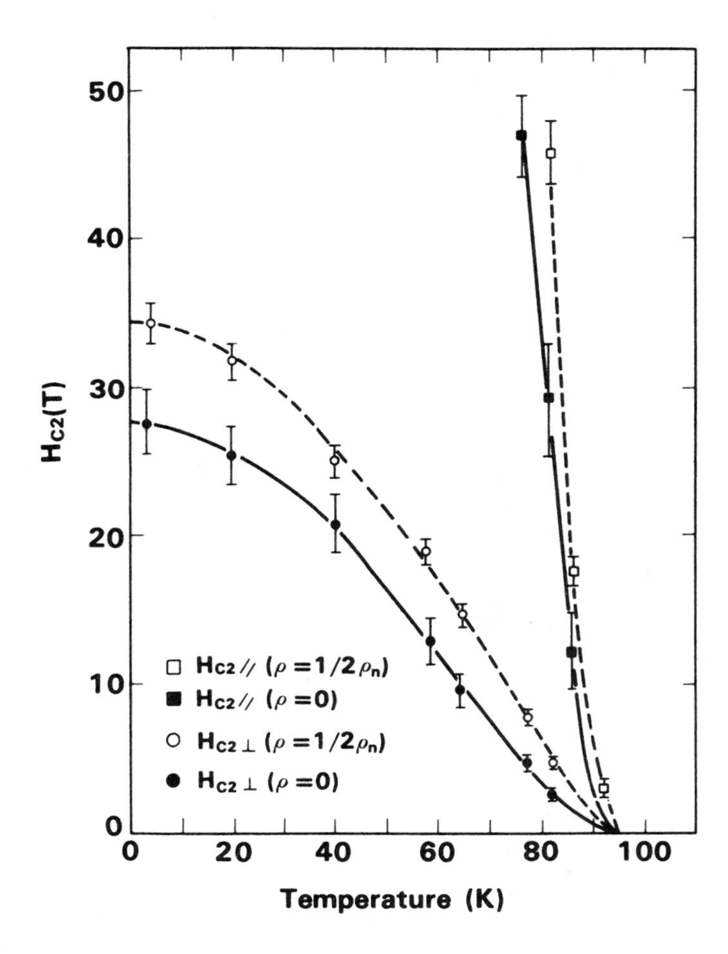

Fig. X.7. Upper critical fields H_{c2} for parallel and perpendicular fields as defined by the A and B constructions in Fig. 6 (ref. 13).

$\rho = 0.50(5)\rho_0$. Both definitions are shown in Fig. 7. For currents in the plane $H_{c2\perp}(T=0)$ is close to 30T, while the extrapolated value of $H_{c2\|}(0)$ is close to 250T. In the anisotropic Landau-Ginsberg theory these values give coherence lengths $\xi_{\|}(0) = 35\text{Å}$ and $\xi_{\perp}(0) = 3.8\text{Å}$. The spacing of the CuO_2-CuO planes in these cuprates is about 3.9Å, so the perpendicular coherence length is very nearly equal to the spacing of the cuprate planes, assuming that both the CuO_2 and CuO planes are metallic.

To understand the curves theoretically one can begin by using the macroscopic anisotropic Ginsberg-Landau equations (Appendix A) which predict the $H_{c2}(T)$ will be linear as $T \rightarrow T_c$ with the coherence lengths proportional to $H_{c2}^{1/2}$, which yields the coherence lengths quoted above.

The anisotropy in $H_{c2}(T)$ can be explained in several ways. To clarify these explanations in specific contexts it is convenient to compare $H_{c2}(T)$ in the layered cuprates with what has been observed in superconductive layered dichalcogenides, both without[14] and with [15] the intercalation of organic molecules to decouple the layers electronically. Some representative data for these analogue compounds are shown in Table 1.

Material	$\epsilon(H_{c2})$	$\delta T_{c\perp}/T_c$	$\delta T_{c\parallel}/T_c$
$NbSe_2$	3.3(2)	$1.7 \cdot 10^{-2}$	$3.5 \cdot 10^{-2}$
TaS_2	6.0(5)		
EBCO	9(1)	$12 \cdot 10^{-2}$	$6 \cdot 10^{-2}$
$TaS_2 + PY_{1/2}$	~50	$4 \cdot 10^{-2}$	$10 \cdot 10^{-2}$

Table 1. Anisotropy $\epsilon = (dH_{c2\parallel}/dT)/(dH_{c2\perp}/dT)$ and positive upward curvature parameters for $EuBa_2Cu_3O_{7-x}$ compared to reference dichalcogenide superconductors.

The anisotropy ratio ϵ of $H_{c2\parallel}/H_{c2\perp}$ is twice as large in the sulfide as in the selenide, and larger again in the cuprate, reflecting increasing ionic character and decreasing ratios of $v_{F\perp}/v_{F\parallel}$ according to band theory. The anisotropy increase on going from the selenide to the sulfide is consistent with nearly constant $v_{F\parallel}$ and decrease by a factor of two in $v_{F\perp}$ with the chalcogen replacement, which is not incompatible with available band-structure calculations.[16] (These calculations for such relatively simple compounds with six atoms/cell contain uncertainties[17] of order at least 0.5 eV for the bands near E_F. This is a useful reminder that the band calculations for the much more complex cuprates described in IV, which are still very preliminary, could be in error by 1 eV or more.) Thus the 50% increase in ϵ from TaS_2 to EBCO shown in Table 1 may not have much significance, except perhaps that a larger increase might have been expected for the cuprate in view of the barrier for c-axis currents provided by the $O^{\square}$ coplanar with Eu.

For $H_{c1} < H < H_{c2}$ one is in the mixed or intermediate state in which flux lines penetrate the sample to generate an array of cylinders of order $\xi_0(T)$ in diameter. For a homogeneous sample which is (nearly) isotropic in the plane perpendicular to the field, the array should be a (close-packed) honeycomb lattice with a spacing determined by the anisotropic Ginsberg-Landau equations (Appendix A) once the flux quantum $\phi_0 = hc/e^*$ is known. The vortex array can be observed directly[18] by cooling the sample in a magnetic field and decorating the surface with fine magnetic particles which are attracted by the flux lines.

Honeycomb short-range order has been observed on the a-b surfaces of single crystals of $YBa_2Cu_3O_7$ when the magnetic particles were deposited at 4.2K. The flux line density scaled with applied field $10g \lesssim H \lesssim 200g$ and corresponded quite accurately to $e^* = 2e$, in accordance with the BCS theory and with spacings previously observed on metallic superconductors. This result agrees well with more accurate but basically equivalent measurements of e^* by the Josephson effect and by flux jumps in cylinders.[18]

Perhaps the most interesting feature of this experiment was the failure to observe short-range flux line order at 77K even though at this temperature $\xi_{o\perp}(T) \sim 100\text{Å} << $ flux line spacing of order 1 μ (with a magnetic particle diameter $\lesssim 0.1\,\mu$). The correlation length of the flux pattern at 4.2K was 2.5 lattice parameters of the fluxoid lattice and the hexagonal autocorrelation function showed at least four well-resolved rings. The failure to observe this pattern at 77K suggests that we are dealing with substantial lateral variations of the energy gap, and these destroy two-dimensional order at temperatures $T_c^* < T_c$, where T_c measured resistively reflects one-dimensional percolation. This problem was discussed before, in connection with sound attenuation measurements (IX.1), which showed an apparent phase transition at $T' \sim 65K$ in $YBa_2Cu_3O_7$. If the heuristic dimensional argument just given is correct, $T_c^* = T' < 77K$, which would indeed explain the absence of the flux lattice at this temperature. With steadily improving single-crystal samples, this would seem to be a promising area for further research.

5. Meissner Effect and Type III Superconductivity

The traditional classification of superconductive diamagnetic regimes in terms of H_{c1} and H_{c2} (Appendix A) arose historically from the Meissner effect for $H < H_{c1}$, that is, the complete or 100% expulsion of magnetic flux from a superconductive sample. Flux penetration in the intermediate regime (mixed state) $H_{c1} < H < H_{c2}$, with zero-electrical resistance, itself already seemed a very novel idea when Abrikosov proposed it in the 1950's. Only very pure simple metals (such as Pb and Sn) have $H_{c2} = H_{c1}$ (no mixed states). The cuprates have carried this evolution one stage further, because so far as I know no high T_c cuprate has been prepared with much more than an 80% Meissner effect. Just as superconductors with $H_{c2} > H_{c1}$ are called type II superconductors, so we may call these high T_c cuprates with only fractional Meissner effects type III superconductors.

What are the general morphological features of type III superconductors? So far not much information is available, because experimentalists have tended to regard an incomplete Meissner effect as an indication of poor sample quality. Because of the presence of non-superconductive (usually non-metallic) phases at vicinal compositions, this was a legitimate concern in early work. However, in subsequent work these unwanted phases were largely eliminated by improved preparation methods, but the fractional Meissner effects remained. In this section we examine chemical and geometrical trends in Meissner (low field, $H < H_{c1}$) regime data as presently available. With more data on better single crystals the nature of type III superconductivity should become clearer, but some features can already be identified.

Just as type I superconductors are characterized by the length scale of the London penetration depth λ and type II similarly by the coherence length ξ, so there is a characteristic length associated with the fractional Meissner effect, which can be called η. It appears that this length depends on sample size and for powdered samples scales approximately with L, where L is the median grain diameter.[19] This is shown by plotting the fractional diamagnetism $f_D(0)$ against median grain diameter in Fig. 8, that is, the ratio $-\chi(0)/4\pi$, where $\chi(0)$ is the static magnetic susceptibility at $T = 0$ and $-1/4\pi$ is the ideal Meissner value. The data points are well fitted by the relation $f_D(0) = 0.8x/(1+x)$, where $x = L/\eta$ and $\eta = 6\mu$. In each

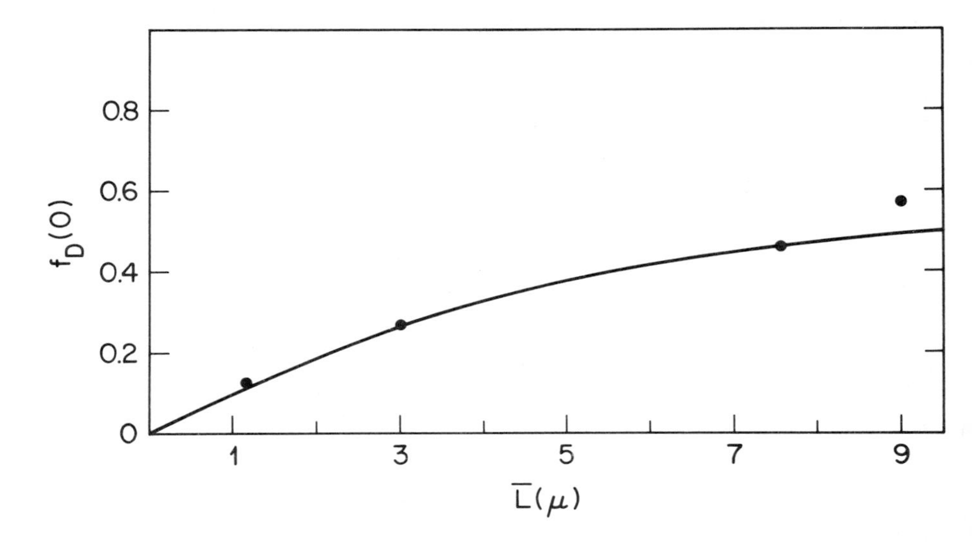

Fig. X.8. Fractional Meissner volume at T = 0 vs. median grain size (Table 1, ref. 18) in powdered $YBa_2Cu_3O_7$. Solid curve: 0.8x/(1+x), see text.

powder sample there is a broad distribution of grain sizes, which means that other relations could be constructed which would fit the data as well or perhaps better, so this expression is mentioned just as a convenient description.

We now need a physical model to explain the origin of this general trend, whose details may vary with sample preparation and which may take a different form for clamped single crystals where $x \gg 1$. The linear trend in unclamped powders with $x \ll 1$ is surprising, because the natural explanation for the monotonic increase of $f_D(0)$ with L is that in larger samples it is easier to relax the elastic stress generated by defects, so that in smaller samples there is a reduced volume fraction of the sample containing the native defects which pin the Fermi energy (IV.5), produce the high T_c through their contribution to $N(E_F)$ (IV.7,8), are responsible for the normal-state transport anomalies (especially the freeze-out of the nominal carrier density $n_H(T)$) (IV.9), and so on. However, effects associated with internal stress involve harmonic long-range (i.e., the η length scale) forces and so ordinarily would scale with x^2 for small x. This classical model would break down in the presence of stress generated by defects pinned at the Fermi energy, and such a quantum-degeneracy probably would produce a linear dependence on x.

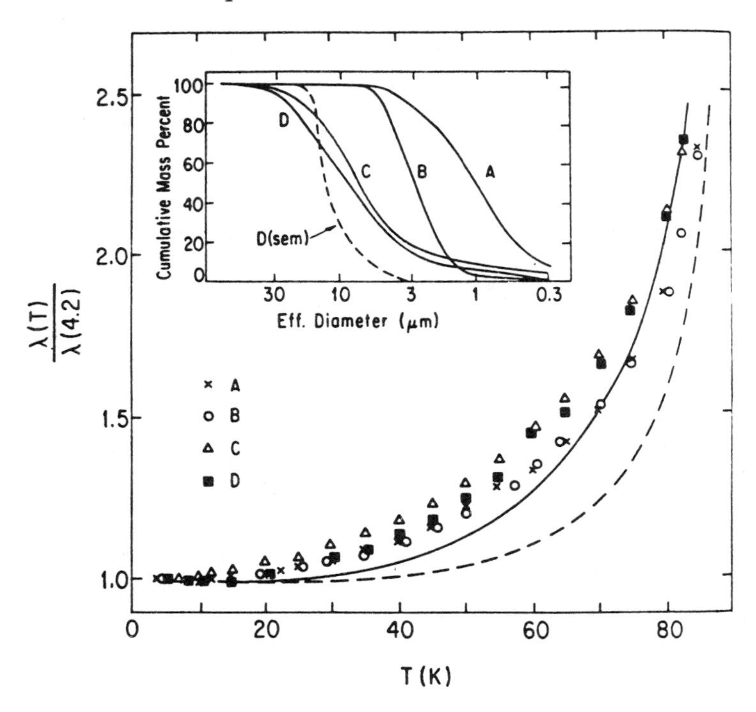

Fig. X.9. Powder size distribution and reduced penetration depth $\lambda(T)/\lambda(4.2)$ in $YBa_2Cu_3O_7$ inferred from magnetic susceptibility measurements (ref. 19).

We will discuss possible meanings of the characteristic length $\eta = 6\mu$ in X.7. Already we can see that a fractional Meissner effect has far-reaching implications for the internal superconductive morphology of powders. Temperature-dependent penetration depths $\lambda(T)$ can either be measured directly from the temperature dependence of the susceptibility,[19] giving the results shown in Fig. 9, or indirectly via muon spin-resonance[20] which is relaxed by interactions with the field gradient in a flux vortex, Fig. 10. These two methods respond differently to the anisotropy of λ (see below), but the susceptibility data are also more sensitive to grading of the superconductive coupling strength or a spatial distribution of $E_g(0)$, the energy gap at $T = 0$. Such a spatial distribution may occur even when $f_D(T < T_c) = 1$, but it most certainly occurs when $f_D(0) < 1$. Moreover such a grading would produce deviations in $\lambda(T)$ from the BCS function which would increase $\lambda(T)$ most in the range near $T = 60K$, for example, if the oxygen-deficient $T_c = 60K$ phase (III.5) of $YBa_2Cu_3O_{6.7}$ were present as a minority constituent of the powder sample. Thus the small deviations from BCS shown in Fig. 9, which are largely absent from Fig. 10, are indirect evidence for gap grading which must be present from general considerations. (Spectroscopic evidence for gap grading was seen in Raman scattering data on single crystals in VI.3.) Finally, even with the complications associated with gap grading, the data in Fig. 10 indicate that we are dealing with electron-phonon coupling in the (strong, dirty) limit.[21]

Because H_{cl} is defined in terms of initial flux penetration, while H_{c2} is defined in terms of final disappearance of zero resistance, the effect of gap grading on H_{cl} tends to be complementary to that for H_{c2}. Moreover, zero resistance is measured perpendicular to the field while flux penetration takes place parallel to the field. Differences in definition of flux penetration thus can change

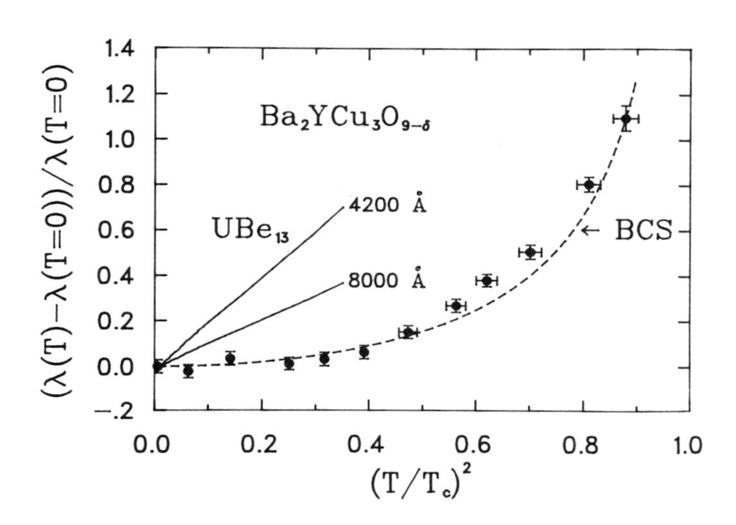

Fig. X.10. Incremental magnetic field penetration depth in $YBa_2Cu_3O_{7-x}$ deduced from muon-spin relaxation. For comparison the same function is shown for the non-s-wave superconductor UBe_{13} (ref. 20).

measured values of H_{cl} by an order of magnitude,[22] and it is not clear that the lower values have not been overly influenced by low-T_c inclusions in a predominantly high-T_c sample. Depending on definition, anisotropy ratios for H_{cl} range from 5 to 10, compared to 10 for H_{c2}. Anisotropy ratios for H_{cl} and H_{c2} are the same in anisotropic Ginsberg-Landau theory, and given the gap grading problem, the H_{cl} data do not seem to be in conflict with this theory. Perhaps the most important point is that all authors agree that $\kappa = \lambda/\xi >> 1$ (typically 30-200).

An anisotropy ratio for H_{cl} of about 10 implies a corresponding anisotropy in $\lambda(T)$ which can explain[19] the discrepancy in $\lambda(0)$ between the susceptibility and μ spin measurements, which range between 0.7 and 0.15 μ respectively. The former estimate should also be corrected because $f_D(0) < 1$, but this correction is difficult to make in the absence of detailed morphological information. Such a correction would appear to reduce the estimated value for $\lambda(0)$, thus also reducing the discrepancy between the two approaches. In any event, the functional form of $\lambda(T)$ is close enough to that given by BCS for s-wave pairing, after allowance has been made for grading, that one can say[20] that the high-T_c cuprates have s-wave pairing and not p- or d-wave pairing, as is thought to be the case for some rare earth intermetallic superconductors (such as UB_{13}). Of course s-wave pairing is much stronger than $\ell \geq 1$ pairing, so with T_c in the cuprates 10^3 times larger than in UBe_{13}, one would be amazed if the cuprate pairing were not s-wave.

Another interesting point is the comparison between $\lambda(0)$ as measured in $YBa_2Cu_3O_7$ compared to $(La, Sr)_2 CuO_4$ powders when both are measured by the same method (muon spin relaxation).[20,23] The value of λ in the latter is about twice as large as in the former, suggesting a smaller carrier concentration in $La_{2-x}Sr_xCuO_4$ which may well be approximately of order x electrons/Cu layer, for $0.1 < x < 0.2$.

6. Twin Boundaries and Giant Flux Creep

For x near 0 and T_c near 90K, samples of $YBa_2Cu_3O_{7-x}$ are orthorhombic, not tetragonal. As a result most samples of $YBa_2Cu_3O_7$ are twinned. The spacing of the twins varies[24] from about 2000Å in polycrystalline ceramics, to 750Å in single crystals, to 350Å in epitaxial films with c axis normal to the $SrTiO_3$ planar substrate. The small spacing in the latter may be attributed to the large lattice mismatch (~2%) between the a axis of the film and the substrate. Twinning limits stress accumulation caused by this mismatch. The free surfaces of grains also facilitate stress relief, and electron channeling studies[25] indicate that these internal stresses (whose thermochemical aspects are discussed in X.7) may be relieved in grains by cracking to form subgrains, misaligned by $\sim 1°$, ranging in size from 3 to 20 μ. (Compare to $L = 6\mu$ in X.5 in discussing fractional diamagnetism.) Finally it is noted that TEM studies seem to give smaller values for twin spacings (200-800Å) than are seen by optical microscopy ($\sim$2000Å) over samples stressed (and damaged) by thinning.[25]

These mechanical aspects of sample geometries are often observed[24] to be correlated with critical current densities J_c. The origin of these correlations can be explained[26] by the assumption that a layer centered on the twin boundary has a transition temperature T_{c1} while the interior of the twin has a different transition temperature T_{c2}. Many of the observed properties are qualitatively independent of the sign of $\delta T = T_{c1} - T_{c2}$ and depend only weakly on its magnitude. For the moment assume that $T_{c2} > T_{c2}$, so that T_c is enhanced in the neighborhood of the twinning plane, as has been observed in Sn and Nb,[27] where $|T_{c1} - T_{c2}|/T_c \sim 0.01$. This is similar to an n-s-n sandwich with the thickness t of the s layer given by $t \sim \xi_o |\delta T_c/T_c|^{1/2}$. Then the Ginsberg-Landau equations shown that $H_{c2\|} \sim (1 - T/T_{c1})^{1/2}$.

Because of the large anisotropy of H_{c2} this functional dependence can be observed only in very homogeneous single crystals or in grain-aligned polycrystalline samples, the latter method being used to produce the convincing result[26] shown in Fig. 11 for $YBa_2Cu_3O_{7-x}$. With $\tau_o = |\delta T_c|/T_c$ one finds more exactly $\tau_o(\tau_o + 2) \sim \xi_{o\|}^4$, so that with $\xi_{o\|} = 34$Å one obtains $T_{c2} = 79$K. This means that $|\delta T_c|/T_c \sim 0.15$, much larger than in Sn or Nb. A possible reason for this large grading is discussed in X.7.

One way of distinguishing the sign of $\delta T = T_{c1} - T_{c2}$ is by a careful measurement of $J_{c\|}(T)$ in zero magnetic field.

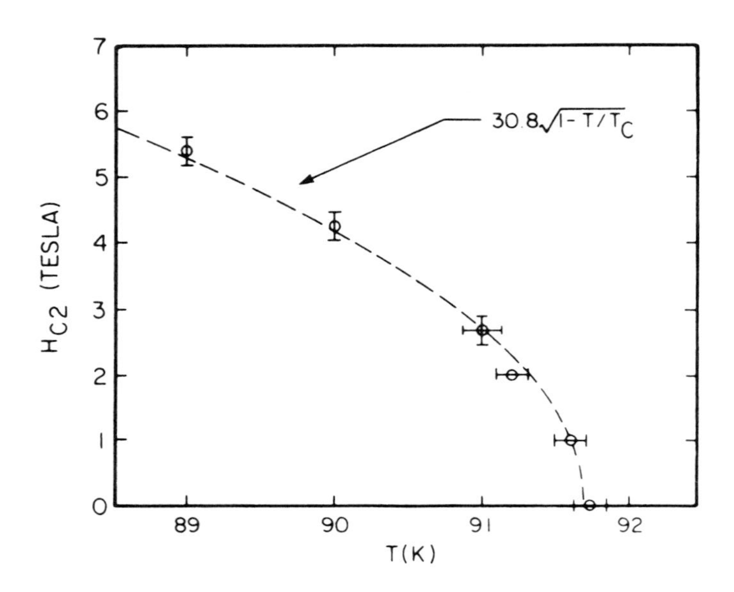

Fig. X.11. Measurement of $H_{c2}(T)$ using a polycrystalline grain-aligned sample of $YBa_2Cu_3O_7$, showing proportionality to $(1-t)^{1/2}$ very near T_c (ref. 26).

If $\delta T < 0$ the twin planes isolate the twin domain interiors by s-n-s junctions and one expects[28] $J_c \propto (T_{c2} - T)^2$. If $\delta T > 0$ the twin planes form a network of thin superconductive films which have $J_c \propto (T_c - T)^{3/2}$. It appears that the latter is observed.[29]

At this writing all resistive measurements using magnetic fields near T_c are being re-examined to analyze the hysteretic effects of flux creep in the mixed state $H_{c1} < H < H_{c2}$, which are very large in single crystals.[30] The relaxation of magnetization is thermally activated, usually giving a quasi-linear dependence on lnt, where $t = T/T_c$. In terms of an activation energy U the relaxation of the magnetization M in a cylinder of radius r is given by $dM/d\ln t = (rJ_c/3c)(kT/U)$ where the measurement is in a critical state such that $J = J_c$. With $H_{c1} < H = 1kg$ parallel to the c axis, this equation is reasonably well satisfied with only small deviations, as shown in Fig. 12. (At low T the factor of T is important, and near T_c one has J_c going to zero.) More quantitatively, the deviations can be fitted by assuming $J_c/U \propto (1-t)^2$ with $U_{o\parallel} = 0.6eV$ and $U_{o\perp} = 0.1eV$. The larger value for H‖c suggests that the twin planes may effectively pin H when it is ‖c and can lie between these planes. The angular average of U_o is 0.15eV, which is more than a factor of 10 smaller than the pinning energy typically found for intermetallic superconductors (such as Nb_3Sn).

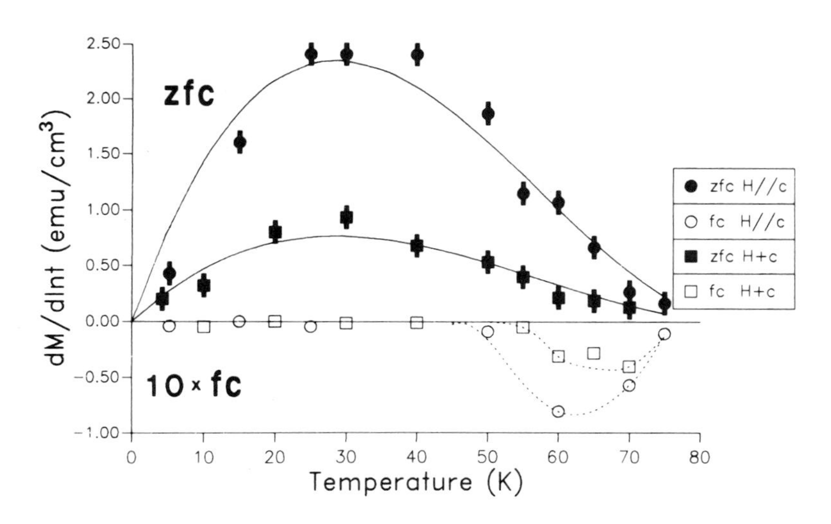

Fig. X.12. Relaxation rate of field-cooled (fc) and zero-field cooled (zfc) for fields ‖ and ⊥ c axis of an $YBa_2Cu_3O_7$ crystal. Solid lines are fits with $J_c/U \propto (1-t)^2$, see text (ref. 30).

The field and temperature dependence of U are probably determined by material-specific interactions. A scaling argument[30] gives

$$U = CH_c^2 \xi \Phi_o / H \tag{1}$$

where C is a pure number but need *not* be of order unity. With Ginsberg-Landau temperature dependences $H_c \propto (1-t)$ and $\xi \propto (1-t)^{-1/2}$, substitution of (1) into the formula for thermally activated critical currents

$$J_c = J_{co}[1 - (kT/U)\ln(Bd\Omega/E_c)] \tag{2}$$

(where d is a distance between pinning centers, Ω is an

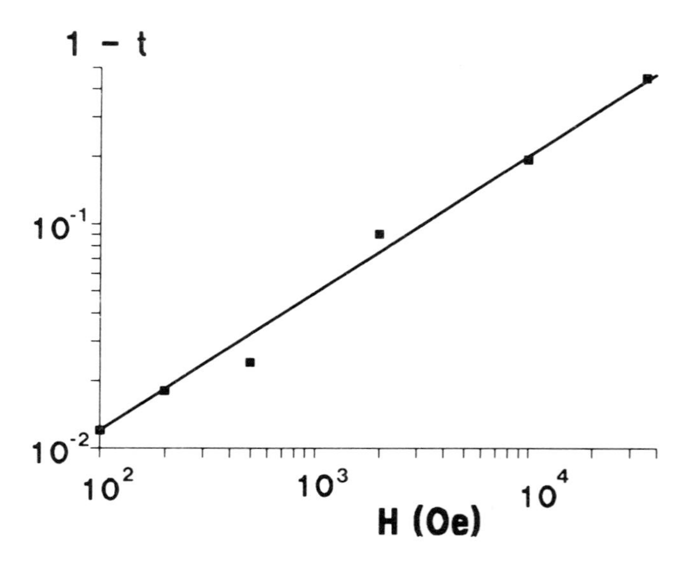

Fig. X.13. Hysteresis or irreversibility line for H||c for an $YBa_2Cu_3O_7$ crystal (ref. 30).

oscillation frequency of a flux line in a pinning well, and E_c a minimum measurable voltage), yields a condition for zero J_c:

$$1 - t \propto H^{2/3}. \qquad (3)$$

This condition describes the slope of the irreversibility line for hysteretic behavior shown in Fig. 13. Fitting C to this line gives $C \sim 5 \cdot 10^{-3}$, which is apparently consistent with flux line pinning by twin planes.[30]

An alternative physical interpretation for U can be obtained by transforming U by the Ginsberg-Landau equations, which shows[31] that $U = 2E_J$, where E_J is the Josephson coupling per vortex line maintaining phase continuity for spacings ξ along it, or $U \propto J_c/H$. Using this Josephson analogy the shot noise resistance associated with thermally activated, field-assisted flux-line motion can be calculated to be[31]

$$R/R_n = [I_o[A(1-t)^{3/2}/2H]]^2 \qquad (4)$$

where I_o is a modified Bessel function.

The scaling properties of (4) immediately explain the field-induced broadening of the resistance transition seen previously in Figs. 5 and 6, and also explain the positive curvature of $H_{c2} \propto (T_c - T)^{3/2}$, instead of $(T_c - T)$, as in the Ginsberg-Landau theory. The actual functional form of (4) gives an excellent one-parameter fit of the field- and temperature-dependence, as shown in Fig. 14.

Detailed numerical analysis of (2) and (4) shows[30,31] that because kT/U is $\sim 10^2$ larger in high T_c cuprate materials than in intermetallics, J_c is severely limited by giant flux

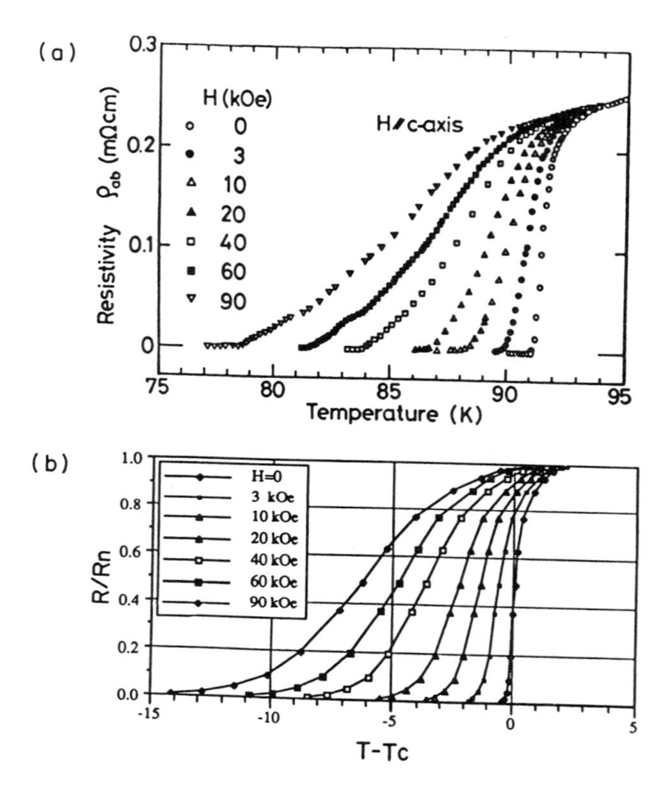

Fig. X.14. (a) Experimental $\rho_{\parallel}(T, H)$ for $YBa_2Cu_3O_7$ crystal, and (b) One-parameter fit to (a) using Eqn. (4) (ref. 31).

creep and its accompanying equivalent resistance. For technological purposes a better understanding of the pinning energy U and its microscopic origin is important. At present it is tempting to try to relate U to some property of the twinning planes.

7. Microscopic Inhomogeneities

While the ternary phase diagram, Fig. X.3, seems complicated enough, it is still only a small part of a macroscopic description of the chemistry and structure of $YBa_2Cu_3O_{7-x}$ in its highest-T_c orthorhombic phase. When the component oxides are first combined to form a polycrystalline sample, usually $x = 0.7$ and the sample is semiconductive. The sample is made metallic and superconductive by sintering at $T \gtrsim 900C$ at an oxygen pressure $P^a_{O_2}$, and is then slowly cooled at a different oxygen pressure $P^b_{O_2}$, transforming from tetragonal to orthorhombic near 700C when $P^a = P^b = 1$ Torr. A more general diagram[32] is shown in Fig. 15. In the orthorhombic phase

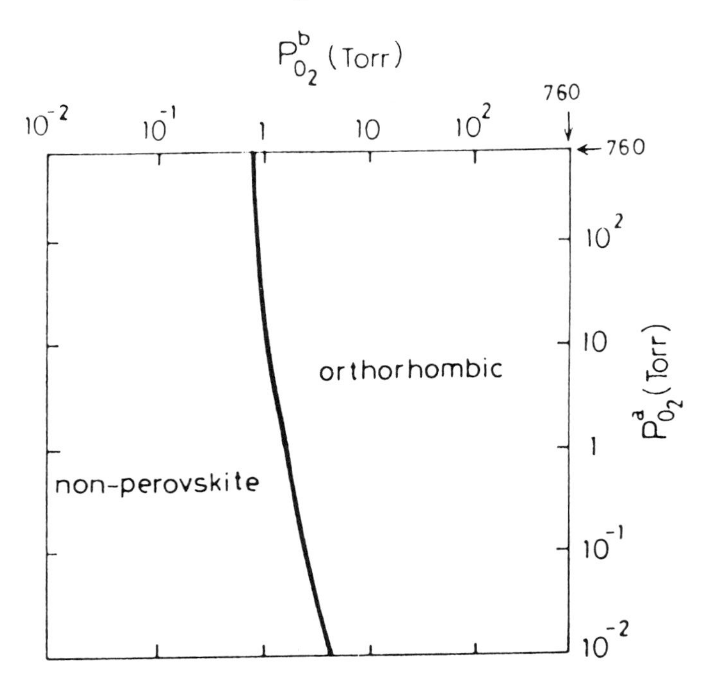

Fig. X.15. Sintering oxygen vs. cooling oxygen partial pressures, showing a clear phase line (ref. 32).

the lattice constants and T_c vary significantly with P^a and P^b.

The internal chemistry and structure of a given sample-polycrystalline, single crystal, or epitaxial film - are still not exhausted for suitable analogues of Figs. 3 and 15, because the oxidation stages are carried out kinetically in the presence of an oxygen gradient. This anion gradient induces [A, B] cation counter-gradients in our pseudoperovskite crystal $A_3B_3O_{9-\delta}$, where $A_3 = YBa_2$ and $B = Cu$. These lead to Cu drift towards the additional oxygen, and Y, Ba drift away. Under extreme conditions new phases nucleate in opposite directions as the sample decomposes,[33] and these new phases can be identified by diffraction.

While the oxidation-induced phase separation just described is observable only under extreme conditions, it is important to realize that to a much smaller degree (not observable by diffraction as the formation of new phases) the [(Y, Ba), Cu] cation counter-ion spinodal diffusion must always take place during oxidation. We saw in III.6 that even a 1% change in Cu site occupancy can change T_c by 20K. Such a small change in site occupancy would be completely undetectable by present diffraction methods, but it would lead to a large grading of T_c.

Let us now imagine a possible scenario for oxygen diffusion in epitaxial films, which are also subjected to annealing in oxygen to increase T_c. Even the best epitaxial films necessarily have twin planes, and the complex atomic arrangements at these planes provide a greater variety of diffusion paths than are available inside a single domain. Thus it seems likely that some of these twinned paths present lower activation energies at the interface for O_2 diffusion than are found inside a single orthorhombic domain, which means higher equivalent P_{O_2} at the twin planes, which become Cu-enriched. This may limit T_c enhancement, but not so much as away from the twin planes, where very unfavorable (Y, Ba) enrichment occurs. Thus a large grading of T_c results, with maximum T_c near (but not necessarily at) the twinning planes. This is consistent with many of the observations discussed in X.6.

At this point it is worth remembering the observation (VI.3) that the softening of the 335 cm^{-1} phonon takes place over a temperature range which is 10 times narrower in an untwinned single crystal of YBCO then in a twinned one. Thus the gap grading is strongly coupled to the twin planes, presumably through microscopic inhomogeneities. In effect, the spinodal waves can be pinned to the twin planes.

When T_c is strongly graded, it is difficult to estimate the filling factor f associated with the high-T_c phase. As an extreme case, suppose that the high-T_c phase is located only at the twin planes which are spaced $\bar{L}$ apart in an epitaxial film. Then the fractional volume f of highest T_c material is about $f = 2\xi_o/\bar{L}$, which with $\xi_o = 35Å$ and $\bar{L} = 350Å$ is only 0.2. When the specific heat jump ΔC_p associated with this volume is normalized to the entire volume, the apparent value of ΔC_p is reduced by about a factor of 5. However, the difference between the weak-coupling (BCS) value of $\Delta C_p/\gamma T_c$ and the maximum value calculated for strong coupling ($\lambda \sim 3.5$) in Fig. II.8 is at most a factor of 7. We can try to increase this filling factor by varying the oxidation procedure, but if the high T_c phase is to remain connected, this approach is severely limited. Taken together with uncertainties in the electronic density of states factor $N(E_F) \propto \gamma$, in band theory, in Fermi-level pinned defect contributions, and in electron-phonon (m^*/m) enhancement factors, these uncertainties produce severe limits on the microscopic information which can be extracted from observed values[34,35] of ΔC_p. The observed variations in $N(E_F)$, as inferred from normal-state susceptibility measurements on different samples,[35] while T_c remains constant, are consistent with a graded gap model.

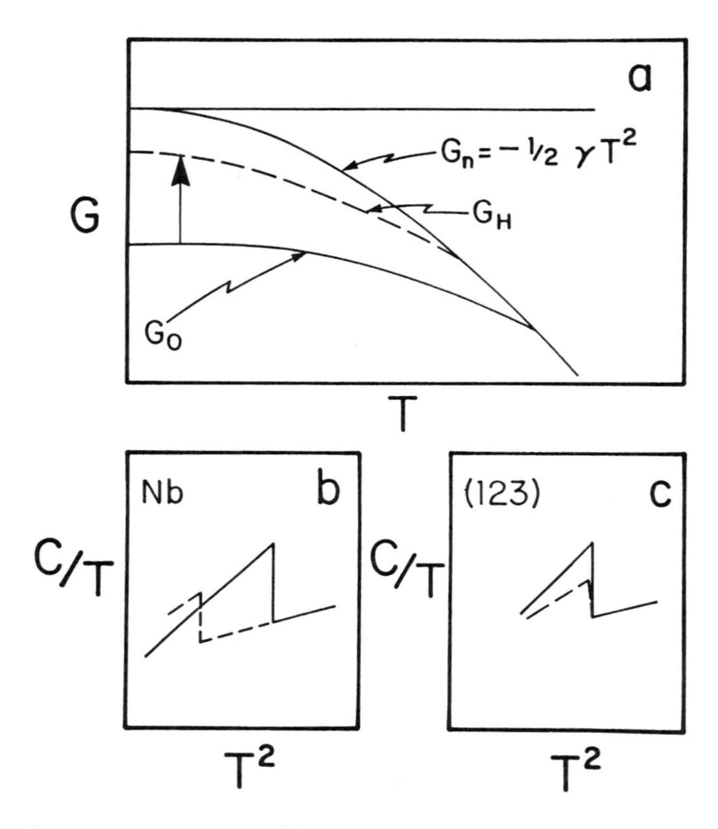

Fig. X.16. Sketches of (a) Gibbs free energy in normal state n, superconducting state G_0, and mixed state G_H; (b) Specific heat jump $\Delta C_p(H)$ in Nb, and (c) in YBCO (ref. 36).

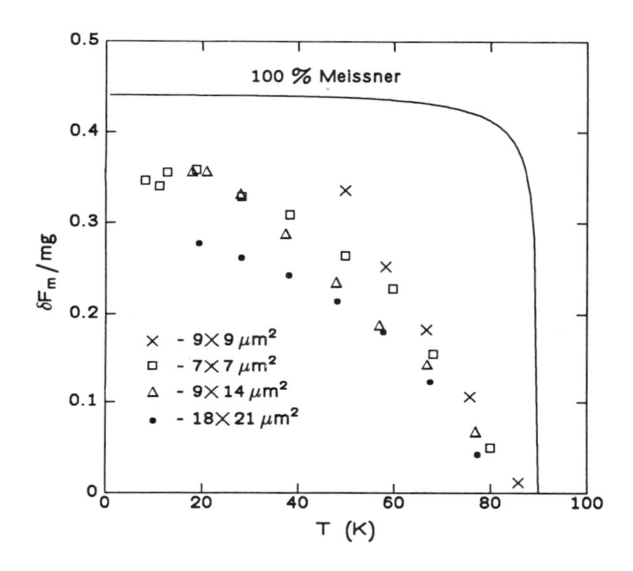

Fig. X.17. Temperature dependence at H= 360g of four grains with areas indicated. The solid line represents 100% Meissner flux exclusion (ref. 37).

Another example of graded gap effects is the variation of the specific heat jump $\Delta C_p(H)$ with magnetic field H for $H > H_{c1}$, that is, in the mixed state.[36] The free energy curves are shown in Fig. 16a, with the behavior of a homogeneous type II superconductor (such as Nb) in Fig. 16b. For fixed field $H_{c1} < H < H_{c2}(O)$, the jump in the specific heat shifts from T_c at $H = 0$ to T defined by $H_{c2}(T) = H$. However, in YBCO the actual behavior is that seen in Fig. 16c (grain-aligned sample). The jump remains at $T = T_c$ but decreases in magnitude as H increases, which seems quite surprisingly different from Fig. 16b. However, in the context of graded gaps, this is just what we would expect. So long as the filling factor f is small enough, the external field will be squeezed into the regions where

$T_c << 92K$ which are normal for $T \sim 90K$. The superconductive regions will have $H = 0$ and so the jump in T_c will remain at 92K. In fact not all the field will be so confined, so some of the superconductive fraction will disappear, $f = f(H)$ will decrease as a function of H. This is exactly what is observed.

It is important to realize that there is nothing strange or mysterious about these observations in a complex oxide. What would have been really surprising would have been to see homogeneous behavior *without* a graded gap! This point becomes especially clear in the context of recent measurements[37] of H_{c1} in single grain particles which are approximately spheroidal with cross-sectional areas as indicated in Fig. 17. The measurements were carried out on individual grains confined by an electromagnetic trap. The restoring force needed to balance gravity is shown in Fig. 17, where it is compared with the force predicted for a homogeneous particle with a normal Meissner effect. The fraction of the particle which remains superconductive is roughly measured by the ratio of H_{c1} (exp) relative to H_{c1} (normal), and it is clear that this decreases rapidly with T. The most natural interpretation of these data is provided by the graded gap model which we have discussed.

REFERENCES

1. S. Jin et al., Appl. Phys. Lett. *52*, 2074 (1988).

2. P. R. Broussard and S. A. Wolf, J. Cryst. Growth, Aug., 1988. See also J. Cryst. Growth, Dec., 1987. Some representative examples: $SrTiO_3$ substrate, evaporation: P. Chaudhari, R. H. Koch, R. B. Laibowitz, T. R. McGuire, and R. J. Gambino, Phys. Rev. Lett. *58*, 2694 (1987); MBE, J. Kwo et al., Phys. Rev. *B36*, 4039 (1987); BaF_2 source, oriented films, P. M. Mankiewich et al., Appl. Phys. Lett. *51*, 1753 (1987); sputtering, H. Adachi, K. Hirochi, K.

Setsune, M. Kitabatake, and K. Wasa, Appl. Phys. Lett. *51*, 2263 (1987); dual sputtering, P. Madakson, J. J. Cuomo, P. S. Yee, R. A. Roy, and G. Seilla, J. Appl. Phys. *63*, 2046 (1988); laser ablation, X. D. Wu et al., Appl. Phys. Lett. *51*, 861 (1987).

3. Y. Hidaka, Y. Enomoto, M. Suzuki, M. Oda and T. Murakami, J. Cryst. Growth *85*, 581 (1987).
4. S. Hirano and S. Takahashi, J. Cryst. Growth *79*, 219 (1986); *85*, 602 (1987).
5. J. Takahashi et al., Phys. Rev. *B37* 9788 (1988); C. Changkang, B. E. Watts, B. M. Wanklyn and P. Thomas, Sol. State Comm. *66*, 441 (1988).
6. R. S. Roth, J. R. Dennis and K. C. Davis, Adv. Ceramic Mat. *2*, 303 (1987).
7. L. F. Schneemeyer et al., Nature *328*, 601 (1987).
8. H. Maeda, Y. Tanaka, M. Fukutomi and T. Asano, Jap. J. Appl. Phys. *27* L209 (1988).
9. M. A. Subramanian et al., Science *239*, 1015 (1988).
10. L. F. Schneemeyer et al., Nature *332*, 422 (1988); S. Nomura et al., Jap. J. Appl. Phys. *27*, L1251 (1988).
11. M. A. Subramanian et al., Nature *332*, 420 (1988).
12. D. St. James, G. Sarma and E. J. Thomas, *Type II Superconductivity* (Pergamon, Oxford, 1969); W. H. Butler, Phys. Rev. Lett. *44*, 1516 (1980).
13. Y. Tajima et al., Phys. Rev. *B37*, 7956 (1988).
14. B. J. Dalrymple and D. E. Prober, J. Low Temp. Phys. *56*, 545 (1984).
15. R. V. Coleman, G. K. Eisenman, S. J. Hillenius, A. T. Mitchell and J. L. Vicent, Phys. Rev. *B27*, 125 (1983).
16. G. Wexler and A. M. Woolley, J. Phys. *C9*, 1185 (1976).
17. N. J. Doran, Physica *99B*, 227 (1980).
18. P. L. Gammel et al., Phys. Rev. Lett. *59*, 2592 (1987).
19. J. R. Cooper, C. T. Chu, L. W. Zhou, B. Dunn and G. Grüner, Phys. Rev. *B37*, 638 (1988).

20. D. R. Harshman et al., Phys. Rev. *B36*, 2386 (1987).

21. J. Rammer, Europhys. Lett. *5*, 77 (1988).

22. T. R. Dinger, T. K. Worthington, W. J. Gallagher, and R. L. Sandstrom Phys. Rev. Lett. *58*, 2687 (1987); A. Umezawa et al. (unpublished).

23. G. Aeppli et al., Phys. Rev. *B35*, 7129 (1987). For magnetic susceptibility, see T. Ishida and H. Mazaki, Jap. J. Appl. Phys. *27*, L199 (1988).

24. C. H. Chen, J. Kwo and M. Hong, Appl. Phys. Lett. *52*, 841 (1988).

25. J. D. Verhoeven, E. D. Gibson, L. S. Chumbley, R. W. McCallum, and H. H. Baker (unpublished Ames).

26. M. M. Fang et al., Phys. Rev. *B37*, 2334 (1988).

27. I. N. Khlustikov and A. I. Buzdin, Adv. Phys. *36*, 271 ((1987).

28. G. Deutscher and K. A. Müller, Phys. Rev. Lett. *59*, 1745 (1987).

29. H. Watanabe et al., Jap. J. Appl. Phys. *26*, L657 (1987); S. B. Ogale, D. Kijkkamp, T. Venkatesan, X. D. Wa and A. Inam, Phys. Rev. *B36*, 7210 (1987).

30. Y. Yeshurun and A. P. Malozemoff, Phys. Rev. Lett. *60*, 2202 (1987).

31. M. Tinkham, Phys. Rev. Lett. *61*, 1658 (1988); M. Tinkham and C. J. Lobb, Solid State Physics *42* (1989).

32. I. Iguchi et al., Jap. J. Appl. Phys. *27*, L992 (1988).

33. P. K. Gallagher, G. S. Grader, and H. M. O'Bryan, Proc. 11th Int. Symp. React. Sol. (Princeton, 1988).

34. N. E. Phillips et al., in *Novel Superconductivity* (Eds. S. A. Wolf and V. Kresin, Plenum, N.Y. 1987), p. 739.

35. A. Junod, A. Bezinge and J. Muller, Physica *C152*, 50 (1988).

36. K. Athreya, O. B. Hyun, J. E. Osterson, J. R. Clem and D. K. Finnemore, Phys. Rev. *B* (1989).

37. G. S. Grader, A. F. Hebard, and R. H. Eick, Appl. Phys. Lett.

XI. Bismates and Thallates

Following the discovery of superconductivity with $T_c > 90K$ in $YBa_2Cu_3O_7$ many scientists searched for other materials with tetragonal crystal structures containing CuO chains or CuO_2 planes. The first new family[1] is based on Bi_2O_2 layers which are lattice-constant matched to CuO_2 planes. As discussed further in Appendix C, the parent compounds[2] for this family are $Bi_4Ti_3O_{12}$, Bi_2MoO_6 and Bi_2WO_6, all of which fall in island C and contain Bi_2O_2 layers, as shown in Fig. 1. These layers are natural templates for CuO_2 planes, but because CuO_2 planes are

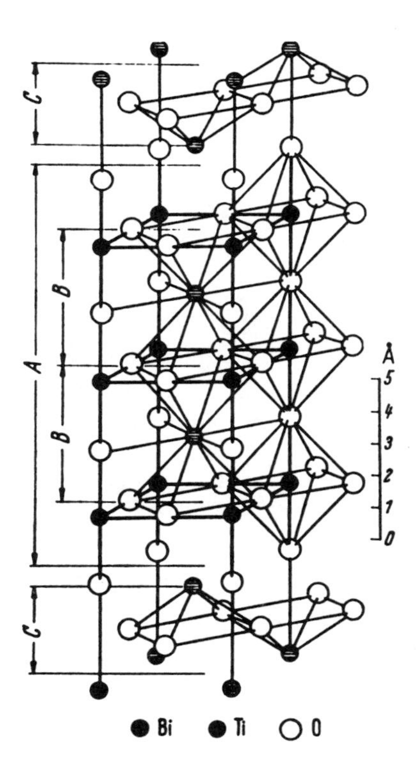

Fig. XI.1. The structure of $Bi_4Ti_3O_{12}$. Layers of perovskite-like $Ti_3Bi_2O_8$ are sandwiched between Bi_2O_2 layers (ref. 2).

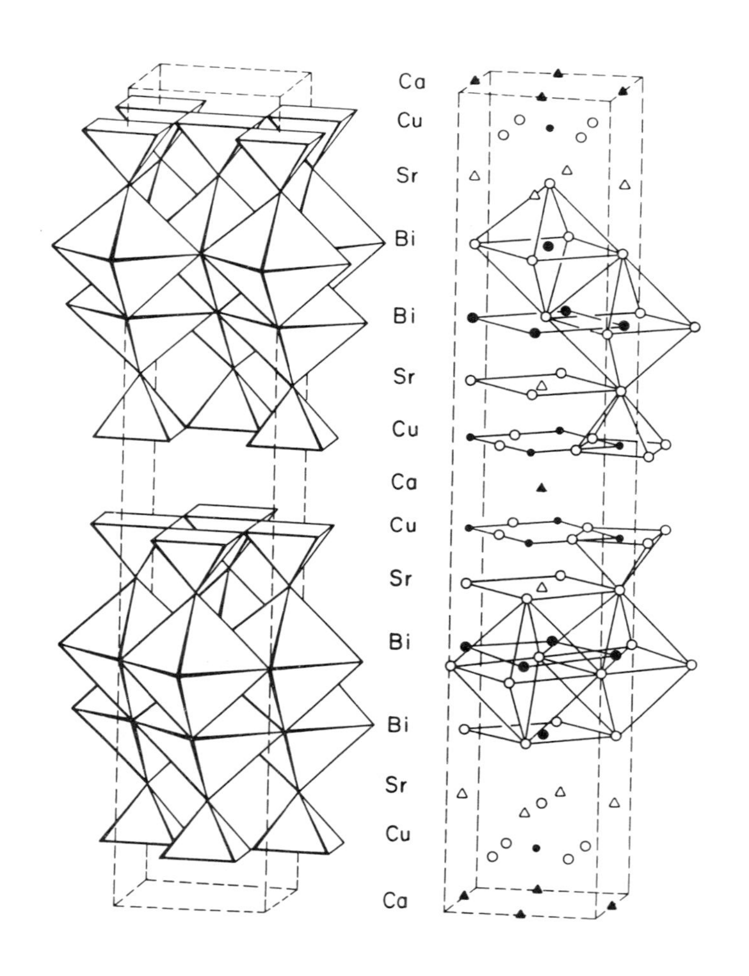

Fig. XI.2. The structure of $Bi_2Sr_2CaCu_2O_8$, sometimes called Bi(2212) (ref. 3).

electron-rich, alkaline earth planes (such as BaO) must added to balance the electron count. However, the epitaxial matching is very delicate as well, because planes of oxygen vacancies $O^{\square}$ also are necessary to balance the stoichiometry. The planar cation order of one of these compounds has been assigned[3] as ($Ca_{0.8}Bi_{0.2}$), Cu, Sr, Bi, Sr, Cu, or formally $Bi_{2.2}Sr_2Ca_{0.8}Cu_2O_{8+\delta}$, as shown in Fig. 2. Every cation plane contains (nearly) coplanar oxygen ions, except for the $Ca_{0.8}Bi_{0.2}$ plane. Actually the cation planes are strongly buckled by modulation displacements,[4] as shown in Fig. 3. Normal to the planes these displacements are 0.2A (Bi, Sr) and 0.3A (Cu). Once again this is dramatic evidence for static distortions which may well be accompanied by dynamical distortions which we have described as fictive phonons (IV.13 and VI.3).

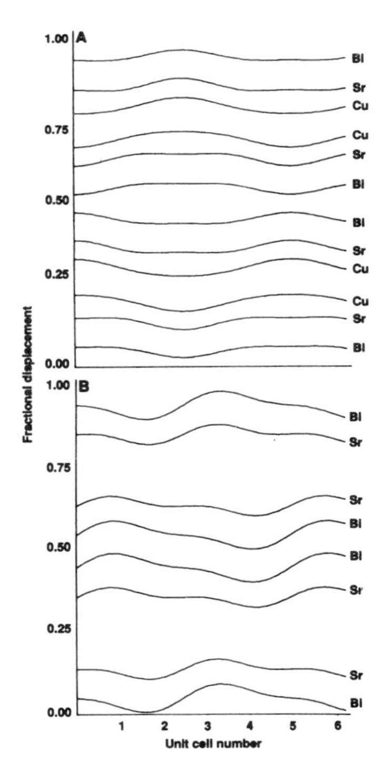

Fig. XI.3. Vertical (c axis, **A**) modulation displacements (along a axis, **B**) of cation layers (relative scale) of Fig. 2 (ref. 4).

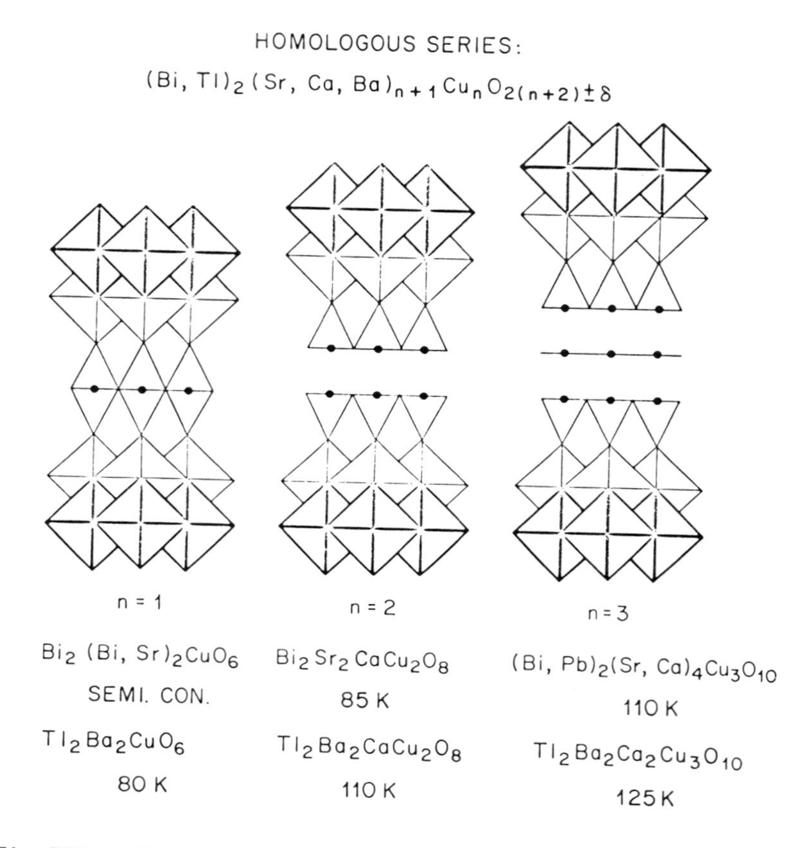

Fig. XI.4. Trends in transition temperature in the homologous tetragonal (Bi, $T\ell$) layer compounds (ref. 8).

In the bismate compound the formal valence of Bi is [3+]. This naturally suggests replacing Bi by $T\ell$, and when this is done similar compounds are obtained.[5,6] The structural differences[7] are small but interesting: the buckling of the CuO_2 sheets decreases from $YBa_2Cu_3O_7$ to $Bi_2Sr_2CaCu_2O_8$ to $T\ell_2Ba_2CaCu_2O_8$. However, at the same time the oxygen atoms in the $T\ell$ plane appear to have moved off center (along the x axis) and $T\ell$ has an apparently large in-plane vibrational amplitude. Whether this is evidence for anharmonicity or (more probably) $T\ell$ vacancies is hard to say, but if the latter, it might be evidence for increasing electronic localization and marginal two-dimensionality in the CuO_2 planes.

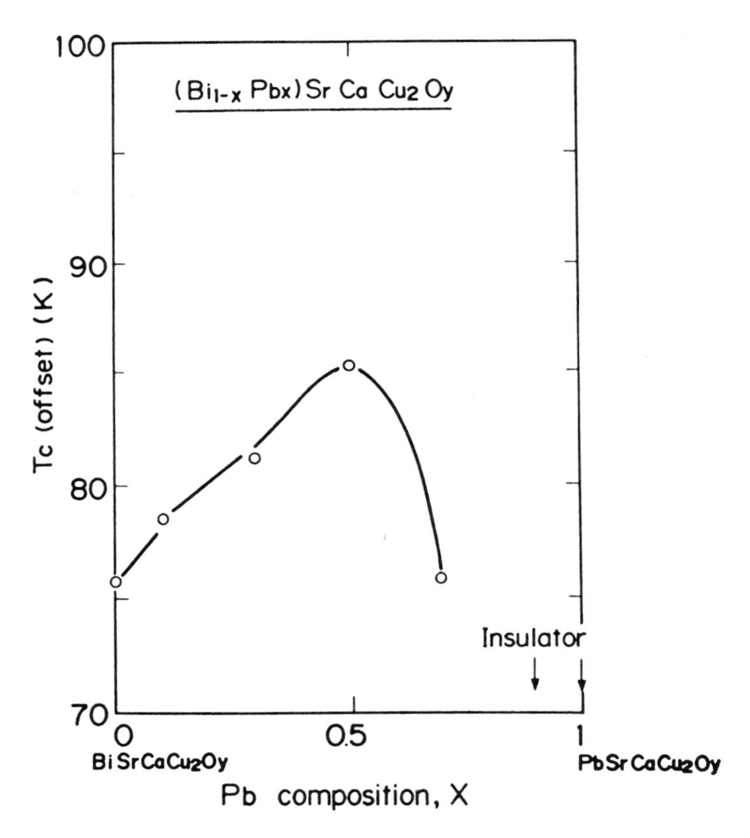

Fig. XI.5. Zero resistivity temperature T_c in Pb-doped bismate compound (ref. 9).

The stacking of superlattices which led to bilayer chains in Y-Ba-Cu-O also appears in the bismate and thallate systems, where it produces grouping of CuO_2 planes separated by Ca^{2+} layers and bounded by BaO layers. The superlattices form an interesting homologous family, shown[8] in Fig. 4 together with onset T_c's. The trends are clear: more grouped CuO_2 layers, or $T\ell$ instead of Bi, increases T_c. However, as we see in Appendix C, the limit $n \longrightarrow \infty$, without BaO boundary layers or $(T\ell, Bi)_2O_2$ layers, is semiconductive. In any event, higher T_c's in this family are achieved by more CuO_2 planes and no CuO planes. This is consistent with the result (III.7) that adding an extra CuO plane to $YBa_2Cu_3O_7$ decreases T_c from 92K to ~80K. It is also consistent with the assignment (VI.3) of the 330 cm^{-1} phonon band to transverse vibrations of O atoms in CuO_2 planes.

The alloy series $(Bi_{1-x}Pb_x)$ $SrCaCu_2O_y$ shows the characteristic increase of T_c with x (Fig. 5) associated with lattice instabilities until a phase transition to an insulating phase occurs.[9] Here again we expect reduction of planar buckling and increased localization in the CuO_2 planes as x increases.

The $T\ell$ compounds are distinguished from the Bi compounds by the fact that the Bi compounds always contain Bi double layers while the $T\ell$ compounds may contain single or double $T\ell$ layers. The latest results[10] can be described by the general formula $(B^{III})_y$ $(A^{II})_2$ Ca_{n-1} Cu_n O_z, where $B^{III} = T\ell$ or Bi, $A^{II} =$ Ba or Sr, and $y = 2$ (Bi) while $y = 1$ or 2 ($T\ell$). For $y = 1$ ($T\ell$), $n = 4$ has been synthesized.[10] The $y = 1$ or 2 ($T\ell$) series have T_c (n) as shown in Fig. 6. Here f_s is the superconductive filling factor associated with the volume of the unit cell filled by CuO_2 planes. A nice fit is obtained to the five points with two straight lines (four parameters). One could argue that the increase in T_c with n reflects increasing lattice instabilities

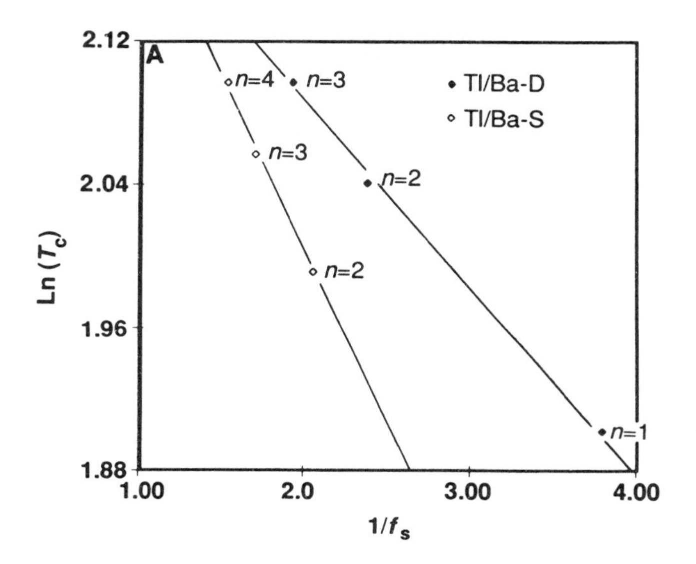

Fig. XI.6. Plots of $\ln T_c$ vs. $1/f_s$ for the two $T\ell$ series, single (S) and double (D), ref. 10.

and increasing concentration of interplanar defects. The fit appears to fail for $n = 5$, where $T_c < 120K$.

Electron microscopic studies show that a variety of phases with various numbers of Ba, Ca and Cu cation layers can form between Bi_2O_2 layers in the same sample.[11] These are referred to as intergrowth phases, and they are essentially mixed phases of the homologous series shown in Fig. 5. The effects of pressure on T_c in the bismates are comparable to other high T_c oxides.[12]

The stacking sequence usually involves double (Bi, Tℓ)O layers, but at least in the case of Tℓ only one layer is needed to produce a high-T_c superconductor. However, in the first report compared to the double-layer compounds, T_c is reduced by ~10K, and there are more indications of flux creep as $T \rightarrow T_c$ in the breadth of the resistive and Meissner transitions as well as the decay of persistent currents,[13] which are similar in Tℓ-Ba-Ca-Cu-O to YBCO.

At present little is known about the concentration of oxygen vacancies $O^{\square}$ in the (Bi, Tℓ) compounds, although one report[14] has it that T_c can be enhanced by annealing in O_2. Perhaps there are fewer $O^{\square}$, because the linear term in T in the specific heat[15] found at low T in $YBa_2Cu_3O_{7-x}$ is too small to measure in the Bi compounds, but is present in the Tℓ compounds with magnitude similar to YBCO, which is consistent with the off-site structural feature of the Tℓ compounds mentioned above. In any event, it is virtually certain that this linear term, thought to be significant in some exotic theories, is not a general feature of high-T_c oxide superconductors. Of course the simplest explanation for this term is that it is associated with a minority metallic phase which is not superconducting, but it could equally well be associated with the two-level defects which produce a "linear" term in C_p in all oxide glasses.

Electronic band structure calculations show that the CuO_2 3d-2p bands found in the other cuprate superconductors are present here as well, as one would expect.[16] The calculations also show that there may (or may not) be a small band gap between the BiO 6s-2p and 6p-2p bands, so that these layers may be semiconductive or semimetallic. At present the effects of defects in the (Bi, Tℓ) layers are likely to be more significant than these band features.

As for the energy gap in these materials, early tunneling data[17,18] with point contacts give E_g/kT_c (bulk) $\sim 5-6$ in $Bi_2CaSr_2Cu_2O_y$ for $T_c = 84K$. At the junction we expect (p. 246) that $T_{c\ell} < T_c$ (bulk), corresponding to an even higher ratio for $E_g/kT_{e\ell}$. Thus it seems that all high-T_c cuprates are in the strong-coupling limit, which of course automatically excludes exotic theories based on exciton- (rather than BCS phonon-) mediated coupling.

Preliminary measurements of maximum critical currents in epitaxial bismate and thallate films indicate large values comparable to those seen in YBCO films,[19] while the critical field anisotropy[20] factor ϵ may be ~ 20 (compared to ~ 10 in

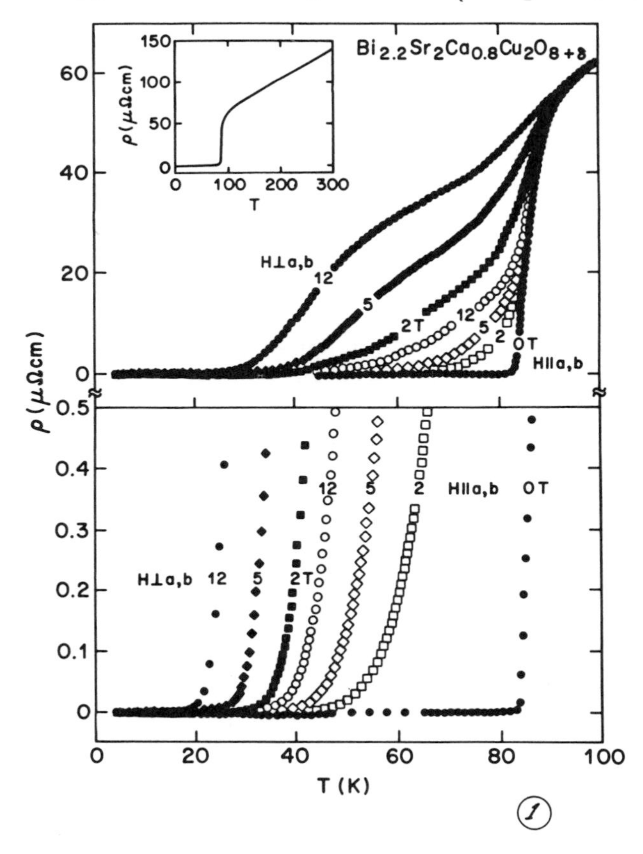

Fig. XI.7. Bismate magnetoresistive superconductive transitions, ref. 22.

YBCO). Because of the increase in the average spacing between groups of CuO_2 planes, this increased anisotropy was to be expected. The effects of flux creep on these quantities remain to be evaluated, but the magnetic broadening of the resistive transition[21] is at least as large as in YBCO.

Measurements of the temperature and field dependence of the resistivity $\rho(T, H)$ in the threshold region $10^{-6} \lesssim \rho/\rho_n \lesssim 10^{-2}$ for $Bi_{2.2}Sr_2Ca_{0.8}Cu_2O_{8+\delta}$ ("2212"), see Fig. 7, show[22] quite different scaling behavior than was found in Eqn. (X.1) for YBCO with $\rho \sim \rho_n/2$. This is not surprising, since near threshold the resistivity is certainly percolative and the appearance of ρ is dominated by the *creation* of resistive links ("Josephson junctions"). As one would expect, $U_{\parallel} \neq U_{\perp}$, but in $\rho = \rho_o \exp(-U_\alpha/kT)$ the prefactor ρ_o is independent of field orientation, suggesting that the important weak links are being created in the ab plane, and that near threshold the percolation is fully three-dimensional regardless of the field orientation. Clearly a variety of scaling "laws" can be expected, depending on the configurational correlations within the defect ensemble whose fluctuations give rise to the formation of weak links. All of this serves to illustrate the importance of defects which both pin the Fermi energy and (together with fictive phonons) can greatly enhance T_c (IV.10). More specifically, we note that the magnetic broadening of the bismate resistivity shown in Fig. 7 shows two regions, which may be associated with axial and planar weak links, respectively, with the threshold behavior associated with thermally activated axial weak links.

Perhaps one of the most remarkable aspects of high-T_c cuprates is the large variety of preparative techniques which have been utilized with comparable success. For the (Bi, $T\ell$) compounds e-beam evaporated 200Å films[20] and dense bulk glasses[22] both yield $T_c \gtrsim$ 80K-110K. Apparently replacing the $T\ell_2O_2$ planes by $T\ell O$ planes[23] reduces T_c drastically to ~ 20K. Stacking faults also reduce[24] T_c.

REFERENCES

1. M. Maeda, Y. Tanaka, M. Fukutoma, and T. Asano, Jap. J. Appl. Phys. *L209* (1988).
2. J. F. Ackerman, J. Sol. State Chem. *62*, 92 (1986).
3. S. A. Sunshine et al., Phys. Rev. *B 38*, 893 (1988).
4. Y. Gao, P. Lee, P. Coppens, M. A. Subramanian and A. W. Sleight, Science *241*, 954 (1988).
5. Z. Z. Sheng and A. M. Hermann, Nature *332*, 55, 138 (1988).
6. R. M. Hazen et al., Phys. Rev. Lett. *60*, 1174 (1988); M. Kikuchi et al., Jap. J. Appl Phys. *27*, L 1050 (1988).
7. M. A. Subramanian et al., Nature *332f*, 420 (1988).
8. L. F. Mattheiss (unpublished).
9. Y. Yamada and S. Murase, Jap. J. Appl. Phys. *27L*, 996 (1988).
10. P. Haldar et al., Science *241*, 1198 (1988); H. Ihara et al., Nature *334*, 510 (1988).
11. S. Ikeda et al., Jap. J. Appl. Phys. *27*, L 999 (1988).
12. K. Kumagai and M. Kurisu, Jap. J. Appl. Phys. *27*, L 1029 (1988).
13. S. S. P. Parkin et al., Phys. Rev. Lett. *61*, 750 (1988); G. S. Grader et al., Appl. Phys. Lett. *53*, 319 (1988).
14. T. Wada et al., Jap. J. Appl. Phys. *27*, L 1031 (1988).
15. R. A. Fisher et al., Phys. Rev. Lett. (1988); K. Kumagai and Y. Nakamura, Physica *C152*, 286 (1988).
16. M. S. Hybertsen and L. F. Mattheiss, Phys. Rev. Lett. *60*, 1661 (1988); D. R. Hamann and L. F. Mattheiss, Phys. Rev. *B38*, 5138 (1988).

17. Y. Chen, T. Hong-jie, Z. Shi-ping, Y. Y.-feng and Y. Qian-Sheng, Mod. Phys. Lett. *B* (1988).

18. Z. Shi-ping, T. Hong-jie, C. Yin-fei, Y. Yi-fen, and Y. Qian-sheng, Sol. State Comm. (1989).

19. Ref. 2, Chap. X.

20. T. Yoshitake, T. Satoh, Y. Kubo, T. Manako and H. Igarashi, Jap. J. Appl. Phys. *27*, L 1094 (1988).

21. D. S. Ginley et al., Appl. Phys. Lett. *53*, 406 (1988).

22. T. T. M. Palstra, B. Batlogg, L. F. Schneemeyer, and J. V. Waszczak Phys. Rev. *B38*, 5102 (1988); Phys. Rev. Lett. *61*, 1662 (1988).

23. D. G. Hinks et al., Appl. Phys. Lett. *53*, 428 (1988).

24. R. Beyers et al., Appl. Phys. Lett. *53*, 432 (1988).

Appendix A. Macroscopic Parametric Relations

Thermodynamic models of the London and Landau-Ginsberg type have been described in several books.[1,2] Here we note a few definitions for anisotropic, uniaxial superconductors appropriate to layered materials[3] such as $NbSe_2$ or high-T_c cuprates. For uniform exponential penetration of the magnetic field, $H < H_{c1}$, the London penetration depth λ in an isotropic material is

$$\lambda_L = c/\omega_{ps} \tag{1}$$

$$\omega_{ps}^2 = \frac{4\pi ne^2}{m^*} \tag{2}$$

where m^* is an effective electron mass which may include band effects and electron-phonon renormalization. When a high density of defect states is present which make a large contribution to $N(E_F)$, their effect should also be included in m^*.

In the uniaxial case $(m^*)^{-1}$ has two components $(m_{\parallel}^*)^{-1}$ and $(m_{\perp}^*)^{-1}$ and relative to the usual band masses these are conventionally corrected in the GL (Ginsberg-Landau) theory for the anisotropy of the energy gap.[3] (Little is known about gap anisotropy and this factor is often neglected in practice.) In the cuprate case the layers may be decoupled and current flow in the normal state perpendicular to the layers is semiconductive: thus the meaning of the gap perpendicular to the layers is unclear, although there is tunneling data on this anisotropy (VIII.3), which indicates a gap anisotropy of about a factor of 2.

The coherence length ξ_o can be defined in terms of the energy gap $E_g = 2\Delta$,

$$\xi_o = \hbar v_F / \pi\Delta \tag{3}$$

and this relation can also be generalized to include $v_{F\parallel} \neq v_{F\perp}$.

Cuprate superconductors have values of $\xi_o < 10^2$A and $\lambda_o > 10^3$ so that

$$\kappa = \lambda_o / \xi_o >> 1 \tag{4}$$

which means that they are type II superconductors with a wide range of mixed states $H_{c1} < H < H_{c2}$. In the uniaxial case, where $\lambda_\parallel$ and $\xi_\parallel$ refer to current flow parallel to the planes and $\kappa_\perp$ refers to $H\perp$ to the planes, these equations are generalized to[3]

$$\kappa_\perp = \lambda_\parallel / \xi_\parallel \tag{5}$$

$$\kappa_\parallel^2 = \lambda_\parallel \lambda_\perp / \xi_\parallel \xi_\perp \tag{6}$$

with an anisotropy ratio ϵ defined by

$$\kappa_\perp = \epsilon \kappa_\parallel \tag{7}$$

$$\xi_\perp = \epsilon \xi_\parallel \tag{8}$$

$$\lambda_{\perp} = \lambda_{\parallel}/\epsilon. \tag{9}$$

In the limit $T \rightarrow T_c$, where H_{c1} and H_{c2} should be proportional to $(T_c - T)$, their slopes define κ:

$$(dH_{c2}/dT)/(dH_{c1}/dT) = 2\kappa^2/\ell n\kappa \tag{10}$$

for $\kappa >> 1$.

All these quantities are derived quantities, which have been tabulated[3] for $NbSe_2$. Evaluation for the cuprates will require corrections for flux creep (X.6).

REFERENCES

1. D. St. James, E. J. Thomas and G. Sarma, *Type II Superconductivity* (Pergamon, Oxford, 1969).
2. M. Tinkham, *Introduction to Superconductivity* (McGraw-Hill, N.Y. 1975; Krieger, Malabar, Fla., 1980, 1985).
3. P. de Trey, S. Gygax and J.-P. Jan., J. Low Temp. Phys. *11*, 421 (1973).

Appendix B. Microscopic Theory

Bardeen has summarized how the BCS theory in the weak-coupling limit[1] is extended to the strong-coupling limit by using Green's function methods to include phonon self energies.[2] This theory has worked very well for three-dimensional metals, where one-electron self-consistent field theory adequately describes electronic properties in the normal state. In high-T_c cuprates electronic conduction takes places primarily in two-dimensional CuO_2 layers, where because of marginal dimensionality the correct description of one-electron states in the presence of defects is unclear. For the reader's convenience we reproduce here Bardeen's summary[2] of the Green's function method, bearing in mind that the entire three-dimensional formalism for the normal state may require significant modifications before it can be applied to the two-dimensional context. However, once the normal state is well understood, I believe that the previous microscopic theory can be applied with only minor changes.

Green's-function method for normal metals.

By use of Green's function methods, Migdal[3] derived a solution of Fröhlich's Hamiltonian, $H = H_{el} + H_{ph} + H_{el-ph}$, for normal metals valid for arbitrarily strong coupling and which involves errors only of order $(m/M)^{1/2}$. The Green's functions are defined by thermal average of time-ordered operators for the electrons and phonons, respectively

$$G = -i(T\psi(1)\psi^{+}(2)) \tag{1a}$$

$$D = -i(T\phi(1)\phi^{+}(2)) \tag{1b}$$

Here $\psi(r,t)$ is the wave-field operator for electron quasi-particles and $\phi(r,t)$ for the phonons, the symbols 1 and 2 represent the space-time points $(r_{1,t_{1)}}$ and (r_2,t_2), and the brackets represent thermal averages over an ensemble.

Fourier transforms of the Green's functions for $H_0 = H_{el} + H_{ph}$ for non-interacting electrons and phonons are

$$G_0(P) = \frac{1}{\omega_n - \epsilon_0(k) + i\delta_k} \tag{2a}$$

$$D_0(Q) = \left[\frac{1}{\nu_n - \omega_0(q) + i\delta} - \frac{1}{\nu_n + \omega_0(q) - i\delta}\right] \tag{2b}$$

where $P = (k,\omega_n)$ and $Q = (q,\nu_n)$ are four-vectors, $\epsilon_0(k)$ is the bare electron quasiparticle energy referred to the Fermi surface, $\omega_0(q)$ the bare phonon frequency and ω_n and ν_n the Matsubara frequencies

$$\omega_n = (2n+1)\pi i k_B T; \quad \nu_n = 2n\pi i k_B T \tag{3}$$

for Fermi and Bose particles, respectively.

As a result of the electron-phonon interaction, H_{el-ph}, both electron and phonon energies are renormalized. The renormalized propagators, G and D, can be given by a sum over Feynman diagrams, each of which represents a term in the perturbation expansion. We shall use light lines to represent the bare propagators, G_0 and D_0, heavy lines for the renormalized propagators, G and D, straight lines for the electrons and curly lines for the phonons.

The electron-phonon interaction is described by the vertex

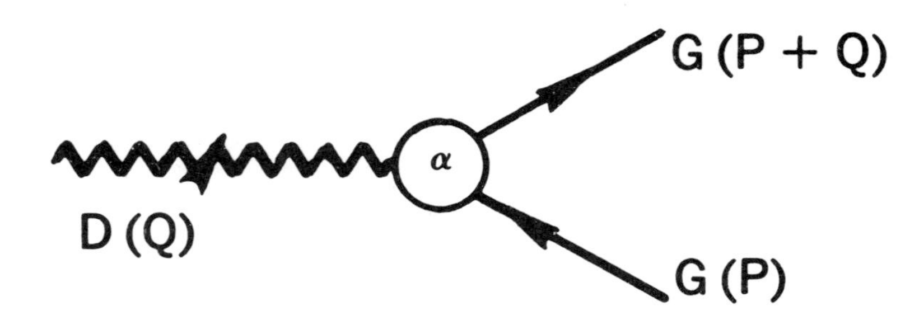

which represents scattering of an electron or hole by emission or absorption of a phonon or creation of an electron and hole by absorption of a phonon by an electron in the Fermi sea. Migdal showed that renormalization of the vertex represents only a small correction, of order $(m/M)^{1/2}$, a result in accord with the Born-Oppenheimer adiabatic approximation. [This may fail in the marginal dimensionality case. JCP] If terms of this order are neglected, the electron and phonon self-energy corrections are given by the lowest-order diagrams provided that fully renormalized propagators are used in these diagrams.

The electron self energy $\Sigma(P)$ in the Dyson equation:

$G_0(P)\,\Sigma(P)\,G(P)$

$G(P)$ = $G_0(P)$ + Σ

is given by the diagram

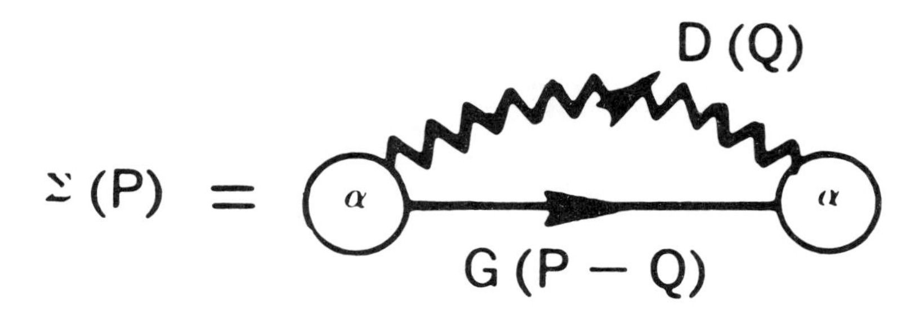

The phonon self-energy, $\pi(Q)$, defined by

D (Q) = D₀ (Q) + D₀ (Q) π (Q) D (Q)

is given by

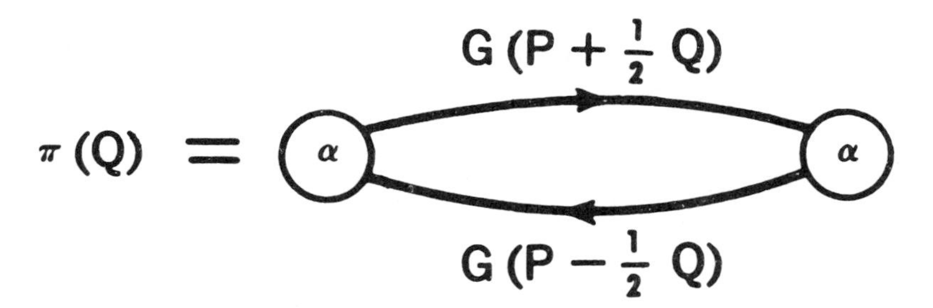

Since to order $(m/M)^{1/2}$ one can use an unrenormalized vertex function $\alpha = \alpha_0$, the Dyson equations form a closed system such that both $\Sigma(P)$ and $\pi(Q)$ can be determined.

The phonon self-energy, $\pi(Q)$, gives only a small renormalization of the phonon frequencies. As to the electrons, Migdal noted that we are interested in states k very close to k_F, so that to a close approximation $\Sigma(k, \omega)$ depends only on the frequency. For an isotropic system,

$$\Sigma(k,\omega) \simeq \Sigma(k_F,\omega) \equiv \Sigma(\omega) \tag{7}$$

The renormalized electron quasi-particle energy, ω_k, is then given by a root of

$$\epsilon(k) = \omega_k = \epsilon_0(k) + \Sigma(\omega_k) \tag{8}$$

In the thermal Green's-function formalism, one may make an analytic continuation from the imaginary frequencies, ω_n, to the real ω axis to determine $\Sigma(\omega)$.

Although $\Sigma(\omega)$ is small compared with the Fermi energy, E_F, it changes rapidly with energy and so can effect the density of states at the Fermi surface and thus the low-temperature electronic specific heat. The mass renormalization factor, m^*/m, at the Fermi surface may be expressed in terms of a parameter λ:

$$m^*/m = Z(k_F) = 1 + \lambda = (d\epsilon_0/dk)_F/(d\epsilon/dk)_f \tag{9}$$

In modern notation, the expression for λ is

$$\lambda = 2\int_0^\infty d\omega \frac{\alpha^2(\omega)F(\omega)}{\omega}, \tag{10}$$

where $F(\omega)$ is the density of phonon states in energy and $\alpha^2(\omega)$ is the square of the electron-phonon coupling constant averaged over polarization directions of the phonons. Note that λ is always positive, so that the Fermi surface is stable if the lattice is stable. Values of λ for various metals range from about 0.5 to 1.5. The parameter λ corresponds roughly to the term $N(0)V_{phonon}$ of the BCS theory.

Nambu-Eliashberg theory

Migdal's theory has important consequences that have been verified experimentally for normal metals, but gave no clue as to the origin of superconductivity. Following the introduction of the BCS theory, Gor'kov showed that pairing could be introduced through the anomalous Green's function

$$F(P) = -i(T\psi_{\uparrow}\psi_{\downarrow}), \tag{11}$$

Nambu showed that both types of Green's functions can be conveniently included with use of a spinor notation

$$\psi = \begin{pmatrix} \psi_{\uparrow}(r,t) \\ \psi_{\downarrow}^{+}(r,t) \end{pmatrix} \tag{12}$$

where $\psi_{\uparrow}$ and $\psi_{\downarrow}$ are wave field operators for up and down spin electrons and a matrix Green's function with components

$$\tilde{G}_{\alpha\beta} = -i(T\psi_{\alpha}\psi_{\beta}^{+}) \tag{13}$$

Thus G_{11} and G_{22} are the single-particle Green's function for up and down spin particles and $G_{12} = G_{21^*} = F(P)$ is the anomalous Green's function of Gor'kov.

These are two self-energies, Σ_1 and Σ_2, defined by the matrix

$$\tilde{\Sigma} - \begin{pmatrix} \Sigma_1 & \Sigma_2 \\ \Sigma_2 & \Sigma_1 \end{pmatrix} \tag{14}$$

Eliashberg noted[4] that one can describe superconductors to the same accuracy as normal metals if one calculates the self-energies with the same diagrams that Migdal used, but with Nambu matrix propagators in place of the usual normal-state Green's function. The matrix equation for $\tilde{G}$ is

$$\tilde{G} = \tilde{G}_0 + \tilde{G}_0 \tilde{\Sigma} \tilde{G} \tag{15}$$

The matrix equation for $\tilde{\Sigma}$ yields a pair of coupled integral equations for Σ_1 and Σ_2. Again Σ_1 and Σ_2 depend mainly on the frequency and are essentially independent of the momentum variables. Following Nambu,[5] one may define a renormalization factor $Z_s(\omega)$ and a pair potential, $\Delta(\omega)$, for isotropic systems through the equations:

$$\omega Z_s(\omega) = \omega + \Sigma_1(\omega) \tag{166}$$

$$\Delta(\omega) = \Sigma_2(\omega)/Z(\omega) \tag{17}$$

Both Z_s and Δ can be complex and include quasi-particle lifetime effects. Eliashberg derived coupled nonlinear integral equations for $Z_s(\omega)$ and $\Delta(\omega)$ which involve the electron-phonon interaction in the function $\alpha^2(\omega)F(\omega)$.

The Eliashberg equations have been used with great success to calculate the properties of strongly coupled superconductors for which the frequency dependence of Z and Δ is important. They reduce to the BCS theory and to the nearly equivalent theory of Bogoliubov based on the principle of "compensation of dangerous diagrams" when the coupling is weak. By weak coupling is meant that the significant phonon frequencies are very large compared with $k_B T_c$, so that $\Delta(\omega)$ can be regarded as a constant independent of frequency in the important range of energies extending to at most a few $k_B T_c$. In weak coupling one may also neglect the difference in quasi-particle energy renormalization and assume that $Z_s = Z_n$."

REFERENCES

1. J. Bardeen, L. N. Cooper and J. R. Schrieffer, Phys. Rev. *108,* 1175 (1957).
2. J. Bardeen, Physics Today, *26,* No. 7, p. 41 (1973).
3. A. B. Midgal, Zh. Eksp. Teor. Fiz. 34, 1438 (1958). English trans. in Sov. Phys. -JETP *7,* 996 (1958).
4. G. M. Eliashberg, Zh. Eksp. Teor. Riz 38, 966 (1960). English trans. in Sov. Phys. -JETP *11,* 696 (1960).
5. Y. Nambu, Phys. Rev. *117,* 648 (1960).

Appendix C. Crystal Chemistry

Historically high-temperature superconductive cuprates have evolved from materials related to the perovskite family ABO_3 which was discussed briefly in III.2. There we noted that the technological interest in these materials originated with the discovery of strong ferroelectricity in $BaTiO_3$ and many other perovskites or pseudoperovskites, such as A_2BCO_6. Many perovskites and related materials which are not ferroelectric are magnetic, such as the ferrite materials based on magnetite (Fe_3O_4) and a number of antiferromagnetics. An extensive 1970 review of perovskite-like structures and magnetic properties[1] contains a summary of Goldschmidt (1927) - Megaw (1946) ionic packing models, and briefly touches on La_2NiO_4 and La_2CuO_4. The formula A_2BX_4 is written as an intergrowth compound $AX \cdot ABX_3$ and the two structures are compared as shown in Fig. 1. The following perspicuous remarks were made: "The A_2BX_4 structure also permits the study of B^{2+} cations in oxides with a smaller B—X—B separation (hence stronger interaction) than is found in the BO compounds with rocksalt structure. The possible significance of this is illustrated by La_2NiO_4. The Ni^{2+} electrons of e_g symmetry appear to be collective in La_2NiO_4, localized in *NiO*" (p. 193). "Many of these compounds [containing Bi_2O_2 sheets sandwiching perovskite-like layers] ... exhibit ferroelectric distortions within the perovskite layers, and they will certainly be important for technical applications in the future" (p. 193). The astute reader will note here that no mention is made of magnetism in the bismates. Indeed I know of no bismate compound which exhibits magnetic properties, and even $BiFeO_3$ is not magnetic but is

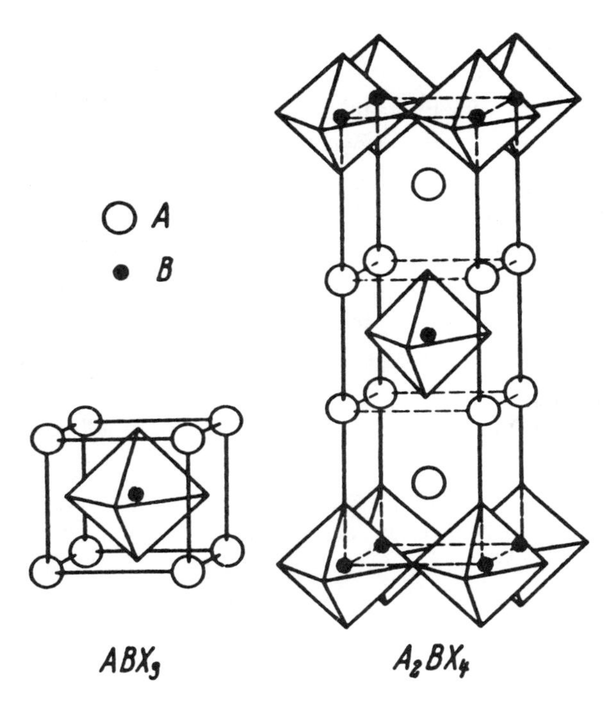

Fig. C1. Comparison between ABX_3 perovskite and $A_2BX_4 = AX \cdot ABX_3$ structures (ref. 1).

ferroelectric (see ref. 3).

While both the ferroelectric and magnetic properties sustained interest in these materials for decades, it was generally believed that the materials were not metallic, except for very heavy elements, e.g., $BaPb_xBi_{1-x}O_3$. (Little attention has been paid to $LiTi_2O_4$, possibly because of preparative difficulties). Indeed, as-formed samples of cuprates are not metallic, and they become so only by doping or by post-deposition oxidation.[2] That such oxidation is readily carried out for bulk samples (because of the openness of the structure and high oxygen mobility) is one of nature's happy accidents.

A balanced discussion[3] of ABX_3, ABX_4 and A_2BX_4 structures shows how these structures can be compared. Broadly speaking the cation ionic radii r_A and r_B discriminate among the various phases moderately well (for fixed X= oxygen or halide). Square planar CuO_4, which is important in the cuprate superconductors, occurs both for $ACuX_4$ and A_2CuX_4. However, in the ABX_4 structures usually either the A or the B coordination configurations are tetrahedral, AX_4 or BX_4. Only in rutile-like compounds do we find both AX_6 and BX_6 octahedra, and these often form chains. Once again, however, in the case of Cu the Jahn-Teller effect favors square-planar coordination in compounds like $CuMoO_4$ and $CuWO_4$. When we examine the structures of these compounds in details we find that the local Cu coordination polyhedra are quite similar to those of La_2CuO_4, as shown in Table I. Nevertheless, the

Compound	(4)	(2)
$MoCuO_4$	1.93-1.98	2.21-2.62
$WCuO_4$	1.96-2.00	2.34-2.45
La_2CuO_4	1.90	2.40

Table I. Coordination lengths (in Å) of CuO_6 octahedra in ABO_4 and A_2BO_4 compounds.

unit cell morphologies differ greatly, because in the $ACuO_4$ structures the layering which is a characteristic feature of high-T_c cuprates is absent. Moreover,, for A= Mo and W, the A coordination configurations are tetrahedral, which explains the absence of layering. Because of the formation of tetrahedra the ABX_4 compounds are less likely than the ABX_3 and A_2BX_4 compounds to served as starting compounds for constructing high-T_c superconductors. More generally we can look for layering in compounds with formulae $(AX)_m\ B_nX_{3n}$ or $(AX)_m\ B_nX_{3n-1}$ and then hope to replace B by Cu. This may also remove some X= O atoms, leaving behind many oxygen vacancies $O^{\square}$. The resulting as-formed material may be semiconductive but by annealing in oxygen we can fill some of the $O^{\square}$. This converts the material to a metal and possibly a high-T_c superconductor. This second stage is actually a doping stage which is analogous to the replacement of La by Sr in $La_{2-x}Sr_xCuO_4$, with the dopant being oxygen anions rather than metallic cations.

While classical ionic radii are useful for separately classifying the various structures assumed by ABX_3, A_2BX_4 and ABX_4 families, these radii so far have not proved to be useful in describing the potentials of these materials as high-T_c superconductors. The key point here is that the semiconductive cuprates should become *metallic* upon doping by cation substitution or anion addition. Traditionally binary *metallic* structures are described by three coordinates: average valence electron number, size differences and electronegativity differences. Here we are dealing with ternary compounds which when doped contain four or even five elements (including O), and our first problem is how we should define size and electronegativity differences when there are more than two elements.

Electronic structure calculations (IV.4) show that in these multinary cuprates holes are present on both the metallic cations and the oxygen anions when they are metallic. Thus it may be possible to discuss the structures and properties of these materials *when they are metallic* using metallic coordinates even for oxygen, even though when they are semiconductive or insulating the structures are better discussed using ionic radii. This is consistent with the viewpoint expressed in IV.5, where we argued that the insulating and metallic phases have entirely different physical properties, not only as regards electrical conductivity, but also as regards magnetism and other electronic properties.

Villars' Theory

For metals many different sets of elemental coordinates, including many different definitions of size and electronegativity, have been proposed. More than 180 such definitions were examined by Villars,[4] and he arrived finally at the average valence electron number, sizes and electronegativities shown in the elemental coordinate periodic table in Fig. 2. In retrospect his choices can be rationalized by theory (both his definitions, of size and electronegativity, have spectroscopic bases, and the valence-number dependence of these bases is explicit, which is not the case for most other definitions). However, in fact his choices were made empirically from studies of the binary crystal structures of some 3000 AB, AB_2, AB_3 and A_3B_5 binary intermetallic compounds. The resulting structural diagrams of intermetallic compounds, including both simple s-p metals and transition (d) and rare earth (d-f) metals, but excluding oxides and halides, are typically 97% successful in separating binary structures. Moreover from his analysis of the ability of ternary alloys to form stable

H 1 2.10[a] 1.25[b]																
Li 1 0.90 1.61	Be 2 1.45 1.08											B 3 1.90 0.795	C 4 2.37 0.64	N 5 2.85 0.54	O 6 3.32 0.465	F 7 3.78 0.405
Na 1 0.89 2.65	Mg 2 1.31 2.03											Al 3 1.64 1.675	Si 4 1.98 1.42	P 5 2.32 1.24	S 6 2.65 1.10	Cl 7 2.98 1.01
K 1 0.80 3.69	Ca 2 1.17 3.00	Sc 3 1.50 2.75	Ti 4 1.86 2.58	V 5 2.22 2.43	Cr 6 2.00 2.44	Mn 7 2.04 2.22	Fe 8 1.67 2.11	Co 9 1.72 2.02	Ni 10 1.76 2.18	Cu 11 1.08 2.04	Zn 12 1.44 1.88	Ga 3 1.70 1.695	Ge 4 1.99 1.56	As 5 2.27 1.415	Se 6 2.54 1.285	Br 7 2.83 1.20
Rb 1 0.80 4.10	Sr 2 1.13 3.21	Y 3 1.41 2.94	Zr 4 1.70 2.825	Nb 5 2.03 2.76	Mo 6 1.94 2.72	Tc 7 2.18 2.65	Ru 8 1.97 2.605	Rh 9 1.99 2.52	Pd 10 2.08 2.45	Ag 11 1.07 2.375	Cd 12 1.40 2.215	In 3 1.63 2.05	Sn 4 1.88 1.88	Sb 5 2.14 1.765	Te 6 2.38 1.67	I 7 2.76 1.585
Cs 1 0.77 4.31	Ba 2 1.08 3.402	La 3 1.35 3.08	Hf 4 1.73 2.91	Ta 5 1.94 2.79	W 6 1.79 2.735	Re 7 2.06 2.68	Os 8 1.85 2.65	Ir 9 1.87 2.628	Pt 10 1.91 2.70	Au 11 1.19 2.66	Hg 12 1.49 2.41	Tl 3 1.69 2.235	Pb 4 1.92 2.09	Bi 5 2.14 1.997	Po 6 2.40 1.90	At 7 2.64 1.83
Fr 1 0.70[a] 4.37[b]	Ra 2 0.90[a] 3.53[b]	Ac 3 1.10[a] 3.12[b]														

Ce 3 1.1[a] 4.50[b]	Pr 3 1.1[a] 4.48[b]	Nd 3 1.2[a] 3.99[b]	Pm 3 1.15[a] 3.99[b]	Sm 3 1.2[a] 4.14[b]	Eu 3 1.15[a] 3.94[b]	Gd 3 1.1[a] 3.91[b]	Tb 3 1.2[a] 3.89[b]	Dy 3 1.15[a] 3.67[b]	Ho 3 1.2[a] 3.65[b]	Er 3 1.2[a] 3.63[b]	Tm 3 1.2[a] 3.60[b]	Yb 3 1.1[a] 3.59[b]	Lu 3 1.2[a] 3.37[b]
Th 3 1.3[a] 4.98[b]	Pa 3 1.5[a] 4.96[b]	U 3 1.7[a] 4.72[b]	Np 3 1.3[a] 4.93[b]	Pu 3 1.3[a] 4.91[b]	Am 3 1.3[a] 4.89[b]								

Fig. C2. Periodic table of metallic valences, sizes and electronegativities, according to Villars (ref. 4).

compounds he was able to construct a definition of size and electronegativity differences which is suitable for discussing simple properties of ternary phases.

In general oxides and halides are insulating and the Villars coordinates may not be suitable for discussing their structures or properties. However, his approach is successful for semiconductive sulfides, selenides, and tellurides, and the cuprate superconductors are essentially doped semiconductors. It is therefore tempting to use his coordinates to attempt to identify regions of elemental configuration space where high T_c superconductors are found. To achieve meaningful results one must define "high T_c" as T_c above a certain temperature T_0 such that T_0 is low enough to provide a large data base but not so low as to include even elemental superconductors such as Pb or Nb. Because the maximum attainable T_c for elemental or pseudoelemental phases is 9.8K, a natural choice for T_0 is 10K. Then it turns out that more than 60 essentially different (which means small changes in composition are excluded, such as isostructural compounds involving replacement of one rare earth by another) high-T_c superconductors are known. This is a robust data base, whose size generally surprises most people who believe themselves knowledgeable. These superconductors are catalogued, together with their Villars coordinates, at the end of this Appendix as Table II. Also some 600 superconductors in the range $1K < T_c < 10K$ are listed, grouped according to crystal structure, for completeness, as Table III.

When we now plot the distributions of these two groups of superconductors using the Villars coordinates $\overline{N}_v$ (average number of valence electrons), $\overline{\Delta X}$ (difference of spectroscopic electronegativities) and $\Delta\overline{r}$ (difference of spectroscopic radii), two results appear.[4] The 600 superconductors with $1K < T_c < 10K$ form a scatter-shot plot, which is not surprising. However, this is not the case for the 60+ superconductors with $T_c > 10K$. These are found to cluster in three small islands which fill less than 1% of the three-dimensional ($\overline{N}_v$, $\overline{\Delta X}$, $\Delta\overline{r}$) volume, as shown in Fig. 3. We call the three islands A, B and C.

In island A we find intermetallic compounds of the A15 (Cr_3Si) family as well as other complex intermetallic structures. Island B is largely dominated by the NbN family, although it also contains borides and carbides. The Chevrel sulfides are found in island C, which spreads with increasing $|\overline{\Delta X}|$ into the oxides including the cuprates. The original version[4] of Fig. 3 was published before the bismates and thallates were discovered, but they are also located in island C. This is a major triumph for this method. One should also note that the ranges of $\Delta\overline{r}$ and ($\Delta\overline{X}$) are narrow (wide), which means that size misfit is critical and charge transfer much less so, for high temperature superconductivity. The narrow range of size misfit is associated with anharmonicity, while the wide range of charge transfer suggests the possibility of defect bands pinned near the Fermi energy. Thus the phenomenology and crystal chemistry of high-temperature superconductivity is fully consistent with the microscopic models of defect states and fictive phonons discussed in IV - VI.

Before we leave the discussion of Fig. 3 there are several very important points to notice. Compared to the total range of values of $\overline{N}_v$, $\overline{\Delta X}$ and $\overline{\Delta r}$, the three islands in this Figure occupy about 1% of the elemental chemical coordinate volume. That some (originally 60, now) 70 ternary, quaternary and quinary high T_c superconductors

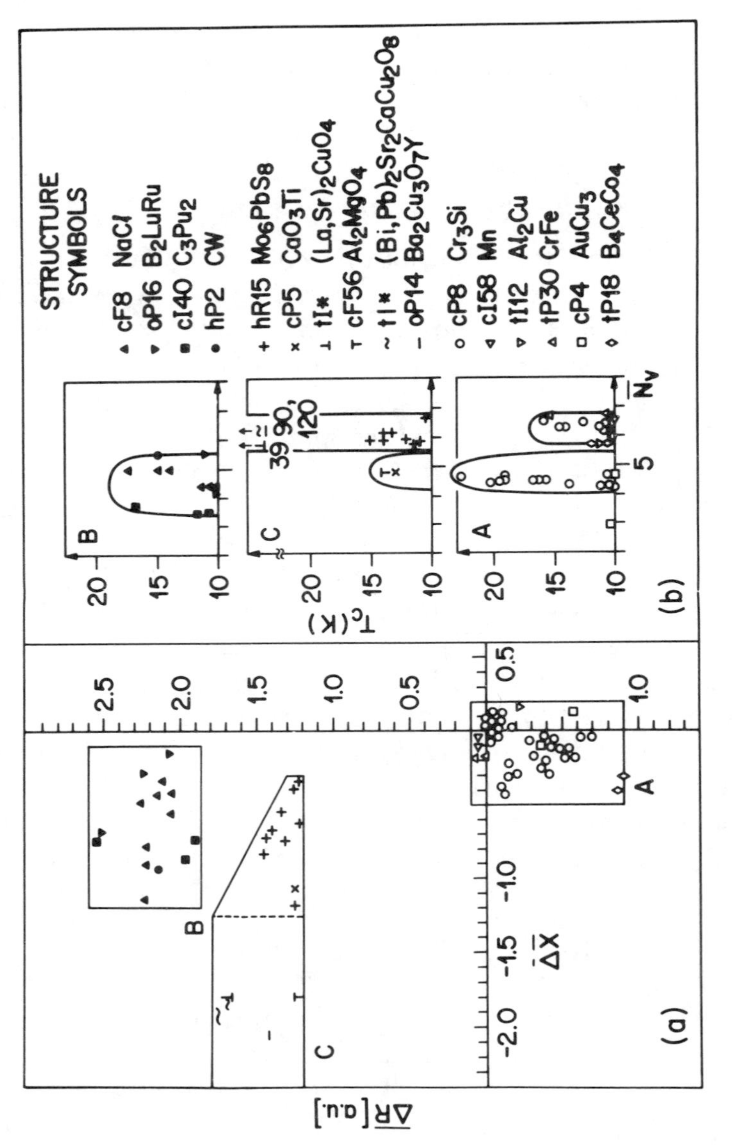

Fig. C3. Occurrence of high-temperature ($T_c > 10K$) superconductivity in the Villars configuration space (ref. 4).

should concentrate in these small volumes is, from a statistical point of view, nearly miraculous. (By comparison, relatively little information is contained in the sparse data base of the isotope effect, Fig. V.4.) The odds against such an event, which transcends the details even of the various complex crystal structures listed on the right hand side of the Figure, occurring by accident are of order 10^{100} to 1. (This rather large number should be compared to the age of the universe, about 10^{16} sec, a small (!) number by comparison.) What conditions must hold for such a miracle to occur?

The first and most obvious condition is that since the same diagrammatic construction works for the cuprate and Chevrel phases (island C) as works for the Nb_3Sn (A-15 and related phases, island A) and NbN (B-1) and related phases, island B), the microscopic mechanism responsible for high-T_c superconductivity in all three systems *must be the same.* Anyone who doubts the force of this statistical argument is invited to plot the coordinates of the 600 low-T_c superconductors with $1K < T_c < 10K$. (There is nothing quite so disillusioning as a scatter-shot plot.) Or, if you like the high-T_c magnon models described in ref. 44 of IV, plot the coordinates of all oxides with Curie temperatures above 10K; again you will obtain a scatter-shot plot, a result all too familiar to those who have attempted to rationalize physical properties using less than optimal global chemical coordinates.

Granted then that Fig. 3 already demonstrates that the same very strong electron-phonon interactions which produce high T_c's in the metallic islands A and B also produce high T_c's in the sulfides and oxides, how does the Villars construction succeed in doing this? For binary compounds his coordinates are successful because they have been very carefully chosen to describe binary crystal structures. For ternaries, however, this procedure is no longer applicable because the number of known ternary structures already exceeds 3000 and is growing rapidly. The real secret to the success of Fig. 3 is that the electronegativity and size difference coordinate prescription is chosen to separate local coordination configurations. This seems to be a rather simple property, but in fact it is both subtle and profound. For such a prescription to be successful, it must be very accurate just on the border-line between different configurations. However, it is just at this borderline that maximum lattice instabilities are found, and, as we have seen repeatedly, it is maximum lattice instabilities that produce high-T_c superconductivity. The same borderline behavior is found for icosahedral quasicrystals, and this method was successful in predicting these materials as well.

We see that the "Villars miracle" can be understood on conceptual grounds as the result of finding the right coordinates for describing crystal structures in general and lattice instabilities in particular. Please note that the high-T_c lattice instability is the most severe kind, namely the structure is most unstable. Weaker instabilities, which transform one phase to another, often accompany superconductivity, and in the simplest cases are recognizable in terms of soft phonons and displacive second-order phase transitions. The weaker correlations of these weaker instabilities are much less convincing than the Villars correlations. Microscopically one expects that at such boundaries, the material may contain a high concentration of point defects. Moreover these defects may well have electronic resonances very close to the Fermi energy (chemical potential), just because the compound almost does not form. At present we have no proof for these remarks, but at the same time full quantum-mechanical calculations of specific defect configurations and electronic energy levels seem unlikely to be feasible in the foreseeable future. Thus the global approach has great practical advantages.

It is certainly not obvious *a priori* that the Villars procedure can succeed against such enormous statistical odds, and indeed the result shown in Fig. 3 is a triumph in pattern recognition that will not soon be equaled by any computer starting, for example, from crystal structures alone! However, granted the obvious *a posteriori* success of the method, there is no room for doubt that it is lattice instabilities and not magnetism that produce high-T_c superconductivity just as much in cuprates as in intermetallic compounds.

New Materials

Before leaving the subject of Villars' diagram, we should mention that his construction can be used to restrict and accelerate searches for new materials. In November, 1987 I realized that La_2CuO_4, the antecedent of the then known high-T_c cuprate superconductors, also was found in island C, and that therefore a search of all ternary oxides in this island could prove rewarding. Learning of the existence of the inorganic data base[5] I contacted I. D. Brown and a joint search was initiated. Although the search was efficient, it was extremely limited in terms of man-hours/week, with the result that our "short list" did not exist until April, 1988, whereas the bismate superconductors with $T_c > 100K$ were announced two months earlier. Selections from our short list[6] are given as Table IV at the end of this section. In fact, $Bi_4Ti_3O_{12}$, Bi_2MoO_6, and Bi_2WO_6, all antecedents of the new bismate materials, fall in island C and appeared on our short list, together with their lattice constants, which showed that all three materials are lattice-constant matched to CuO_2 planes. It is true that our work was completed two months too late, but previously crystallographic bibliographies had not been searched for material properties at all. Villars' diagrams (together with the fact that these data bases are on tape) make such a search relatively easy and very efficient. Recent advances in bibliographic languages and formatting now make searches of this type possible in a few days.[6]

Ferroelectrics

The same diagrammatic approach used to discuss high-T_c superconductivity also succeeds for high-T_c ferroelectric Here for ferroelectrics the Curie temperature T_c is said to be high when $T_c > 500K$. There are 50 such ternary compounds, and they are confined to an even smaller range of $\overline{N_v}$ and $\overline{\Delta X}$ than the ternary superconductors, Fig. 4. Also shown in this diagram is the outline of island C from the high-T_c superconductor diagram (Fig. 3). Comparison of the two diagrams suggests that high-T_c superconductivity and high-T_c ferroelectricity are mutually exclusive. This conclusion is consistent with empirical chemical trends in $MBa_2Cu_3O_7$ soft phonon anomalies (VI.3, p. 213) where we saw that increasing the ionic radius of M stiffens the infrared active (ferroelectric) mode, but softens the Raman (340 cm^{-1} superconductive) mode.

Diagrammatic analysis enables one to identify dramatic and specific parallels between high T_c ferroelectrics and superconductors. The ferroelectric with the highest known Curie temperature is monoclinic $La_2Ti_2O_7$, $T_c = 1770K$. Also tetragonal $Bi_4Ti_3O_{12}$ is ferroelectric, $T_c = 950K$. These

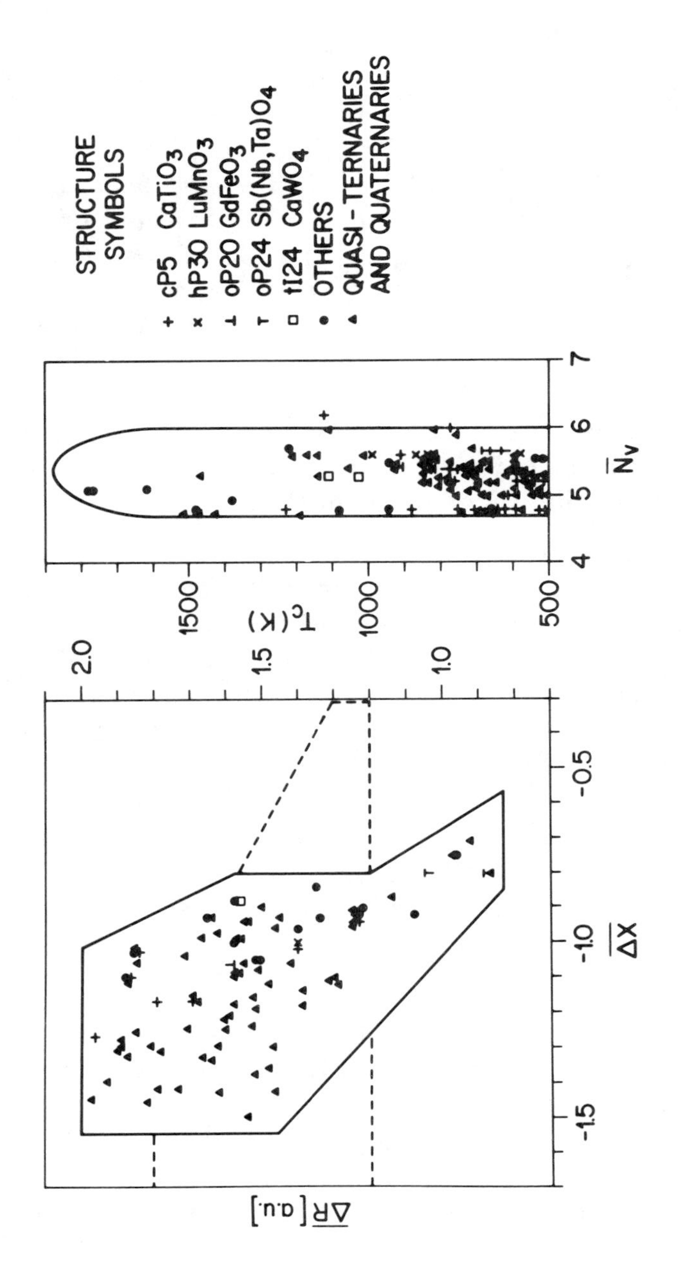

Fig. C4. Villars plot for high T_c ternary ferroelectric oxides (ref. 7).

are the two highest T_c ferroelectric titanates, and *both* are converted to high T_c superconductors by replacing Ti by an appropriate mixture of Cu and alkaline earths (Ca, Sr and/or Ba). [We are certain that $T\ell_4Ti_3O_{12}$ would be similar to $Bi_4Ti_3O_{12}$, but that it has not been prepared because $T\ell$ is so toxic.] Other high T_c titanates include $Bi_4PbTi_4O_{15}$, $T_c = 840K$; cubic $PbTiO_3$, $T_c = 760K$, and a wide variety of pseudoternary perovskites with mixtures of Pb, Bi, and alkaline earths, all with similar or lower T_c's. Very few tungstates and molybdenates have been tested for ferroelectricity, but T_c $(WBi_2O_6) = 1220K$. Thus the ternary ferroelectrics with the highest Curie temperatures are the ones most suitable for doping with Cu and alkaline earths to derive high T_c superconductors.

When we next turn to a closer look at the bonding structures of specific cuprates, we are likely first to use the bond-valence method,[8] which among crystal chemists is the preferred way of discussing bond lengths in ceramic solids, especially those with irregular coordination figures.[9] This method works quite well[6] for $YBa_2Cu_3O_6$ and confirms that Cu(1) is Cu(I), that is monovalent, and Cu(2) is Cu(II). However, the method cannot encompass Jahn-Teller effects for partially filled valence bonds (such as the $Cud_{x^2-y^2} - Op_{x,y}$ σ antibonding band), so that the results are less successful in describing the specifically metallic effects which are central to the structure of superconductive phases.

The bond valence method has provided the most widely successful explanation[8] for the distortions of ABO_3 perovskite structures, which were discussed in detail by SCF methods for $BaBiO_3$ in IV.2. In the ideal structure two bond lengths, AO and BO, should be satisfied, but only the cubic lattice constant is adjustable, Fig. 5(a). An extra

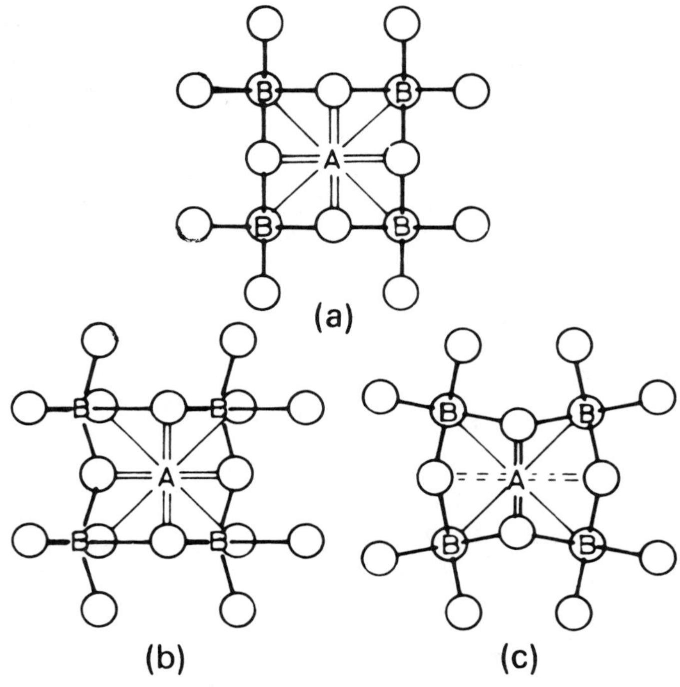

Fig. C5. ABO_3 perovskite structures: a, undistorted, b, tetragonal distortion by Cartesian displacement of B ion, c, distorted by rocking of B ion coordination complex (ref. 8).

adjustable parameter is introduced either by the (ferroelectric) tetragonal distortion, Fig. 5(b), or by a (superconductive) rocking distortion, Fig. 5(c), as in $BaBiO_3$. Which pattern is selected is dictated by the sign of ΔV_B as calculated by this technique (Fig. 6).

Again we see from this method some hints which provide support for the fictive phonon model. The method, strictly speaking, predicts only static distortions in insulators and does not include in the bond energy the contribution of metallic electrons, or the possibility of retaining during quenching a larger number of metallic electrons to reduce bond frustration. However the bonding contribution alone satisfies several general theorems.[10] For instance, it can be shown that any deviation of the bond lengths from their average value will increase the average bond length. This is just the situation for the large amplitude vibrations of the cuprates which are present at high temperatures where oxidation and metallization are the critical "final" touches responsible for high T_c's in the cuprates. When the sample

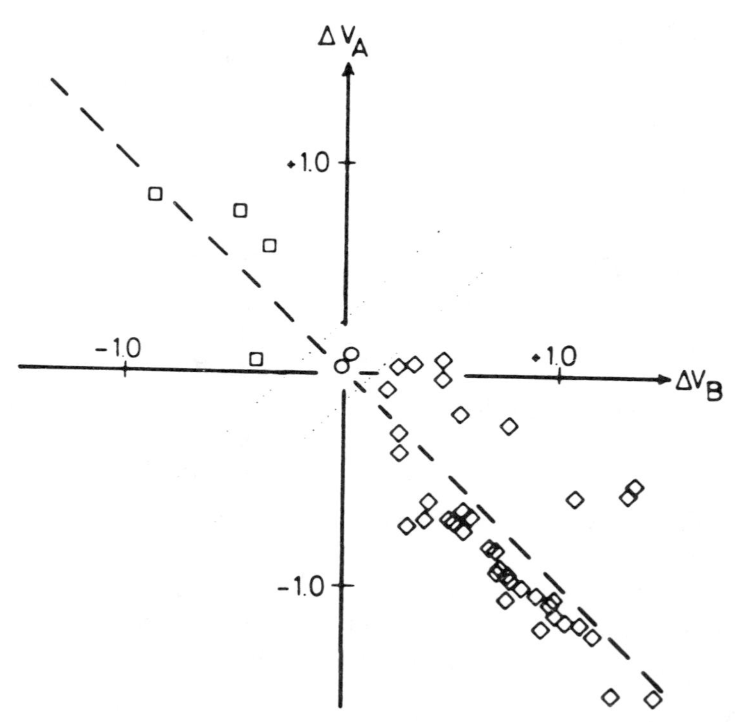

Fig. C6. Experimental and theoretical valence deviations of perovskites. Circles: undistorted; squares, distorted as in 4b, and diamonds, as in 4c (ref. 8).

is quenched, purely mechanical considerations suggest that the presence of additional oxygen (above that required for Cu^{2+}) will produce bond stress which can be reduced by weakening the bonds by retaining their large vibrational amplitudes, which increases average bond lengths (Fig. VI-15).

Madelung Potentials

In discussing lattice vibrations, we saw in VI.1 that an ionic model (including compressible oxygen ions) gives quite a good account of the lattice dynamics without adjustable parameters. Thus we might expect to gain some structural insights directly from calculating Madelung potentials.[11] When this is done for Cu_2O, CuO, Cu_4O_3 and $KCuO_2$ one finds that these Madelung potentials are generally self-consistent in the sense that at the ith site

$$\phi_i = -\alpha q_i \tag{1}$$

with $\alpha =$ 12.5, 12.5 and 11.4V for Cu(I, II, III) and for O(II): 11.7V. Similar calculations for Ag_3O_4 lead to the conclusion that it may already be metallic, which might explain why the highest T_c's so far observed[12] for Ag-O compounds have been for the clathrates Ag_7O_8X (X= NO_3, F and BF_4, $T_c =$ 1.0, 0.3, and 0.15K). Or these low T_c's might reflect the nearly cubic clathrate structure, but higher T_c's were obtained in the similar Chevrel compounds. Note that the clathrate structure can be regarded as molecular fragments which have been separated from a network by too large an internal stress (too large frozen-in vibrational amplitudes).

When these calculations are applied to $YBa_2Cu_3O_{6,7}$ and the results compared to Cu_2O, etc., the main conclusion is that the O Madelung potentials and charges are reduced, i.e., the O2p level (which is well below Cu 3d in CuO) is brought up nearly equal to Cu 3d, as was also found in self-consistent band calculations (IV.4). This shift is traced to the presence of large positive ions such as Ba. Correspondingly some of the Ba-O bond lengths are compressed, and this generates the anharmonicities discussed in V.4. These anharmonicities are important to the fictive phonon mechanism for high T_c's (IV.13 and VI.3), so the overall picture is one of a chemically, electronically and mechanically marginally stable structure.

Dramatic evidence for the marginal stability of these structures has been obtained by the synthesis[13] of a pseudo-ternary antecedent or parent cuprate compound, $Ca_{1-x}Sr_xCuO_2$, which is stable in a layered defect (oxygen vacancy) structure only for $x = 0.15 \pm 0.01$. Both $CaCuO_2$ and $SrCuO_2$ are orthorhombic, and Cu is only two-fold coordinated.[14] In tetragonal $Ca_{0.86}Sr_{0.14}CuO_2$ there are CuO_2^{2-} planes separated by planes of Ca^{2+}. Structurally this compound represents the $n = \infty$ limit of the bismates and thallates (XI) with formula $A_2B_2Ca_{n-1}Cu_nO_{4+2n}$ with $A = Bi$ or $T\ell$, $B = Sr$ or Ba. The synthesis of small single crystals of this material confirms the electronic and mechanical marginal stability of these materials.

REFERENCES

1. J. B. Goodenough and J. M. Longo, in Landott-Börnstein, III/4a (Ed. K.-H. Hellwege and A. M. Hellwege, Springer, New York, 1970), p. 126. See also J. B. Goodenough, Prog. Sol. State Chem. (Ed. H. Reiss, Pergamon, Oxford, 1971), Vol. *5*, 145, esp. p. 273.
2. C. Michel and B. Raveau, Rev. Chim. Min. *21*, 407 (1984).
3. O. Muller and R. Roy, *The Major Ternary Structural Families* (Springer, New York, 1974).
4. P. Villars and J. C. Phillips, Phys. Rev. *B37*, 2345 (1988).
5. G. Bergerhoff, R. Hundt, R. Sievers, and I. D. Brown, J. Chem. Inf. *23*, 66 (1983); D. Altermatt and I. D. Brown, Acta Cryst. *B41*, 240 (1985).
6. I. D. Brown, J. C. Phillips, and K. Rabe (unpublished); John Rogers, Canadian Inst. Scien. Tech. Infor., Nat. Resch. Council (Canada); private comm.; also J. R. Rogers and G. H. Wood, Crystallographic Data Bases (Int. Union Cryst., 1987), p. 96.
7. P. Villars, J. C. Phillips, K. Rabe and I. D. Brown (unpublished).
8. I. D. Brown, in *Structure and Bonding in Crystals* (Ed. M. O'Keeffe and A. Navrotsky, Academic Press, New York, 1981) Vol. *2*, p. 1.
9. M. O'Keeffe and S. Hansen, J. Am. Chem. Soc. *110*, 1506 (1988).
10. I. D. Brown, Chem. Soc. Rev. *7*, 359 (1978).
11. M. O'Keeffe,
12. M. B. Robin et al., Phys. Rev. Lett. *17*, 917 (1966).
13. T. Siegrist, S. M. Zahurak, D. W. Murphy and R. S. Roth, Nature *334*, 231 (1988).
14. C. L. Teske and H. Müller-Buchsbaum, Z. f. Anorg. Allg. Chem. *379*, 113, 234 (1970).

Table II

Group A

Structure Type	Compound	$T_c(^\circ K)$	Ref.	$\overline{N}_v$	$\overline{\Delta X}$	$\overline{\Delta R}$
Cr_3Si	$GeNb_3$	23.2	6	4.75	-0.02	-0.60
Cr_3Si	$GaNb_3$	21	11	4.5	-0.16	0.53
Cr_3Si	Nb_3Si	19	6	4.75	-0.02	-0.67
Cr_3Si	$AlNb_3$	19	11	4.5	-0.19	-0.54
Cr_3Si	$SnNb_3$	18	11	4.75	-0.07	-0.44
Cr_3Si	SiV_3	17.2	6	4.75	-0.12	-0.50
Cr_3Si	$PbTa_3$	17	6	4.75	-0.01	-0.35
Cr_3Si	GaV_3	16.8	6	4.5	-0.26	-0.36
Cr_3Si	$AuTa_3$	16.0	6	6.5	-0.37	-0.06
CrFe	$MoTc_3$	15.8	6	6.75	-0.12	0.03
Cr_3Si	Mo_3Re	15	6	6.25	0.06	-0.02
Cr_3Si	Mo_3T_c	15	6	6.25	0.12	-0.03
Cr_3Si	InV_3	13.9	6	4.5	-0.29	-0.19
Cr_3Si	$OsMo_3$	12.7	6	6.5	-0.04	-0.03
Cr_3Si	Mn_3Si	12.5	38	6.25	-0.03	-0.40
CrFe	$Mo_{0.3}T_{c_{0.7}}$	12.0	38	6.7	-0.14	0.04

CrFe	$AlNb_3$	12	38	4.5	-0.19	-0.54
Cr_3Si	AlV_3	11.8	6	4.5	-0.29	-0.37
B_4CeCo_4	B_4LuRh_4	11.7	30	5.61	-0.40	-0.76
Cr_3Si	$AuNb_3$	11.5	6	6.5	-0.42	-0.05
Cr_3Si	ReW_3	11.4	6	6.25	0.13	-0.02
Al_2Cu	$RhZr_2$	11.3	6	5.65	0.19	-0.20
B_4CeCo_4	B_4Rh_4Y	11.3	29	5.61	-0.31	-0.95
Cr_3Si	GeV_3	11	6	4.75	-0.11	-0.43
Cr_3Si	Nb_3Pt	10.9	6	6.25	-0.06	-0.03
Cr_3Si	Mo_3Ru	10.6	6	6.5	0.01	-0.05
Mn	$NbTc_3$	10.5	39	6.5	-0.07	0.05
$AuCu_3$	$InLa_3$	10.4	7	3	0.14	-0.51
Cr_3Si	$RhTa_3$	10	6	6	0.02	-0.13
Mn	$Nb_{0.18}Re_{0.82}$	10	39	6.64	-0.01	0.02
CrFe	$Ir_{0.4}Nb_{0.6}$	10	38	6.6	-0.12	-0.10
$AuCu_3$	Nb_3Pb	9.6	8	4.75	-0.05	-0.33
Cr_3Si	$IrMo_3$	9.6	6	6.75	-0.03	-0.04
Cr_3Si	$InNb_3$	9.2	6	4.5	-0.20	-0.35
Cr_3Si	Nb_3Tl	9	6	4.5	-0.17	-0.26

Group B

ClNa	NNb	17.3	24	5	-0.82	2.22
C_3Pu_2	$C_{0.66}Th_{0.13}Y_{0.21}$	17.0	44	3.6	-0.71	2.62
MoN	MoN	14.8	32	5.5	-0.91	2.18
ClNa	CMo	14.3	24	5	-0.43	2.08
ClNa	NTa	14	24	5	-0.91	2.25
ClNa	CNb	11.5	24	4.5	-0.34	2.12
C_3Pu_2	C_3Y_2	11.5	44	3.6	-0.76	1.84
C_3Pu_2	C_3La_2	11.0	44	3.6	-0.81	1.95
ClNa	NZr	10.7	24	4.5	-1.15	2.28
ClNa	CTa	10.3	24	4.5	-0.43	2.15
ClNa	CW	10	24	5	-0.58	2.09
B_2LuRu	B_2LuRu	10	19	4.25	-0.70	2.57

Group C

$T\ell_2Ba_2CaCu_2O_8$		120		5.46	-1.98	1.79
$Bi_2Sr_2CaCu_2O_8$		110		5.73	-1.90	1.73
$Ba_2Cu_3O_7Y$	$Ba_2Cu_3O_7Y$	90		6.31	-2.06	1.45
$Ba_2Cu_3O_6Y$	$Ba_2Cu_3O_6La$	80		6.34	-2.24	1.57
K_2NiF_4	$(La, Sr)_2CuO_4$	39		5.82	-1.85	1.65
K_2NiF_4	$(La, Ba)_2CuO_4$	36		5.82	-1.85	1.66
K_2NiF_4	$(La, Ca)_2CuO_4$	18		5.82	-1.85	1.64
Mo_6PbS_8	$Mo_6Pb_{0.9}S_{7.5}$	15.2	30	5.88	-0.68	1.40
Mo_6PbS_8	$Mo_3S_4Sn_{0.6}$	14.2	18	5.84	-0.71	1.31
Mo_6PbS_8	$I_2Mo_6S_6$	14.0	31	6.14	-0.35	1.21
Mo_6PbS_8	$Br_2Mo_6S_6$	13.8	31	6.14	-0.31	0.99
Al_2MgO_4	LiO_4Ti_2	13.7	30	4.72	-1.79	1.27
CaO_3Ti	$BaBi_{0.2}O_3Pb_{0.8}$	13.2	30	4.84	-1.08	1.28
Mo_6PbS_8	Mo_6Se_8Tl	12.2	25	5.79	-0.63	1.21
Mo_6PbS_8	$Mo_6O_2PbS_6$	11.7	31	5.82	-0.81	1.48
Mo_6PbS_8	$LaMo_6Se_8$	11.4	25	5.79	-0.72	1.45
Mo_6PbS_8	$Cu_{1.8}Mo_6S_8$	10.8	30	6.55	-1.07	1.29

Table III

Pearson symbol	Structure Type
Compound(s)	T_c Ref. [Sec. Ref.]

mC12 Bi_2Pd

Bi_2Pd	1.70	38[43]

mC24 GaTe

GaTe	0.17	38[43]

oP4 AuCd

IrMo	8.8	38[43]	$Nb_{0.9}Rh_{1.1}$	3.07	38[43]

IrMo	1.85	38[43]	MgZn	0.9	11

oP8 BFe

CuLa	5.85	13[43]	SiTh	2.41	38[43]

oP8 MnP

GeIr	4.7	14	BiRh	2.06	14
AuGa	1.2	11	GeRh	0.96	14
PdSi	0.93	14	SiPt	0.9	11
SiPd	0.9	11	PtSi	0.88	14
AuIn	0.6	11	AsRh	0.58	14
PdSn	0.41	14	GePt	0.40	14

oP12 IrTa

$Ir_{1.15}Nb_{0.85}$	4.6	38[43]

oP12 Co_2Si

PRhTa	4.41	4	NbPRh	4.08	4
HfIrSi	3.50	23	PRuZr	3.46	30
IrSiY	2.70	23	IrSiZr	2.04	23
PRhZr	1.55	9	Pd_2Sn	0.41	14[43]

oP16 Al_3Ni

$CoLa_3$	4.01	2	Bi_3Ni	4.06	44[43]

oP16 Ge_3Rh_5

Ge_3Rh_5	2.1	11[43]

oP16 $CaLiSi_2$

Bi_3Rh	3.2	14[43]

oP16 CFe_3

RhY_3	0.65	2[43]	CFe_3	1.30	38[43]
PPd_3	0.75	14			

oP16 B_2LuRu

B_2LuRu	9.86	19	B_2RuY	7.80	19
B_2LuOs	2.66	19	B_2OsY	2.22	19
B_2OsSc	1.34	19			

oP28 $CaFe_2O_4$

P_4Rh_5	1.22	38[43]

oC8 BCr

BNb	8.25	38[43]	BTa	4.0	1

PtTh	0.44	2	IrTh	0.37	2
RhTh	0.36	2			

oC12 Si_2Zr

Ge_2Sc	1.31	38[43]	Ge_2Zr	0.30	38[43]

oC16 $BCMo_2$

$BCMo_2$	7.0	30

oC16 $CeNiSi_2$

$LaRhSi_2$	3.42	5	$IrLaSi_2$	2.03	5

oC20 $PdSn_4$

Pb_4Pt	2.80	14	$PtSn_4$	2.38	38[43]

oC32 BiPd

BiPd	3.7	14[43]

oC108 B_4LuRh_4

B_4LuRh_4	6.2	30	B_4Rh_4Tm	5.4	30
B_4ErRh_4	4.3	29	B_4HoRh_4	1.4	29

oI36 FeSiTi

NbReSi	5.1	4	ReSiTa	4.4	4
RuSiTa	3.15	4	NbRuSi	2.65	4

oI40 $Co_3Si_5U_2$

$La_2Rh_3Si_5$	4.45	5	$Ir_3Si_5Y_2$	2.83	5
$Rh_3Si_5Y_2$	2.70	5			

oI44 Sn_5Ti_6

Nb_6Sn_5	2.8	38[43]

tP4 AuCu

$BaBi_3$	5.69	39[43]	BiLi	2.47	38[43]
BiNa	2.25	38[43]			

tP4 OPb

OSn	3.81	38[43]

tP10 Pd_4Se

Pd_4Se	0.42	16[43]

tP14 Hg_5Mn_2

Ga_5V_2	3.55	39[43]

tP18 B_4CeCo_4

B_4Rh_4Tm	9.8	29	B_4ErRh_4	8.7	29
B_4NdRh_4	5.3	29	B_4Rh_4Th	4.3	29
B_4Rh_4Sm	2.7	29	B_4ErIr_4	2.1	29
B_4HoIr_4	2.0	29	B_4Ir_4Tm	1.6	29

tP30 CrFe

Ir_2Nb_3	9.8	38	$T_{c_3}W_2$	7.9	38
MoRe	7.8	39	Mo_3Ru_2	7.0	39
$IrMo_3$	6.8	39	Fe_3Re_2	6.55	38
Re_3V	6.26	38	$Mo_{0.62}Os_{0.38}$	5.65	39
$NbRe_3$	5.27	38	$Re_{0.52}W_{0.48}$	5.2	39
RuW	5.12	39	$Ir_{0.28}W_{0.72}$	4.46	38
Nb_3Rh_2	4.04	38	$Nb_{0.62}Pt_{0.38}$	4.01	38
OsW_2	3.81	39	NbRe	3.8	39
$Cr_{0.42}Re_{0.58}$	2.50	39	Rh_2Ta_3	2.35	39
Cr_3Ru_2	2.10	39	Nb_3Os_2	1.85	39
$Pt_{0.3}Ta_{0.7}$	1.5	39	Re_3Ta_2	1.4	39
$Ir_{0.35}Ta_{0.65}$	1.2	38			

tP32 PTi_3

Nb_3P	1.8	11	$SiZr_3$	0.5	11
$GeZr_3$	0.4	11	$SiNb_3$	0.3	11
$AsNb_3$	0.3	11	Nb_3Sb	0.2	11

tP38 $Co_4Sc_5Si_{10}$

$Ge_{10}As_4Y_5$	9.06	27	$Rh_4Sc_5Si_{10}$	8.54	22
$Ir_4Sc_5Si_{10}$	8.46	22	$Co_4Sc_5Si_{10}$	5.0	22
$Lu_5Rh_4Si_{10}$	3.95	27	$Ir_4Lu_5Si_{10}$	3.9	27
$Ir_4Si_{10}Y_5$	3.10	27	$Ge_{10}Lu_5Rh_4$	2.79	27
$Ge_{10}Ir_4Y_5$	2.62	22	$Ge_{10}Ir_4Lu_5$	2.60	27
$Ge_{10}Rh_4Y_5$	1.35	27			

tP40 $Fe_3Sc_2Si_5$

$Fe_3Lu_2Si_5$	6.1	29	$Fe_3Sc_2Si_5$	4.52	29
$Fe_3Si_5Y_2$	2.4	29	$Re_3Si_5Y_2$	1.76	29
$Fe_3Si_5Tm_2$	1.3	29			

tI2 In

CdHg	1.77	39[43]

tI4 Sn

Si	6.7	11	Ge	5.3	11
GaSb	4.2	11	Sn	3.7	11
AlSb	2.8	11	InSb	2.0	11

tI6 C_2Ca

C_2Y	3.88	38[43]	C_2Lu	3.33	44
C_2La	1.66	44			

tI6 $MoSi_2$

Bi_2Pd	4.25	39[43]	Hg_2Mg	4.0	15[43]

tI8 Al_3Ti

Al_3Nb	0.64	11

tI10 Al_4Ba

$LaRh_2Si_2$	3.90	5	Rh_2Si_2Y	3.11	5
Ir_2Si_2Y	2.60	5	Ir_2Si_2Th	2.14	5

tI10 $BaNiSn_3$

$LaRhSi_3$	2.7	5	$IrLaSi_3$	2.7	5
$RhSi_3Th$	1.76	5	$IrSi_3Th$	1.75	5

tI12 Al_2Cu

$CoZr_2$	6.30	39	$AuPb_2$	4.42	38
$CuTh_2$	3.49	39	BTa_2	3.12	38

$AuTh_2$	3.08	39	Pb_2Pd	2.95	14
Pb_2Rh	2.66	14	$AgTh_2$	2.26	39
$AgIn_2$	2.46	38	AgTh	2.2	11
$NiZr_2$	1.52	44			
Al_2Cu	1.02	38	$PdTh_2$	0.85	2
$GaZr_2$	0.38	11	$GaHf_2$	0.21	11
$AlTh_2$	0.1	11			

tI12 Si_2Th

Ge_2Y	3.8	13	Si_2Th	3.2	11
Ge_2La	2.6	11	$LaSi_2$	2.5	11

tI12 LaPtSi

IrSiTh	6.50	4	$Rh_{0.86}Si_{1.04}Th$	6.45	4
GeLaPt	3.53	4	LaPtSi	3.48	4
GeIrLa	1.64	4			

tI12 Mo_2N

Mo_2N	5.0	38[43]

tI16 BMo

BMo	0.5	38[43]

tI26 $Mn_{12}Th$

$Be_{13}W$	4.1	39[43]

tI28 MnU_6

FeU_6	3.86	38[43]	MnU_6	2.32	38[43]
CoU_6	2.29	38[43]			

tI32 Ni_3P

Mo_3P	5.31	40	PW_3	2.26	40

tI72 B_4LuRu_4

$B_4Rh_{3.4}Ru_{0.6}$	8.38	33	B_4Ru_4Sc	7.2	29
B_4LuRu_4	2.0	29	B_4Ru_4Y	1.4	29

tI140 Rh_2Y_3

Ir_2Y_3	1.61	2[43]	Rh_2Y_3	1.48	2[43]

hP2 CW

CMo	9.3	32	CRu	2.00	38[43]

hP2 Mg

Tc	7.8	11	La	4.9	13[43]
Y	2.5	11	Re	1.7	11
Lu	1.0	11	Zn	0.85	11

Os	0.7	11	Zr	0.6	11
Ca	0.52	11	Ru	0.5	11
Ti	0.4	11			

hP3 AlB_2

BMo_2	4.74	38[43]	Si_2Th	2.4	11
Hg_2Na	1.62	39[43]			

hP3 CdI_2

CW_2	5.2	38[43]	$PdTe_{2.3}$	1.85	16

hP4 AsNi

BiNi	4.25	39	$Pd_{1.1}Te$	4.07	16
Bi_2Pd_3	4.0	38	$BiPd_{0.45}Pt_{0.55}$	3.7	38[43]
$BiNi_{0.5}Rh_{0.5}$	3.0	38[43]	BiPt	2.4	14

BiRh	2.2	38	PtSb	2.1	14
$Bi_{0.1}PtSb_{0.9}$	2.05	38	$Bi_{0.5}NiSb_{0.5}$	2.0	38[43]
$AsNi_{0.25}Pd_{0.75}$	1.6	38	$Bi_{0.5}PtSb_{0.5}$	1.5	38[43]
PdSb	1.5	14	BiCu	1.40	38[43]
AuSn	1.25	38	Pd_3Sn_2	0.64	14
PtTe	0.58	16[43]	PtSn	0.37	38

hP4 SZn

GaN	5.85	38[43]

hP6 $CaCu_5$

B_2LuOs_3	4.62	29	Ir_5Th	3.93	2
B_2Ru_3Y	2.85	29	B_2LaRh_3	2.82	29
Ir_5La	2.13	2[43]	B_2Ir_3Th	2.09	29
B_2Ru_3Th	1.79	29	B_2Ir_3La	1.65	29
Rh_5Y	1.56	2[43]			

hP6 $InNi_2$

$BiIn_2$	5.6	39[43]

hP6 MoS_2

$NbSe_2$	7	38[43]

hP8 Ni_3Sn

$AlLa_3$	5.57	13	$CoLa_3$	4.28	13
Hg_3Li	1.7	39[43]	$CaHg_3$	1.6	15[43]

hP8 $AsNa_3$

$HgMg_3$	0.17	15[43]

hP9 Bi_2Pt

Bi_2Pt	0.16	14[43]

hP9 Fe_2P

HfPRu	9.9	4	AsOsZr	8.0	4
POsZr	7.44	29	OsPZr	7.4	4
HfOsP	6.1	4	AsHfRu	4.9	4
AsHfOs	3.2	4	$AsPd_2$	1.70	14
PRuTi	1.3	35	PRuTi	1.3	4
OsPTi	1.2	4			

hP12 CuS

CuS	1.62	38[43]

hP12 B_2BaPt_3

$B_2Ba_{0.67}Pt_3$	5.60	29[43]	$B_2Pt_3Sr_{0.67}$	2.78	29[43]
$B_2Ca_{0.67}Pt_3$	1.57	29[43]			

hP12 $MgZn_2$

Re_2Zr	6.8	39	$LaOs_2$	6.50	44

$HfRe_2$	5.61	39			
Os_2Y	4.7	13	Os_2Sc	4.6	13
$Nb_{0.5}V_{1.5}Zr$	4.3	38	Re_2Sc	4.2	38
$LuOs_2$	3.49	13	Os_2Zr	3.0	39
$HfOs_2$	2.69	39	Ru_2Y	2.42	38
$Ir_{1.5}Os_{0.5}Y$	2.40	39	OsReY	2.00	39
Ru_2Zr	1.84	39	Re_2Y	1.83	39
Ru_2Y	1.52	44	Ir_2Sc	1.03	44
Ru_2Sc	1.67	13	$LuRu_2$	0.86	13

hP12 $LaRu_3Si_2$

$LaRu_3Si_2$	7.60	29	Ru_3Si_2Th	3.98	29
Ru_3Si_2Y	3.51	29			

hP14 Fe_3Te_3Tl

Mo_3Se_3Tl	4.0	37

hP14 Nb_3Te_4

Nb_3Se_4	2.0	11	Nb_3Te_4	1.8	11

hP16 Mn_5Si_3

$C_{0.6}Mo_{4.8}Si_3$	7.6	37	Ga_3Zr_5	3.8	11
Hg_3Mg_5	0.48	15[43]			

hP20 Fe_3Th_7

B_3Ru_7	2.58	38[43]	La_7Rh_3	2.58	2
Ir_3La_7	2.24	2	Rh_3Th_7	2.15	39
Ni_3Th_7	1.98	39	Fe_3Th_7	1.86	39
Co_3Th_7	1.83	39	Ir_3Th_7	1.52	39
Os_3Th_7	1.51	39	Pt_3Th_7	0.98	2
Pt_3Y_7	0.82	2	Rh_3Y_7	0.32	13

hP24 $MgNi_2$

$HfMo_2$	0.05	38

hP24 $CeNi_3$

$CeIr_3$	3.34	2[43]	$LaRh_3$	2.60	2[43]
Rh_3Y	1.07	44			

hR4 NbS_2

NbS_2	6.3	38[43]

hR6 Au_5Sn

Au_5Sn	1.1	38[43]

hR12 Be_3Nb

Ir_3Y	3.50	2[43]	Ir_3La	2.46	2[43]

hR15 Mo_6PbS_8

Mo_6PrSe_8	9.2	34	Mo_6S_8Yb	9.2	30

$Ag_{1.6}Mo_{6.4}S_8$	9.1	34	$Cu_2Mo_6O_2S_6$	9	31
Mo_6S_8Tl	8.7	25	$Mo_6Na_2S_8$	8.6	34
Mo_6NdSe_8	8.2	34	$InMo_6Se_8$	8.1	25
$Mo_6Pb_{1.2}Se_8$	6.75	28	$Mo_6Se_8Tm_{1.2}$	6.3	34
$Mo_6Se_8Yb_{1.2}$	6.2	34	$Lu_{1.2}Mo_6Se_8$	6.2	34
$Mo_{5.25}Nb_{0.75}Se_8$	6.2	31	$ErMo_6Se_8$	6.2	10
$Ho_{1.2}Mo_6Se_8$	6.1	34	IMo_6Se_7	7.6	31
$LaMo_6S_8$	7.1	34	$Br_3Mo_6Se_5$	7.1	31
$BrMo_6Se_7$	7.1	31	$Mo_6Se_8Sm_{1.2}$	6.8	34
$Mo_6Se_8Sn_{1.2}$	6.8	34	Mo_6Se_8	6.3	34
$Ag_{1.2}Mo_6Se_8$	5.9	34	$Cu_2Mo_6Se_8$	5.9	34
$Dy_{1.2}Mo_6Se_8$	5.8	34	Mo_6Se_8Tb	5.7	10
$Cl_3Mo_6Se_5$	5.7	31	$GdMo_6Se_8$	5.6	10
$Li_2Mo_6S_8$	4.2	34	Mo_6PrS_8	4.0	34
$Mo_{6.6}S_8Zn_{11}$	3.6	34	Mo_6NdS_8	3.6	30
S_8ScMo_6	3.6	34	$GdMo_{6.6}S_8$	3.5	34
$Mg_{1.14}Mo_{6.6}S_8$	3.5	34	$Mo_4Re_2Te_8$	3.5	31
$K_2Mo_{15}S_{19}$	3.32	31	$Mo_6S_8Y_{1.2}$	3.0	34
$Mo_6S_8Sm_{1.2}$	2.9	34	$Ba_2Mo_{15}Se_{19}$	2.75	31
$I_2Mo_6Te_6$	2.6	31	$K_2Mo_{15}Se_{19}$	2.45	31
$ErMo_6S_8$	2.2	10	$DyMo_6S_8$	2.1	10
$Mo_6S_8Tm_{1.2}$	2.1	34	$Lu_{1.2}Mo_6S_8$	2.0	34

$Ho_{1.2}Mo_6S_8$	2.0	34	Mo_6S_8Tb	2.0	10
Mo_6S_8	1.85	25	$Ir_{1.2}Mo_6S_8$	1.85	34
$Cl_2Mo_6Te_6$	1.7	31	$Mo_4Ru_2Te_8$	1.7	31
Mo_6Te_8	1.7	31	$Mo_{4.66}Rh_{1.33}Te_8$	1.7	31
$Mo_{5.25}Ta_{0.75}Te_8$	1.7	31	$Mo_{5.25}Nb_{0.75}Te_8$	1.7	31

hR20 P_3Pd_7

$PPd_{2.3}$	1.00	14[43]

cP2 ClCs

BaHg	2.32	15[43]	CaHg	1.6	15[43]
HgMg	1.39	15[43]	LaZn	1.04	13
AgLa	0.94	13	AgLa	0.9	38[43]
CuY	0.33	13	AgLu	0.33	13
AgY	0.33	13	YZn	0.33	13

cP4 $AuCu_3$

La_3Tl	8.86	7	$LaSn_3$	6.55	8

$GaLa_3$	5.84	13[43]	$NaPb_3$	5.62	7
Pb_3Th	5.55	7	Pb_3Y	4.72	7
$Bi_{0.26}Tl_{0.74}$	4.15	8	$LaPb_3$	4.10	7
Sn_3Th	3.33	12	Hg_3Zr	3.28	8
Ga_3Lu	2.3	12	Ga_3Lu	2.3	8
$InMg_3$	2.2	38[43]	$CaTl_3$	2.0	8
$LaTl_3$	1.63	8	Al_2Ge_2U	1.60	26
Tl_3Y	1.52	8	Al_2Si_2U	1.34	26
Al_3Yb	0.94	8	$ThTl_3$	0.87	8
Ga_2Ge_2U	0.87	26	$AlZr_3$	0.73	8
In_3Lu	0.2	11	Ru_3U	0.15	26

cP6 NbO

NbO	1.25	38[43]

cP8 Cr_3Si

$IrMo_3$	9.6	6	$InNb_3$	9.2	6

Nb_3Tl	9	6	Mo_3Pt	8.8	6
$SnTa_3$	8.35	6	GeV_3	8.2	6
$GeTa_3$	8.0	6	Mo_3Os	7.2	39
SnV_3	7.0	11	$SbTi_3$	5.80	6
$SnTi_3$	5.80	6	$IrTi_3$	5.40	6
OsV_3	5.15	6	Cr_3Os	4.68	6
$BiNb_3$	4.5	6	V_3Sn	3.7	38
$SnTa_2V$	3.7	39	Cr_3Ru	3.43	6
AuV_3	3.22	6	$BiZr_3$	2.85	6
PtV_3	2.83	38	$SnTaV_2$	2.8	39
Nb_3Rh	2.64	6	Nb_3Sb	2.2	6
$GeMo_3$	1.75	6	IrV_3	1.71	6
Mo_3Si	1.70	6	$IrNb_3$	1.7	39
$SiMo_3$	1.4	11	$GaMo_3$	1.20	6
Nb_3Os	1.05	6	$AuZr_3$	0.92	6
$SnZr_3$	0.92	6	SbV_3	0.80	6
NiV_3	0.78	6	Cr_3Ir	0.78	6
$PbZr_3$	0.76	6	$SbTa_3$	0.72	6
$AlMo_3$	0.6	11	$PtTi_3$	0.58	6
$AuTa_{4.3}$	0.55	38[43]	$PtTa_3$	0.4	6
RhV_3	0.38	6	Cr_3Rh	0.3	6
AsV_3	0.20	6			

cP12 Si_2Sr

LaRhSi	4.35	5

cP32 H_3U

$AuZn_3$	1.28	38[43]

cP40 $Pr_3Rh_2Sn_{13}$

$CaRh_{1.2}Sn_{4.5}$	8.7	29	$Rh_{14}Sn_{4.6}Yb$	8.6	29
$Ca_3Ir_4Sn_{13}$	7.1	41	$Ca_3Co_4Sn_{13}$	5.9	29,41
Os_xSn_yTh	5.6	42	$Ir_4Sn_{13}Sr_3$	5.1	41,30
$Rh_4Sc_3Sn_{13}$	4.5	41	$Rh_4Sn_{13}Sr_3$	4.3	41,29
$LuRh_{1.2}Sn_{4.0}$	4.0	29	$Ge_{13}Os_4Y_3$	3.9	29
$LaRu_{1.5}Sn_{4.5}$	3.9	30	$Ge_{13}Lu_3Os_4$	3.6	29
$La_3Rh_4Sn_{13}$	3.2	41,29	$Rh_4Sn_{13}Y_3$	3.2	41
$Ir_4Lu_3Sn_{13}$	3.2	41	$Co_4La_3Sn_{13}$	2.8	41
Ir_xSn_yTh	2.6	42	$Ir_4La_3Sn_{13}$	2.6	41,29
Os_xSn_yY	2.5	29	Co_xSn_yYb	2.5	42

$Ge_{13}Lu_3Ru_4$	2.3	29	Rh_xSn_yTm	2.3	42
$La_3Pb_{13}Rh_4$	2.2	41	$Ir_4Sn_{13}Y_3$	2.2	41
$Ca_3Ge_{13}Rh_4$	2.1	41	Ir_xSn_yYb	2	42
$Ge_{13}Rh_4Sc_3$	1.9	41	Rh_xSn_yTh	1.9	29
$LuOs_xSn_y$	1.8	42	$Ca_3Ge_{13}Ir_4$	1.7	41
$Ge_{13}Ru_4Y_3$	1.7	29	Os_xScSn_y	1.5	42
Co_xLuSn_y	1.5	42	$Ge_{13}Ir_4Sc_3$	1.4	41
$HoOs_{1.2}Sn_{2.5}$	1.4	30	$Os_{1.5}Sn_{2.6}Tb$	1.4	30
$ErRh_{1.1}Sn_{3.6}$	1.36	29	$ErOs_{1.1}Sn_{2.7}$	1.3	30
$Ru_{1.1}Sn_{3.1}Y$	1.3	29	Os_xSn_yTm	1.1	30
$Ir_4Sc_3Sn_{13}$	1.1	41			

cP64 $Pd_{17}Se_{15}$

$Rh_{17}S_{15}$	5.8	14[43]

cI2 W

Nb	9.2	11	V	5.4	11

Ta	4.5	11	$Nb_{0.26}U_{0.74}$	1.85	39[43]
Pd_3Te	0.76	16[43]	U	0.2	18[43]
Mo	0.9	11			

cI16 CoU

CoU	1.70	44

cI28 P_4Th_3

La_3Se_4	8.6	11	La_3S_4	8.1	11
$C_{10}Sc_{13}$	8.50	44			
La_3Te_4	5.3	11	As_4La_3	0.6	11[43]

cI34 Fe_4LaP_{12}

$LaRu_4P_{12}$	7.2	30

cI46 $Pd_{16}S_7$

$Pd_{2.2}S$	1.63	14[43]

cI58 Mn

$MoRe_3$	9.89	39	$T_{c_6}Zr$	9.7	39
$Nb_{0.18}Re_{0.82}$	9.7	39	Re_3W	9.0	39
$MoRe_4$	7.85	38	Re_3Zr	7.40	38
Re_3Ta	6.78	39	$Re_{24}Ti_5$	6.6	38
$Hf_{0.14}Re_{0.86}$	5.86	39	Al_5Re_{24}	3.35	38
NbOs	2.86	39	$NbOs_2$	2.52	39
Nb_3Pd_2	2.47	39	$Re_{24}Sc_5$	2.2	38
OsTa	1.95	39	$Ag_{3.3}Al$	0.34	38[43]

cI120 Bi_4Rh

Bi_4Rh	2.70	38[43]

cF4 Cu

La	6.06	13[43]

cF8 ClNa

NV	8.5	24	NHf	8.4	24
HfN	8.4	32	NTi	5.5	24
C_3MoRe	3.8	3[43]	C_2ReW	3.8	3[43]
Bzr	3.4	1	BHf	3.1	1
SeTh	1.7	11	In_4SbTe_3	1.5	38[43]
LaN	1.35	38[43]	TeY	1.02	38[43]
OTi	0.68	38[43]	STh	0.5	11
GeTe	0.41	38[43]			

cF12 CaF_2

Ga_2Pt	1.8	11	$AuGa_2$	1.6	11
PRh_2	1.3	14	$CoSi_2$	1.22	38[43]
Al_2Pt	0.5	11	$AuIn_2$	0.2	11
Al_2Au	0.1	11			

cF16 BiF_3

Pd_2SnY	4.92	36	$PbPd_2Y$	4.05	36
$LuPd_2Sn$	3.11	36	Pd_2ScSn	2.25	36
Pd_2SnTm	1.77	36	Pd_2SnYb	1.79	36

cF24 $AuBe_5$

$CeIr_5$	1.82	2[43]	Au_5Ca	0.38	38[43]

cF24 Cu_2Mg

V_2Zr	8.8	39	$LaOs_2$	6.5	13
Ir_2Th	6.50	39	$CaRh_2$	6.40	39
Rh_2Sr	6.2	39	$CaIr_2$	6.15	39
$BaRh_2$	6.0	38	Ir_2Sr	5.7	39
$Ce_xGd_yRu_{0.66}$	5.20	38	$CeRu_2$	4.90	39
Bi_2Cs	4.75	39	Bi_2Rb	4.25	38
Pt_2Y	4.1	38	Ir_2Zr	4.10	39
Bi_2K	3.58	39	Ru_2Th	3.5	39

Al_2La	3.23	13	Ir_2Lu	2.89	2
$NbSn_2$	2.60	38[43]	IrOsY	2.60	39
Ir_2Y	2.18	13	W_2Zr	2.16	39
Ir_2Sc	2.07	13	Hf_2Rh	1.98	38
Au_2Bi	1.84	39	$LaRu_2$	1.63	13
Pt_2Y	1.57	44			
CeCo	1.5	38	$Lu_{0.275}Rh_{0.725}$	1.27	2
Au_2Pb	1.18	38	$LaMg_2$	1.05	13
Al_2Lu	1.02	13	Al_2Sc	1.02	13
Pt_2Sr	0.7	38	$La_{0.28}Pt_{0.72}$	0.54	2
$LuRu_2$	0.86	44			
Ir_2La	0.48	13	$Ni_{1.5}V_{0.5}Zr$	0.43	39
Al_2Y	0.35	13	Rh_3Sc_7	0.32	13

cF96 $NiTi_2$

$CoTi_2$	3.44	38[43]	$CoHf_2$	0.50	38[43]
$CoZr_2$	6.30	44			

cF1832 Al_3Mg_2

Al_3Mg_2	0.84	39[43]

m**

O_5ReTi	5.71	38	BiRu	5.7	38
$HgSn_6$	5.1	38	$BiIn_5$	4.1	38
Bi_3Sn	3.77	38	Bi_3Mo	3.7	38
$AgBi_2$	3.0	38	AgBi	2.78	38
Bi_2Ir	2.2	38[43]	Bi_4Mg	1.0	38
Bi_3Fe	1.0	38	Bi_3Zn	0.87	38
BiCo	0.5	38			

o**

Bi_3Ni	4.06	39[43]

t**

Bi_2Pb	4.25	38	Nb_3Ru_2	1.2	39

h**

SiTa	4.38	39	La_3Y	2.50	39

Ag_4Ge	0.8	11	$AsRh_{1.4}$	0.56	14
Al_3Tl	0.75	38	Ag_4Sn	0.1	11

cF*

$CIrMo_3$	3.2	3[43]	COs_2W_3	2.9	3[43]
CIr_2W_3	2.1	3[43]	CIr_2Mo_3	1.8	3
CMo_3Pt_2	1.1	3	CPt_2W_3	1.2	3[43]

No structural information

PPb	7.8	38	OV_3Zr_3	7.5	38
Mo_3Os	7.20	38	$K_{0.33}O_3W$	6.3	37
OReTi	5.74	38	$N_{0.34}Re$	5	38
Ir_3Th	4.71	2	$NbSe_{0.25}Te_{0.75}$	4.4	38
$Bi_{0.45}PdSb_{0.55}$	3.7	38	CTc	3.85	38
$O_{0.14}Rh_{0.287}Ti_{0.573}$	3.37	38	Pt_5Th	3.13	2
BW_2	3.10	38	$Na_{0.33}O_3W$	3.0	37
PbS_3Ta	3.0	37	Si_2W_3	2.84	38

BRe_2	2.80	38	$Pd_{2.5}Se$	2.3	14
Pd_2Se	2.20	38	$BiRe_2$	2.2	38
$OsTl_{0.3}W$	2.14	38	$O_{0.105}Pd_{0.285}Zr_{0.61}$	2.09	38
As_3Pd_5	1.9	38	$GeNb_2$	1.90	38
$Ba_{0.13}P_3W$	1.9	38	$CaHg_5$	1.7	15
$Pd_{0.9}Pt_{0.1}Te_2$	1.65	38	$LaRh_5$	1.62	2
$Ni_{0.1}Pd_{0.9}Te_2$	1.30	38	$Bi_{0.5}Pb_{0.5}Pt$	1.29	38
Rh_5Th	1.07	2	Rh_2Se_5	1.04	38
$PdSi_2$	0.93	38	Pt_2Y_3	0.90	2
Ge_7Ir_3	0.87	14	$IrZn_2$	0.78	38
Au_5B	0.7	38	Au_5Ba	0.7	38[43]
$Pd_{6.5}Se$	0.66	16	$Ba_{0.1}O_3Sr_{0.9}Ti$	0.6	38
$Ca_{0.31}O_3Sr_{0.7}Ti$	0.6	38	$LuRh_5$	0.49	2
$HgMg_2$	0.48	15	As_2Pd_5	0.46	14
O_3SrTi	0.39	38	La_3Sb_4	0.2	11

REFERENCES

1. O. I. Shulishova and I. A. Shcherbak, Izvestiya Akademii Nauk SSSR, Neorganicheskie Materialy, *3* No 8 (1967) 1495.
2. T. H. Geballe, B. T. Matthias, V. B. Compton, E. Corenzwit, G. W. Hull, Jr. and L. D Longinotti, Physical Review, *1A* (1965) 119.
3. A. C. Lawson, Journal of the Less-Common Metals, *23* (1971) 103.
4. G. V. Subba Rao, K. Wagner, Geetha Balakhrishnan, J. Janaki, W. Paulus and R. Schöllhorn, Bulletin of Materials Science, *7* (1985) 215.
5. B. Chevalier, P. Lejay, B. Lloret, Wang- Xian- Zhong, J. Etourneau and P. Hagenmuller, Annales de Chimie, *9* (1984) 191.
6. H. D. Wiesinger, Physica Status Solidi *41A* (1977) 465.
7. W. Rong-Yao, Physica Status Solidi, *94A* (1986) 445.
8. W. Rong-Yao, L. Qi- Guang and Z. Xiao, Physica Status Solidi, *90A* (1985) 763.
9. R. Müller, R. N. Shelton, J. W. Richardson, Jr. and R. A. Jacobson, Journal of the Less-Common Metals, *92* (1983) 177.
10. R. N. Shelton, Journal of the Less- Common Metals, *94* (1983) 69.
11. I. M. Chapnik, Journal of Materials Science Letters, *4* (1985) 370.
12. W. Rong- Yao, Journal of Materials Science Letters, *5* (1986) 87.
13. T. F. Smith and H. L. Luo, Journal of Physics and Chemistry of Solids, *28* (1967) 569.
14. Ch. J. Raub, W. H. Zachariasen, T. H. Geballe and B. T. Matthias, Journal of Physics and Chemistry of Solids, *24* (1963) 1093.

15. T. Claeson and H. L. Luo, Journal of Physics and Chemistry of Solids, *27* (1966) 1081.

16. Ch. J. Raub, V. B. Compton, T. H. Geballe, B. T. Matthias, J. P. Maita and G. W. Hull, Jr., Journal of Physics and Chemistry of Solids, *26* (1965) 2051.

17. M. Marezio, P. D. Dernier, J. P. Remeika, E. Corenzwit and B. T. Mattias, Materials Research Bulletin, *8* (1973) 657.

18. R. N. Shelton, A. C. Lawson and D. C. Johnston, Materials Research Bulletin, *10* (1975) 297.

19. H. C. Ku and R. N. Shelton, Materials Research Bulletin, *15* (1980) 1441.

20. H. Barz, Materials Reseach Bulletin, *15* (1980) 1489.

21. G. P. Espinosa, A. S. Cooper, H. Barz and J. P. Remeika, Materials Research Bulletin, *15* (1980) 1635.

22. G. Venturini, M. Meot- Meyer, E. McRae, J. F. Mareche and B. Rogues, Materials Research Bulletin, *19* (1984) 1647.

23. W. Xian- Zhong, B. Chevalier, J. Etourneau and P. Hagenmuller, Materials Research Bulletin, *20* (1985) 517.

24. You-Xiang Zhao and Shou-an He in High Pressure in Science and Technology eds. C. Homan, R. K. MacCrone and E. Whalley, North Holland, *22* (1983) 51.

25. J. M. Tarascon, F. J. DiSalvo, D. W. Murphy, G. Hull and J. V. Waszczak, Physical Review, *29B* (1984) 172.

26. H. R. Ott, F. Hulliger, H. Rudigier and Z. Fisk, Physical Review, *31B* (1985) 1329.

27. H. D. Yang, R. N. Shelton and H. F. Braun, Physical Review, *33B* (1986) 5062.

28. R. Flückiger and R. Baillif, in Topics in Current Physics, eds. O. Fischer and M. B. Maple, Springer Verlag, *32* (1982) 113.

29. D. C. Johnston and H. F. Braun, in Topics in Current Physics, eds. M. B. Maple and O. Fischer, Springer Verlag, *34* (1982) 11.

30. G. V. Subba and Geetha Balakrishnan, Bulletin of Materials Science, *6* (1984) 283.

31. R. Chevrel and M. Sergent, in Topics in Current Physics eds. O. Fischer and M. B. Maple, Springer Verlag, *32* (1982) 25.

32. Z. You- xiang and He Shou-an, Solid State Communications, *45* (1983) 281.

33. D. C. Johston, Solid State Communications, *24* (1977) 699.

34. O. Fischer, Applied Physics, *16* (1978) 1.

35. G. P. Meisner and H. C. Ku, Applied Physics, *A31* (1983) 201.

36. M. J. Johnson, Ames Lab, Ames IA (USA), Report 1984, IS-T-1140.

37. R. N. Shelton, in Superconductivity in d- and f- Band Metals 1982 eds. W. Buckel and W. Weber, Kernforschungszentrum Karlsruhe, (1982) 123.

38. E. M. Savitskii, V. V. Baron, Yu. V. Efimov, M. I. Bychkova and L. F. Myzenkova in Superconducting Materials, Plenum Press, (1981) 107.

39. B. W. Roberts in Intermetallic Compounds ed. J. H. Westbrook, John Wiley and Sons, Inc., (1967) 581.

40. R. D. Blaugher, J. K. Hulm and P. N. Yocom, Journal of Physics and Chemistry of Solids, *26* (1965) 2037.

41. G. Venturini, M. Kamta, E. McRae, J. F. Mareche, B. Malaman and B. Roques, Materials Research Bulletin, *21* (1986) 1203.

42. G. P. Espinosa, A. S. Cooper and H. Barz, Materials Research Bulletin, *17* (1982) 963.

43. P. Villars and L. D. Calvert, Pearson's Handbook of Crystallographic Data for Intermetallic Phases, Vols. 1-3, ASM, Ohio (1985).

44. J. K. Hulm and R. D. Blaugher, in Superconductivity in d- and f- Band Metals, ed. D. H. Douglass, American Institute of Physics, *4* (1972) 1.

Compound	Year	$\overline{N}_v$	$\overline{\Delta X}$	$\overline{\Delta r}$	$\overline{a}$(A)	c(A)
La_2NiO_4	58	5.71	−1.45	1.73	3.86	12.65
Ba_2SnO_4	59	4.57	−1.46	1.65	4.13	13.27
$Bi_4Ti_3O_{12}$	49	5.47	−1.05	1.50	5.43	32.84
Bi_2MoO_6	56	5.78	−0.88	1.34	5.61	12.99
Bi_2WO_6	69	5.78	−0.94	1.35	5.45	16.43
La_2CuO_4	73	5.86	−1.84	1.65	5.38	13.17
La_2CoO_4	80	5.57	−1.48	1.64	5.49	12.55
$CaCuO_2$	70	6.25	−2.24	1.58	3.52	12.23

Table IV

Table IV. Antecedent or parent ternaries for high-T_c superconductors. Because La_2CuO_4 and La_2NiO_4 fall in island C of the Villars diagram, Fig. C-3, it seems possible that other ternaries which fall in island C may be antecedents as well. We define this island by the coordinate ranges: $4.00 \leq \overline{N}_v \leq 7.25$, $\overline{\Delta X} < -0.2$, $1.2 \leq \overline{\Delta r} \leq 1.80$, and restrict the search to oxides. This gave more than 1500 compounds. However, this number can be further reduced quite easily. Suppose we accept only those compounds which potentially have square planar or uniaxially stretched octahedral layers, i.e., are tetragonal. This leaves only 80 tetragonal oxides in island C. About half of these are ABO_4 compounds where either A or B is tetrahedrally coordinated, which cannot form layer structures. The remaining 40 compounds must be examined individually, but if we further impose the condition that the basal lattice constant be compatible with CuO_2 square planar structures ($a = 3.85A$ or $3.85\sqrt{2} = 5.43A$) or AgO_2($a = 4.04A$ or $5.72A$) then the list becomes very short. Here are the compounds, the year first reported, with Villars coordinates and lattice constants. The last compound given, $CaCuO_2$, falls in island C, but is orthorhombic and (unlike $Ca_{1-x}Sr_xCuO_2$ with $x = 0.15$ (1), see ref. 12), does not contain CuO_2 planes.

First Author Index

A

B

Subject Index

Q

R

S

T

V

W

Z